GENETICS OF COMMON DISEASES

future therapeutic and diagnostic possibilities

UCL Molecular Pathology series

Editor: **D.S. Latchman**
Department of Molecular Pathology, University College London,
The Windeyer Building, 46 Cleveland Street, London W1P 6DB

From Genetics to Gene Therapy

Autoimmune Diseases: focus on Sjögren's syndrome

Apoptosis and Cell Cycle Control in Cancer

Ischaemia: preconditioning and adaptation

Genetics of Common Diseases: future therapeutic and diagnostic possibilities

GENETICS OF COMMON DISEASES

future therapeutic and diagnostic possibilities

Ian N.M. Day
Department of Medicine, University College London Medical School, The Rayne Institute, London, UK (address from October 1997: Wessex Human Genetics Institute, Southampton University Hospital, Southampton, UK)

Steve E. Humphries
Department of Medicine, University College London Medical School, The Rayne Institute, London, UK

BIOS SCIENTIFIC PUBLISHERS

First published 1997

A CIP catalogue record for this book is available from the British Library.

ISBN 1 85996 041 3

BIOS Scientific Publishers Ltd
9 Newtec Place, Magdalen Road, Oxford OX4 1RE, UK
Tel. +44 (0) 1865 726286. Fax +44 (0) 1865 246823
World-Wide Web home page: http://www.Bookshop.co.uk/BIOS/

DISTRIBUTORS

Australia and New Zealand
Blackwell Science Asia
54 University Street
Carlton, South Victoria 3053

India
Viva Books Private Limited
4325/3 Ansari Road
Daryaganj
New Delhi 110002

Singapore and South East Asia
Toppan Company (S) PTE Ltd
38 Liu Fang Road, Jurong
Singapore 2262

USA and Canada
BIOS Scientific Publishers
PO Box 605, Herndon
VA 20172-0605

Typeset by Banbury Pre-Press Centre, Banbury, UK.
Printed and bound in Great Britain by Biddles Ltd, Guildford and King's Lynn, UK.

Contents

Contributors **xi**

Abbreviations **xiii**

Preface **xv**

1 Perspectives in human linkage studies. *S. Povey* **1**
Pairwise linkage analysis of Mendelian traits 2
Finding the cause of a disease which shows clear Mendelian inheritance 3
After a LOD score of +3, what next? 5
Narrowing the search by searching for critical recombinants 5
Linkage disequilibrium 7
Linkage analysis in more complex situations 8
Linkage in complex traits 10
Why are we doing this? 15
References 16

2 Twin research: nature versus nurture in common diseases. *K.O. Kyvik* **19**
Introduction 19
History 19
Biology of twinning 20
Demography 21
Twin studies 21
Assumptions 22
The concept of concordance rates 24
Heritability 27
Twin studies and common diseases 28
Other types of twin studies 29
Conclusion 31
References 31

3 **The molecular basis of genetic variation: mutation detection methodologies and limitations.** *E. Spanakis and I.N.M. Day* **33**
Introduction 33
Current theories of mutation 34
Parallel development of theory with methodology 42
Methods of mutation detection 47
Limitations of current methods for mutation detection 58
Our approaches 59
Conclusions 62
References 63

4 **Finding susceptibility genes for schizophrenia.** *M.J. Owen and M.C. O'Donovan* **75**
Introduction 75
Molecular genetic approaches to schizophrenia 76
Candidate gene association studies 81
Conclusions 85
References 86

5 **Approaches to determining the genetic basis of non-insulin-dependent diabetes mellitus.** *M. McCarthy* **91**
Non-insulin-dependent diabetes mellitus (NIDDM) 91
Overview of approaches to dissect NIDDM 96
Choosing a candidate gene 97
Exploring candidate genes 113
Assessing genotype–phenotype correlations 114
Understanding the biology of complex traits 118
References 120

6 **Population-scale genotype assays: *APOE* gene in Alzheimer's dementia and coronary risk.** *I.N.M. Day, D. Palamand and E. Spanakis* **129**
Introduction: availability of population-scale gene testing 129
Background 132
Some future possibilities 145
References 147

7 **Genetic tests for coronary artery disease risk: the fibrinogen and stromelysin genes as examples.** *S.E. Humphries, H. Montgomery, S. Ye and A. Henney* **151**
Introduction 151
Molecular biology of variability 152
Sequence polymorphism at the fibrinogen locus, and variability in plasma fibrinogen levels 154
Fibrinogen and the acute phase response 157
Acute phase stimulation of fibrinogen by exercise 159
Genes determining development of atherosclerosis and plaque rupture 160
Stromelysin genotype and progression of atherosclerosis 161
Conclusion 166
References 167

8 **HLA genes and rheumatoid arthritis susceptibility.** *W. Ollier* **171**
Introduction 171
Epidemiology of RA 172
Defining the genetic basis of RA 172
HLA associations with RA 173
The shared epitope hypothesis 174
Is HLA associated with RA severity or susceptibility? 175
Is HLA-encoded RA susceptibility due to a single or multiple genetic factors? 176
How much of the genetic contribution is due to HLA? 178
References 179

9 **Practical benefits from understanding the genetics of chronic diseases.** *R.R. Williams, S.C. Hunt, P.N. Hopkins, L. Wu and S. Stephenson* **185**
Introduction 185
From genes to treatment and prevention 185
Improved diagnosis 192
More successful practical applications: MED PED FH, – a practical approach to early diagnosis, treatment and prevention 193
More public awareness and financial support are needed 195

MED PED social issues: cost and confidentiality 198
Predictions for the future 198
References 200
Appendix 202

10 **Specific approaches to pulmonary emphysema and its therapy.** *N.A. Kalsheker* **203**
Genetics of pulmonary emphysema 203
α_1-Antitrypsin deficiency 204
Structure of α_1-antitrypsin 205
α_1-Antitrypsin gene structure and mRNA species 206
Genetic mutations associated with disease 209
Therapeutic trials with α_1-antitrypsin 210
Gene therapy 212
Gene transfer techniques 212
Results of administration 213
Gene therapy and α_1-antitrypsin deficiency 213
Response to therapy 214
Conclusions 215
References 215

11 **Gene therapy for neurological diseases: *quo vadis*? Achievements and expectations of, and challenges for, the brave new technology.** *P.R. Lowenstein, J. Jaszai and M.G. Castro* **219**
Introduction 219
Challenges towards clinical implementation: the central questions and the development of a viable clinical strategy 220
Gene therapy tools: their status circa June 1997 226
How to achieve cell type-specific expression 230
Conclusions 232
References 234
Bibliography 239

12 **Impact of genomics on the discovery and development of modern medicines.** *D.S. Bailey and G.I. Johnston* **241**
Introduction 241
The traditional route to pharmaceutical discovery 242

Supporting the discovery process through genomics 244
Realizing new paradigms for drug discovery and development through genomics 253
References 257

13 **Oligonucleotides and their future potential as therapeutic agents.** *K.R. Fox* **263**
Introduction 263
Triplex recognition motifs 265
Current strategies 269
Applications 276
Outstanding problems 278
References 279

14 **Protein engineering of therapeutic antibodies: use of antibodies for immunosuppression and treatment of leukaemias.** *M. Clark* **285**
Introduction 285
Choice and specificity of antibodies 287
Immunogenicity 288
Biological properties 290
Clinical use 296
References 301

15 **Getting to know your genes privately.** *P.G. Debenham* **307**
Who knows about the genetics revolution? 307
Commercial genetic diagnostics is already with us 308
Medical genetic diagnostics: the perception of private tests 310
CF carrier testing service: the component parts 312
Guidelines for future commercial genetic tests available directly to the public 315
What genetic diagnostics will be provided next? 316
Presenting the private possibilities 319
Genetics for diagnostics or for what? 320
References 321

Index **323**

Contributors

Bailey, D.S. Incyte Pharmaceuticals, 3174 Porter Drive, Palo Alto, CA 94304, USA

Castro, M.G. Molecular Medicine Unit, Stopford Building, University of Manchester School of Medicine, Department of Medicine, Oxford Road, Manchester M13 9PT, UK

Clark, M. Department of Pathology, University of Cambridge, Tennis Court Road, Cambridge CB2 1QP, UK

Day, I.N.M. Department of Medicine, University College London Medical School, The Rayne Institute, 5 University Street, London WC1E 6JJ, UK. Address from October 1997: Wessex Human Genetics Institute, Southampton University Hospital, Southampton SO16 6YD, UK

Debenham, P.G. University Diagnostics Ltd, South Bank Technopark, 90 London Road, London SE1 6LN, UK

Fox, K.R. Division of Biochemistry and Molecular Biology, School of Biological Sciences, University of Southampton, Bassett Crescent East, Southampton SO16 7PX, UK

Henney, A. Wellcome Trust Centre for Human Genetics, University of Oxford, Windmill Road, Oxford OX3 7BN, UK

Hopkins, P.N. University of Utah, Room 161, Cardiovascular Genetics Research Clinic, 410 Chipeta Way, Salt Lake City, UT 84108, USA

Humphries, S.E. Department of Medicine, University College London Medical School, The Rayne Institute, 5 University Street, London WC1E 6JJ, UK

Hunt, S.C. University of Utah, Room 161, Cardiovascular Genetics Research Clinic, 410 Chipeta Way, Salt Lake City, UT 84108, USA

Jaszai, J. Molecular Medicine Unit, Stopford Building, University of Manchester School of Medicine, Department of Medicine, Oxford Road, Manchester M13 9PT, UK

Johnston, G.I. Pfizer Central Research, Ramsgate Road, Sandwich, Kent CT13 9NJ, UK

Kalsheker, N.A. Clinical Chemistry, Department of Clinical Laboratory Sciences, University Hospital, Queen's Medical Centre, Nottingham NG7 2UH, UK

Kyvik, K.O. The Danish Twin Register, Department of Genetic

Epidemiology, Institute of Community Health, Odense Universitet, Winsløwparken 15, DK-5000 Odense, Denmark

Lowenstein, P.R. Molecular Medicine Unit, Room 1.302, Stopford Building, University of Manchester School of Medicine, Department of Medicine, Oxford Road, Manchester M13 9PT, UK

McCarthy, M. Department of Metabolic Medicine, Imperial College School of Medicine, St Mary's Hospital, London W2 1PG, UK

Montgomery, H. Department of Medicine, University College London Medical School, The Rayne Institute, 5 University Street, London WC1E 6JJ, UK

O'Donovan, M.C. Department of Psychological Medicine, University of Wales College of Medicine, Heath Park, Cardiff CF4 4XN, UK

Ollier, W. Epidemiology Research Unit, University of Manchester Medical School, Oxford Road, Manchester M13 9PT, UK

Owen, M.J. Neuropsychiatric Genetics Unit, Tenovus Building, University of Wales College of Medicine, Heath Park, Cardiff CF4 4XN, UK

Palamand, D. Department of Medicine, University College London Medical School, The Rayne Institute, 5 University Street, London WC1E 6JJ, UK

Povey, S. MRC Human Biochemical Genetics Unit, The Galton Laboratory, University College London, Wolfson House, 4 Stephenson Way, London NW1 2HE, UK

Spanakis, E. Department of Medicine, University College London Medical School, The Rayne Institute, 5 University Street, London WC1E 6JJ, UK

Stephenson, S. University of Utah, Room 161, Cardiovascular Genetics Research Clinic, 410 Chipeta Way, Salt Lake City, UT 84108, USA

Williams, R.R. University of Utah, Room 161, Cardiovascular Genetics Research Clinic, 410 Chipeta Way, Salt Lake City, UT 84108, USA

Wu, L. University of Utah, Room 161, Cardiovascular Genetics Research Clinic, 410 Chipeta Way, Salt Lake City, UT 84108, USA

Ye, S. Wellcome Trust Centre for Human Genetics, University of Oxford, Windmill Road, Oxford OX3 7BN, UK

Abbreviations

ACE	angiotensin-converting enzyme
ADA	adenosine deaminase
ADCC	antibody-dependent cell-mediated cytotoxicity
AFBAC	affected family-based controls
AIDS	acquired immune deficiency syndrome
APZ	adipocyte fatty acid binding protein
ARMS	amplification refractory mutation system
ASO	allele-specific oligonucleotides
C&S	culture and sensitivity
C/EBP	CCAAT enhancer-binding protein
CAD	coronary artery disease
CDGE	constant denaturant gel electrophoresis
CDK	cyclin-dependent kinase
CDR	complementarity-determining region
CF	cystic fibrosis
CFTR	cystic fibrosis transmembrane regulator
CHD	coronary heart disease
COAD	chronic obstructive airways disease
CSF	cerebrospinal fluid
CT	computed tomography
D	dietary intervention
DC	diet plus cholestyranine
DGGE	denaturing gradient gel electrophoresis
DRG	dorsal root ganglion
DZ	dizygotic
EDTA	ethylenediamine tetraacetic acid
EMSA	electrophoretic mobility shift assay
eNos	endothenal nitric oxide synthase
EST	expressed sequence tags
FDB	familial defective apoB
FH	familial hypercholesterolaemia
FWR	framework residue
GGPD	glucose-6-phosphate deficiency
GIP	gastric inhibitory polypeptide
GK	Goto-Kakizaki
GLP1	glucagon-like peptide-1
GLPR	glucagon-like peptide receptor
GLUT2	glucose transporter 2
GLUT4	glucose transporter 4
GPI	glycosyl-phosphatidylinositol
GRA	glucocorticoid remediable aldosteronism
GRR	genotypic relative risk
HA	heteroduplex analysis
H-DGGE	high throughput DGGE
HDL	high-density lipoprotein
HDN	haemolytic disease of the newborn
HFFFZ	human fetal foreskin fibroblasts
HLA	human leukocyte antigen
HNF	hepatocyte nuclear factor
HRR	haplotoid relative risk
HSV-1	herpes simplex virus type 1
IDDM	insulin-dependent diabetes
IDL	intermediate density lipoprotein

INS	insulin
IRS1	insulin receptor substrate-1
K-cell	killer cell
LADA	latent autoimmune diabetes of adulthood
LDL	low density lipoprotein
LDLR	low density lipoprotein receptor
LGL	large granular lymphocytes
LOD	logarithm of the odds
MADGE	microplate array diagonal gel electrophoresis
MAWS	mean absolute width of coronary segments
ME	military exercise
MED PED -FH	medical pedigrees with familial hypercholesterolaemia
MEN-II	multiple endoneoplasia
MI	myocardial infarction
MIEHCMV	major immediate-early human cytomegalovirus
MinAWS	minimum absolute width of coronary segments
MMP	metalloproteinase
MODY	maturity-onset diabetes of the young
MTHFR	methylene tetrahydrofolate reductase
MZ	monozygotic
NF	neurofibromatosis
NIDDM	non-insulin-dependent diabetes mellitus
NK-cell	natural killer cell
NPY	neuropeptide Y
PCR	polymerase chain reaction
PDGF	platelet-derived growth factor
PMD	programmable melting display
PODGE	profiling mismatch and perfect match oligonucleotide dissociation gel electrophoresis
PPARG	peroxisome proliferator-activated receptor g
RA	restriction analysis
RED	repeat expansion detection
REG	regenerating gene
RF	rheumatoid factors
RSV	Rous sarcoma virus
SCA-1	spinocerebellar ataxia
SCID	severe combined immunodeficiency
SMC	smooth muscle cells
SSCP	single-stranded conformational polymorphism
SUR	sulphonylurea receptor
TDT	transmission disequilibrium test
TDT	transmission distortion test
TGF-b	transforming growth factor-b
TGGE	temperature gradient gel electrophoresis
TNF	tumour necrosis factor
UC	usual care
UCP	uncoupling protein
UZ	unknown zygosity
VAPSE	variations that are likely to affect protein structure or expression
VNTR	variable number of tandem repeats

Preface

University College London Medical School hosted the Fifth Annual Molecular Pathology Symposium on 10 December 1996. This book represents the contributions to that meeting made by a set of distinguished scientists and clinicians whose work pertains closely to the furtherment of our understanding of the genetic components of common diseases, and potential future approaches to their management based on this new genetic understanding. A wide spectrum of topics is covered, the first half of the book (Chapters 1–8) representing the morning sessions and devoted to the principles of analysis of genetic disease, and to specific examples of applications to diseases which are major sources of morbidity and mortality in the UK and in the acultural (Western) world in general. The second half of the book (Chapters 9–15), derived from the afternoon contributions, represents a range of approaches and potential approaches to management, based on methods and principles which utilize genetic knowledge and techniques.

The first and second halves of the meeting were introduced by keynote presentations from Professor Sue Povey and Professor Roger Williams respectively, and for principles of analysis and progress toward management respectively, both convey a fascination of the history of progress in disease genetics and also the excitement of what may be possible in the next few years. In the past decade, very many single gene disorders have been characterized by the new methods available, and for some, novel approaches such as gene therapy are already in trial. However, common diseases are much more complex in their aetiology and although they can be shown to have genetic components, progress in understanding the molecular pathogenesis has lagged far behind. Systematic progress of the Human Genome Project, and the systematic strategies applied by pharmaceutical companies in search of new therapeutics, may converge to generate a whole new stratum of preventive medicine. At the level of applied reasearch, this is already evident in the massive investments which existent pharmaceuticals, venture capitalists, new genomics companies, and even private investors are prepared to make in this field. The examples set out by the contributors to this volume illustrate clearly the potential and demonstrate the sorts of detail to be addressed in analysing pathogenesis, developing new therapeutic modalities and in implementation in the population setting. We hope that this Symposium volume will inspire others to enter this new but rapidly growing field of research and development.

We acknowledge with gratitude the diligence of our contributors in adhering to the strict schedule imposed, and the help of Dr Divya Palamand with preparing the index, to ensure a volume which is up-to-date at the time of publication.

Ian N.M. Day and Steve E. Humphries

1

Perspectives in human linkage studies

Sue Povey

Genetic linkage is found when two genes are close together on one chromosome, and do not sort themselves out randomly when passed from one generation to the next. There have been dramatic changes in perspective since Penrose, who subsequently became Galton Professor of Human Genetics here at University College London, first suggested the sib-pair method of analysis in 1935. *Table 1.1*, taken from his paper, shows part of his analysis of 50 sib-pairs for the inheritance of eye colour and the ABO blood group antigens. He proposed that if these two characters were linked those sibs alike for one character would be more likely than random to be similar for the other character as well. He suggested that this method could be helpful only in the study of common traits and also that, even in the most favourable circumstances, at least 100 sib-pairs would be required. Sixty years later the methods in use bear a striking, if not always recognized, similarity to this approach! However, in 1935 practical progress in human linkage analysis was virtually impossible, not only because of the small size of human families, but also because of the almost complete absence of suitable genetic markers. Complex mathematical methods were required to extract the maximum of information out of the minimum data, and the possibility of understanding the inheritance of any human traits which were not determined by single genes seemed extremely remote.

Genetics of Common Diseases: future therapeutic and diagnostic possibilities,
edited by I. Day and S. Humphries.

Table 1.1. Agglutinogen A and blue eyes

	Like	Unlike	Total
Like	22	11	33
Unlike	11	6	17
Total	33	17	50

Data from Penrose, 1935.

Since, until the last two or three years, the vast majority of work on genetically determined diseases in humans concentrated on those determined by single genes of major effect, it is worth considering these briefly before proceeding to the complex traits which are the main subject of this volume.

1.1 Pairwise linkage analysis of Mendelian traits

Much of the early theoretical work, developing maxium likelihood as the method of analysis, was carried out in the Galton Laboratory by J.B.S. Haldane, and later by Cedric Smith (still an active member of the Department in 1996). Linkage may be simply defined as the non-random segregation of genes from parents to children. Usually under consideration is a disease-causing gene, which is transmitted on average to half the children, and a marker gene, which the parent possesses in two alternative, normal, but distinguishable, forms. If the disease and marker genes are unlinked, a child who inherits the disease gene will have an equal chance of inheriting either form of the marker; if they are linked, an

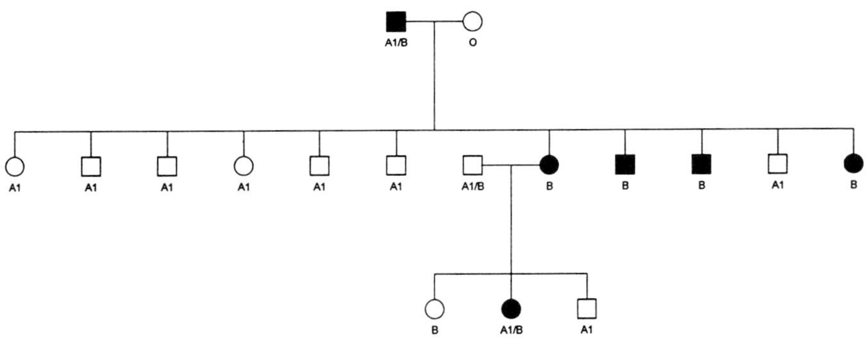

Figure 1.1. Nail-patella syndrome and ABO blood group.

affected child will have a greater likelihood of receiving one form of the marker than the other. If this is observed in a sufficient number of children the segregation is clearly non-random.

One of the earliest positive results was that of the ABO blood group linked to the nail-patella syndrome, found by Renwick and Lawler in 1955 (and a disease gene still uncloned at the time of writing). *Figure 1.1* illustrates part of one of their pedigrees, showing the segregation of nail-patella syndrome with the B allele in the family. At about that time Newton Morton introduced the concept of logarithm of the odds (LOD) scores. He suggested that families could be tested sequentially and the results added together, and that linkage could be regarded as proven when a LOD score of +3 was reached (Morton, 1955). At this point the results obtained would be 1000 times more likely if the two genes under consideration were linked than if they were segregating randomly. The part of the pedigree shown in *Figure 1.1,* considered alone, gives a LOD score of more than +3, for the hypothesis of very close linkage with no recombination, but of course most families would not be as informative as this.

With the important proviso that the mode of inheritance of the disease as a Mendelian trait must be clear before the analysis is begun, the validity of a LOD score of +3 has stood the test of time. Empirically it has been found that only 3% of LOD scores of +3 will turn out to have been false positives. The reason this figure is not 1 in 1000 is because the prior chance of any two loci being linked is about 1 in 50. (Winning the lottery is about 1000 times more likely if 1000 tickets are purchased rather than one, but is by no means a guarantee of winning because of the low prior probability!) It is interesting that if the inheritance of the disease is clearly Mendelian (i.e. there is really a gene there to find), there is probably no need to require a higher LOD score than +3 even if a large number of markers have been tested. This is because each negative result slightly increases the prior probability that the next marker to be tried will be linked.

1.2 Finding the cause of a disease which shows clear Mendelian inheritance

Only in the early 1980s was it gradually realized that linkage analysis was not an esoteric pursuit of only academic interest but might actually lead directly to the understanding of the causes of diseases, some of which had already defeated many years of biochemical investigation. The first 'disease gene' to be identified by what was at first called 'reverse genetics', but which is now known as 'positional cloning', was that causing the sex-linked form of chronic granulomatous disease, cloned in

Table 1.2. Properties of an easily mappable 'disease' gene

- Mendelian inheritance
- Homogeneous – *all families have mutation in same gene*
- Fully penetrant – *all persons carrying mutation get disease*
- No phenocopies – *all people with disease have mutation*
- Enough scorable children – *20 to prove linkage; 200 for close mapping*

1986 (Royer-Pokora *et al.*, 1984). (It is perhaps worth recording the characteristics of a 'disease gene' which make it definitely mappable, and these are shown in *Table 1.2*.)

For many years in the history of linkage analysis a very highly significant figure was the amount of person-hours per publishable result. For example, the relatively straightforward task of finding a linkage for cystic fibrosis (CF), a disease showing most of the features in *Table 1.2*, took at least 100 person-years of work, and was reported in 1985. Nowadays a similar problem could certainly be tackled by a single person in a few months at most.

Four resources can be identified as critical in this revolution and are listed in *Table 1.3*. The discovery of highly polymorphic markers which can be typed by polymerase chain reaction (PCR), mostly in the form of the microsatellites known as dinucleotide repeats which are widely distributed over the genome, was first reported by Weber (1990) and greatly extended by the group headed by Jean Weissenbach (1992). The second has been the availability worldwide of deoxyribonucleic acid (DNA) from a set of 40 reference families [the Centre d'Etudes Polymorphism Humaine (CEPH) families] chosen only on the basis of large numbers of children and available grandparents. These samples were perpetuated as permanent cell-lines. Through the far-sightedness of Jean Dausset they have been made available to many workers without

Table 1.3. Landmarks in resources for linkage analysis

- Microsatellite markers
- CEPH families
- Reliable genetic maps
- Fluorescent genotyping

any cost except the requirement that at the time of publication the primary data should be made available to CEPH. These two initiatives allowed the development of the third resource, the construction and availability of reliable genetic maps. On a global scale this was again carried out most notably by the group of Jean Weissenbach at Genethon [see Dib *et al.* (1996) for what is probably the last whole genome genetic map to appear in paper format], but many details have been resolved in workshops devoted to single chromosomes. [For summaries the reader is referred to the Genome Database at http://gdbwww.gdb.org or the United Kingdom (UK) mirror site at the Human Genome Mapping Project (HGMP) Resource Centre: http://www.hgmp.mrc.ac.uk.] The fourth resource is the technological advance in carrying out this genotyping using automated DNA sequencers and fluorescent labelling, and the development of panels of appropriate primers, pioneered in the UK by the group of John Todd in Oxford (Reed *et al.,* 1994). As a result there are several centres where 'genome scanning' (i.e. testing of about 300 evenly spaced highly informative markers in, say, 60 individuals) can be done within a few weeks and a linkage found. This approach is open to anyone with an automated DNA sequencer. However, to solve a single particular problem it may be now more efficient to collaborate with a larger centre; for example, the newly set-up Linkage Hotel at the Medical Research Council (MRC), HGMP Resource Centre, near Cambridge, UK.

1.3 After a LOD score of +3, what next?

With the resources outlined above, the finding of a positive LOD score is by no means the highest hurdle to be surmounted before the cause of the disease is identified. In pinpointing the position of the gene more precisely several approaches may be used, and two of these will be briefly discussed because of their relevance to the search for genes causing common disorders.

1.4 Narrowing the search by searching for critical recombinants

After the initial finding of linkage between a disease gene and a mapped marker, the region of interest can be further defined by utilizing recombination events within the families in which surrounding genes have separated from the disease gene by recombination. *Figure 1.2* shows the interpretation of typing of markers on 9q34 in a family with tuberous sclerosis (taken from Povey *et al.,* 1994) which allows the definition of the

TSC1 candidate region as lying in between *ABL* and *D9S298*. This type of analysis relies on the phenomenon of interference; that is, the ability of one cross-over event to inhibit other nearby events. This means that the possibility of double recombination events can be ignored over short genetic distances.

In the family shown in *Figure 1.2* there are three informative recombinant events; that is, chromosomes which have undergone recombination in the previous generation and can therefore give information about the order of genes within the region under consideration. Two of these informative chromosomes give some information about the localization of the disease; individual 2 showing that the *TSC1* is proximal to the marker *D9S298*, and, with less certainty (because of some clinical doubt in this particular person), that *TSC1* is distal to the gene *ABL*. The other recombinant event, on a chromosome inherited from the father, confirms that the marker *DBH* (the gene coding for dopamine-β-hydroxylase) is proximal to the anonymous DNA segment *D9S10*. Any of these pieces of information may be absolutely critical to the search for the gene. The existence of one recombinant individual will determine where all the effort is placed. A critical recombinant can save years of work, but a wrong interpretation can waste an enormous amount of time and money. Most investigators recheck the

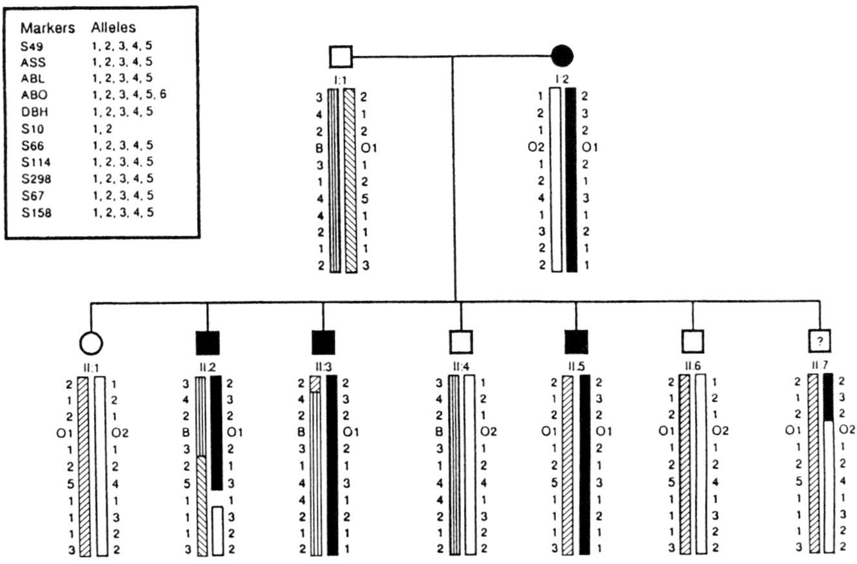

Figure 1.2. Family giving positional information on chromosome 9. Reproduced from Povey *et al.* (1994) and reproduced by kind permission of the editors of the *Annals of Human Genetics.*

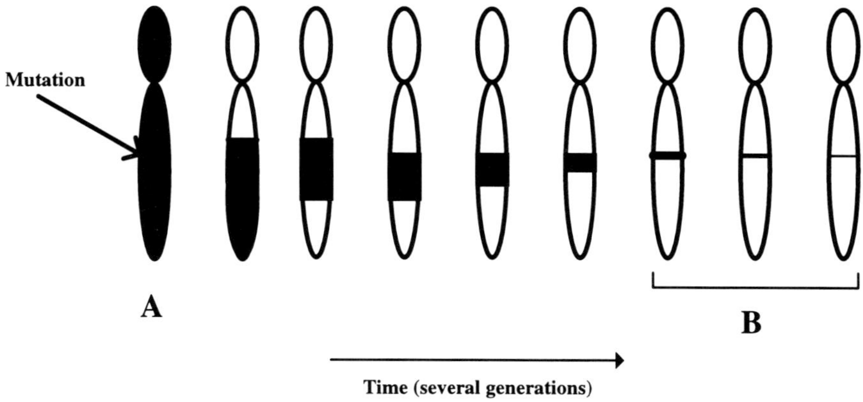

Figure 1.3. A diagrammatic view of chromosomes showing linkage disequilibrium in the affected present-day descendents (B) of a founder individual (A) in whom a mutation occurred at the position shown. After several generations of random recombination events all affected individuals still retain a small piece of the ancestral chromosome.

DNA results on their most critical recombinants; experience suggests that a hard critical look at the diagnosis at this point is also worthwhile!

Linkage, as described here, may well be successful in narrowing the region down to about 1% recombination or roughly 1 million base pairs. Most of the work in finding the genes is still to do at this point unless one of several possible short cuts is found. Perhaps in one patient the disease is caused by a large genomic rearrangement which is easily detectable. Or perhaps the Human Genome Project has already identified an excellent candidate gene in the region. This happens increasingly frequently and is indeed the whole *raison d'être* for the Human Genome Project, but is beyond the scope of this discussion of linkage. The third short cut may be the finding of linkage disequilibrium, something which needs to be considered in more detail because it is extremely relevant to the search for truly complex disorders and especially in isolated populations.

1.5 Linkage disequilibrium

It is clear that particularly for recessive disorders, and especially in isolated populations, a high proportion of cases of disease may be caused by a mutation which occurred only once and has spread as a founder effect throughout the population. *Figure 1.3* represents the idea of the inheritance of the original chromosome on which the mutation happened through many generations. The decendants who share the mutated gene will also share a part of the ancestral chromosome. The amount shared will depend on how many generations have occurred since the original mutation. Information can be obtained from all single cases as well as families, and the smallest region in which all genetic markers are the same must contain the gene. In some populations with strong founder effects, such as Finland, the candidate region for some diseases has been narrowed down to as little as 60 kb this way (Hastbacka *et al.*, 1994). The pinpointing of the causative mutation (as distinct from a normal polymorphism present by chance in all individuals who have inherited this piece of chromosome) may require mutation screening in the candidate region in patients obtained from outside the founder populations.

More than 200 monogenic disorders have now been identified from the initial starting point of linkage. In the majority of these, clues in the form of structural rearrangements, obvious candidate genes or linkage disequilibrium were available, although for some of the major efforts, for example cystic fibrosis (CF) and Huntington's disease, most of the work was done the hard way, and the value of linkage disequilibrium is seen more clearly in retrospect. One of the first genes to be cloned which was achieved entirely by hard work apparently unrelieved by any luck was that of the breast cancer susceptibility gene *BRCA1*, and it is undoubtedly significant that this included industrial partnership (i.e. serious money) (Miki *et al.*, 1994). *BRCA1* fulfils hardly any of the criteria of a mappable gene as defined in *Table 1.2* and is probably responsible for not more than 5% of cases of breast cancer. Nevertheless, because there are families in which *BRCA1* behaves as a gene of major effect and with high penetrance, the identification of *BRCA1* is the first step in understanding the genetic contribution to the whole spectrum of this common disease.

1.6 Linkage analysis in more complex situations

As with *BRCA1*, progress has been made in many situations where there is some evidence for Mendelian inheritance, but the situation is far from ideal from a linkage point of view. One of the most frequent problems is locus heterogeneity – two genes each independently able to cause clinically indistinguishable disease. This was recognized by Newton Morton (1956) who reported that some but not all forms of elliptocytosis were linked to the rhesus blood group. A statistical method for dealing with heterogeneity, the admixture test, was proposed by C.A.B. Smith in 1963 and later further developed by Jurg Ott and incorporated into the widely used HOMOG programmes (Ott, 1991). It is apparent that heterogeneity cannot be coped with simply by separating the families with positive and negative evidence for linkage, since completely unlinked families can be separated into those with small positive and small negative LOD scores and spurious high LOD scores generated. The most convincing defence against heterogeneity is to have large enough families so that each one tested alone can generate a LOD score of more than +3. Such families are of course more likely to be found in a mild condition such as elliptocytosis.

If all families are small, it is still possible to analyse the results with reference to the possibility of this being two or even three loci, with clinically indistinguishable phenotypes, each independently capable of causing the disease, but the confidence limits of this approach are inevitably very wide. For example, we have no doubt that tuberous sclerosis can be caused by *TSC1* on chromosome 9 or *TSC2* on chromosome 16, but our confidence was greatly increased by two large families, showing significant linkage respectively to chromosome 9 and to chromosome 16. Until all mutations are identified it remains difficult to exclude the existence of a third locus which may be causing disease in some of the smaller families. However, even the existence of one large positive family helps the situation dramatically, since this justifies taking all those families negative for the first region and looking in the rest of the genome. One of the most successful demonstrations of positive results in the face of heterogeneity and small families has been the application of homogyzosity mapping in consanguineous families with pseudohypoaldosteronism, where testing of only 12 affected children led to identification of loci in two different positions, each fortunately containing an excellent candidate gene (Strautnieks *et al.*, 1996).

One of the most difficult problems is any lack of certainty between the complete correlation between genotype and phenotype. If a diagnosis of affected is certain, but unaffected is less certain (another way of saying penetrance is incomplete), this can be dealt with in two ways, by regarding all unaffecteds as 'unknown' with respect to disease status or by factoring in some estimate of penetrance. The former approach is, of course, safer but loses power.

A more intractable problem is the existence of phenocopies which may be a polite name for misdiagnoses, or a recognition that the disease is not entirely monogenic and that clinically indistinguishable disease can have other causes. If this is common, as in breast cancer, the statistics are much more complicated, and the practical result is that no one piece of information (e.g. about the position of the gene) can be regarded as absolute, and that therefore, unlike the case of the truly monogenic disorders, no single piece of evidence is ever critical.

1.7 Linkage in complex traits

The current intellectual challenge to linkage is, of course, those diseases which can be recognized as running in families, but in which a clear Mendelian mode of inheritance is not present. Although in rare families there may be single genes of large effect which predispose to disease, these are swamped by the numbers in which the disease is truly multifactorial. There has been an absolute deluge of theoretical papers about approaches to looking at complex traits. [For somewhat controversial guidelines and many references, see Lander and Kruglyak (1996).] The two methods which have been tried can be divided into the linkage approach and the candidate gene approach. The work of John Todd's group on insulin dependent diabetes myelitus (IDDM) illustrates the first (Davies *et al.*, 1994).

The risk of the sib of a child with CF also having CF is about 500 times that of the general population; the risk of a sib of a child with IDDM having IDDM is about 15 times that of the general population. There is clear familial clustering. John Todd's argument was that of Penrose (1934/5). At any locus, full sibs have a 50% chance of inheriting the same allele from their parent. If a locus makes a significant contribution to susceptibility to a disease, two affected sibs will share inheritance of the same allele at that locus on more than 50% of occasions. The number of sibships required to demonstrate linkage with a significance equivalent to a LOD score of +3 depends on two factors: the increased risk of disease conferred by that allele over the risk conferred by other alleles – a figure referred to by Risch as the genotypic relative risk (GRR); and the population frequency of the allele. Recently Risch and Merikangas (1996) have calculated the number of affected sib-pairs required to establish linkage for different values of these parameters. Some of the results are shown in *Table 1.4*, and make rather depressing reading, since unless one gene has a fairly strong effect unrealistic numbers of sib-pairs are required. An additional practical problem is that although, theoretically, parental genotypes are not necessary, in practice the typing of parents dramatically improves the accuracy of the results, and to find affected sib-pairs with parents available is difficult in some diseases and impossible in others.

It is clear that this approach is not going to succeed if there are many genes each of tiny effect, especially since most studies have only started

Table 1.4. Number of families needed for identification of a disease gene (with 80% power)

		Number of families required		
Genotypic risk ratio	Frequency of disease allele	Linkage	Association	
		Sib-pairs	Singletons	Sib-pairs
(γ)	(ρ)	(N)	(N)	(N)
4.0	0.01	4 260	1 098	235
	0.10	185	150	48
	0.50	297	103	61
2.0	0.01	296 710	5 823	1 970
	0.10	5 382	695	264
	0.50	2 498	340	180
1.5	0.01	4 620 807	19 320	7 776
	0.10	67 816	2 218	941
	0.50	17 997	949	484

Data from Risch and Merikangas (1996).

with 100 sib-pairs (usually already quite hard to collect). Many studies will turn up false positives which are not repeatable.

Nevertheless the general power of the method was shown by the fact that the already known contributions of loci in the regions of the major histocompatability complex (HLA) and the insulin gene were readily detected by linkage studies and sib-pairs, and that at least some of the other loci suggested by the first study (Davies *et al.*, 1994) do indeed seem to be genuine. This type of analysis is not, of course, confined to sibs. The likelihood of identity by descent at any marker locus can be calculated for any pair of affected relatives, and, with suitable precautions, additional affected members of the kinship can be incorporated in the calculations. By the method of sib-pair analysis, doing a total genome screen and then identifying a reasonably dense set of markers in the indicated region, a gene of reasonably strong effect may be localized to within a few centimorgans (cM). Complex multipoint linkage analysis may be required to estimate the size of the candidate region, since once again no single recombination event can give definitive information.

John Todd was one of the first people bold enough to embark on a total genome screen to look for linkage in what was clearly recognized at the outset to be a complex trait (i.e. one in which no single major gene is predominant). Many other people have used an alternative approach – that of the candidate gene. Any genetically determined variation in a candidate gene can be tested to see if the distribution of this variation is different in people with and without the disease in question. Intuitively, it is more appealing to look at polymorphisms in coding regions which may have a direct effect on the function of the gene. However, within a gene a completely silent polymorphism may by chance be associated (i.e. in linkage disequilibrium) with the causative mutation. This type of study is clearly fraught with pit-falls, the most obvious of which is population stratification. For example, it is likely that in travellers on the London Underground one might detect an association between the delta 508 mutation of the CF gene, *CFTR*, and blue eyes, which would mean only that both the delta 508 mutation and blue eyes are more common in people of European origin than in people from elsewhere. The problem is therefore that of choosing exactly comparable control chromosomes. The ideal solution appears to be to use as controls those chromosomes present in the parent but not transmitted to the child with the disease. Richard Spielman has devised a test for association called the transmission distortion test (TDT) which is really this idea viewed from another direction (Spielman *et al.*, 1993). This tests a hypothesis that possession of a particular allele increases susceptibility to a disease. If this is the case any parent who is heterozygous for the susceptibility allele will transmit that allele to the affected offspring more than 50% of the time. This test can be applied to families with only a single case of the disease as long as least one parent is available. It is of course necessary to test all available parents, not to ascertain them by occurrence of the postulated susceptibility allele in the affected offspring. However, if a particular variation in a plausible candidate gene is being tested, deviation from 50% transmission will be substantial even if the gene has a relatively small effect.

Can this approach be applied to genomic screening, with no particular hypothesis in mind? The difficulty here is that of interpreting the level of significance when multiple loci have been tested; transmission distortion will certainly occur by chance for some loci. Unlike the situation of Mendelian inheritance, correction for multiple tests is required because there is no certainty that there is actually a gene with sufficient effect to be found, so that previous negative tests do not improve the prior probability of the next locus tested being the right one.

In their recent paper, Risch and Merikangas (1996) presented figures based on the daunting assumption that one has tested each parent and affected child for five polymorphisms at each of 100 000 loci. They estimated the number of families needed to achieve the same power and the same posterior probability of error as in the linkage analysis. The results are shown on the righthand side of *Table 1.4*. Clearly in this situation fewer families are required for definitive results than by the linkage approach. However, does the idea of one million tests on each sample have any place except in nightmares? There are two positive answers to these questions, both to my mind equally amazing.

The first is that, although the sequencing of the first human genome (a patchwork derived from many different individuals) continues to be a very major commitment, the identification of variation from this 'standard' will be very much easier. Other people will describe very impressive increases in the throughput of samples which can now be examined for the presence of particular variation, many thousands of bases being screened in a day. The methods described by Dr Spanakis and Dr Day (this volume) will soon be available in a relatively cheap and user-friendly form to any laboratory. In contrast, chip technology is a more distant and much more expensive prospect although its application to the sequencing of mitochondrial variation has been recently described (Chee *et al.*, 1996). Since as far as I am aware this frequently mentioned but as yet not fully evaluated methodology is not described elsewhere in this volume, a very brief description of the method used by Chee *et al.* may be appropriate here.

The first surprise is that the 'chip' referred to is not electronic, but is a piece of glass to which a very large number of oligonucleotides have been bonded. For a human sequence 15 oligomers are synthesized which represent bases 1–15, 2–16, 3–17 and so on. These are placed in a linear array on the glass. Each oligomer is also synthesized with three different base substitutions at one of the positions (say position 7 in each oligomer), so that the total number of oligonucleotides arrayed on the glass is 4 x L ('L' being the length of the sequence in bases). For the mitochondrial genome of 16 kb, 69 000 oligomers were arrayed on a glass chip measuring 1.2 cm^2.

A fluorescently labelled probe was prepared from the DNA of the person to be tested, and hybridized to the glass chip. After appropriate washing a fluorescence microscope was used to capture the image of the chip, with fluorescent labelling occurring only over the oligonucleotides for which a perfect match was found. A computer could then read off the sequence. This should be possible for any region where the 'standard' sequence is known and only occasional relatively straightforward changes are found. At present this system can scan 16 kb in an afternoon in a number of samples – not more than that achievable by other methods. Clearly the potential power is awesome. However, it remains to be seen which technology will be most useful in practice.

Another finding, which to me is at least as exciting as the wonders of technology, is that the use of linkage disequilibrium as a tool may not be confined to isolated populations, but may be perfectly applicable at least to European populations as a whole. An early suggestion of this came from John Todd's group (Copeman *et al.*, 1995) whose linkage analysis in sib-pairs had suggested a possible susceptibility locus for diabetes on chromosome 2. They noticed that a particular allele of a highly polymorphic marker was associated with diabetes, both in the study populations from Oxford and the study populations from Sardinia. There seems no likelihood that the variation *per se* is causing susceptibility to diabetes. The conclusion is that a single mutation in some ancestral European founder has not yet been completely separated by recombination from a chance nearby polymorphic allele. The region of chromosome 2 can be recognized in today's Europeans as conferring susceptibility to diabetes.

These data, in agreement with other work discussed at meetings, suggest that linkage disequilibrium may be a powerful tool with which to search for genetic contribution to complex traits. It may not be necessary initially to test the exact causative polymorphism or even another polymorphism in the same gene. A polymorphic marker 100 kb away (perhaps further) may also give a positive result. The exact distance over which disequilibrium occurs (which will certainly not be identical in all chromosome regions) has profound implications for the number of markers needed for a genome search by TDT. Only 30 000 evenly spaced markers would be needed to provide one in every hundred kb, and this is a realistic goal which is now being pursued in several large groups. There are several reasons why the polymorphisms being sought are now di-allelic ones, ideally within coding regions. One is that these genetic markers are less mutable than variable numbers of tandem repeats (VNTRs), preserving linkage disequilibrium for more generations. Another is that tests for signficance of linkage disequilibrium are much easier in two-allele than multi-allele systems. The third is that, as described above, simple sequence changes may eventually be the easiest to detect in an entirely automated way.

The extent of linkage disequilibrium has a profound effect, not only on the search for complex traits by genomic association screening, but also on the more precise pinpointing of areas suggested by linkage analysis. It seems likely that association studies, realistically probably in the form of TDT tests on those small families with only one affected member not suitable for linkage analysis, may be a powerful means of narrowing a region (as demonstrated on chromosome 2). The final proof that a commonly occurring variant is actually causally related to a disease may be difficult and controversial. The understanding of the functional significance of the mutations found, not only in isolation but in relation to genetic variation at other loci, will be the next challenge.

So what is the current state of linkage analysis in early 1997? For those diseases showing clear Mendelian inheritance, even where several independent loci may be involved, the way forward is relatively straightforward. For the commoner diseases, described as complex traits, it seems clear that we already have the basic knowledge of the human genome and human population genetics needed to identify genes of moderate effect, which increase the risk of developing a disease say four times above the population average. The problem is that before starting on such a quest it is difficult to be sure whether there is such a gene, or whether the genetic predisposition is made up of many genes each of very small effect.

1.8 Why are we doing this?

The early work on linkage was undoubtedly driven by curiosity. From one of the earliest discoveries, that of the linkage of myotonic dystrophy with the gene determining secretion of blood group substances, it was suggested that linkage might allow predictive testing. However, the possibility that linkage might lead to positional cloning of a disease-causing gene seemed a fantasy until the mid-1980s, the first actual example being achieved in 1986. In contrast the work on complex disorders has been led not from ivory towers but by clinicians and clinical scientists, each with a passionate interest in finding the causes and possible methods of cure or prevention of some specific disease affecting patients under their care. This has led to a very exciting and fast-moving situation described in more detail by many other contributors. For the 'single gene' diseases, identification of the exact cause of the disease has produced definable benefits to the families involved, although so far examples of successful therapy as a direct result are relatively few. Definitive diagnosis and predictive testing first had a dramatic effect in two diseases, thalassemia and Tay-Sachs disease, where mutations were identified by classical means. The prevalence of both these diseases in high risk populations has been substantially reduced. Currently tests for two of those identified by linkage and positional cloning, Duchenne muscular dystrophy and CF, are on the verge of having some population impact in the UK. For the complex traits, the goal is treatment, either prophylactic or therapeutic. How far this will actually translate into reality is not yet proved.

One's view of almost any mutation may depend upon one's perspective. For example, consider deficiency of glucose-6-phosphate dehydrogenase (G6PD), a mutation which gives some resistance to malaria and also produces severe illness (favism) if broad beans are eaten. In a country where malaria was prevalent, G6PD deficiency might become the norm and favism could be regarded as an example of poisoning. In a non-malarious area, heavily dependent on broad beans in the diet, favism could be regarded entirely as a genetic disease. For those genes which we are beginning to uncover as contributory to complex traits, we have not begun to understand the complexity of their interactions with each other and with the environment, and the way in which this may change in time, either in the short term or on an evolutionary timescale.

Already we can identify a few combinations of genetic variants at different loci, each one individually 'normal', which together confer some specific risk in the presence of some specific environmental challenge. But in no case are the full implications of such variation understood. Perhaps fortunately, for most of us the future seems likely to remain unknowable!

References

Chee M, Yang R, Hubbell E *et al.* (1996) Accessing genetic information with high-density DNA arrays. *Science* **274,** 610–614.

Copeman JB, Cucca F, Hearne CM *et al.* (1995) Linkage disequilibrium mapping of a type-1 diabetes susceptibility gene (*iddm7*) to chromosome 2q31-q33. *Nature Genet.* **9** (1), 80–85.

Davies JL, Kawaguchi Y, Bennett ST *et al.* (1994) A genome-wide search for human type-1 diabetes susceptibility genes. *Nature* **371** (6493), 130–136.

Dib C, Fauré S, Fizames C *et al.* (1996) A comprehensive genetic map of the human genome based on 5 264 microsatellites. *Nature* **380,** 152–154.

Hastbacka J, Delachapelle A, Mahtani MM *et al.* (1994) The diastrophic dysplasia gene encodes a novel sulfate transporter – positional cloning by fine-structure linkage disequilibrium mapping. *Cell* **78** (6), 1073–1087.

Lander ES and Kruglyak L. (1996) Genetic dissection of complex traits. *Nature Genet.* **12** (4), 357–358.

Miki Y, Swensen J, Shattuck-Eidens D *et al.* (1994) A strong candidate for the breast and ovarian cancer susceptibility gene *BRCA1*. *Science* **266,** 66–71.

Morton NE. (1955) Sequential tests for the detection of linkage. *Am. J. Hum. Genet.* **7,** 277–318.

Morton NE. (1956) The detection and estimation of linkage between the genes for elliptocytosis and the Rh blood type. *Am. J. Hum. Genet.* **8,** 80–96.

Ott J. (1991) *Analysis of Human Genetic Linkage* (revised edition). John Hopkins University Press, Baltimore.

Penrose LS. (1934/5) Detection of autosomal linkage. In: *Annals of Eugenics* (ed. RA Fisher), p.137. Cambridge University Press, London.

Povey S, Burley MW, Attwood J *et al.* (1994) Two loci for tuberous sclerosis: one on 9q34 and one on 16p13. *Ann. Hum. Genet.* **58,** 107–127.

Reed PW, Davies JL, Copeman JB *et al.* (1994) Chromosome-specific microsatellite sets for fluorescence-based, semiautomated genome mapping. *Nature Genet.* **7** (3), 390–395.

Renwick JH and Lawler SD. (1955) Genetic linkage between the ABO and nail-patella loci. *Ann. Hum. Genet.* **19,** 312–331.

Risch N and Merikangas K. (1996) The future of genetic studies of complex human diseases. *Science* **273,** 1516–1517.

Royer-Pokora B, Kunkel L, Monaco AP, Goff SC, Newburger PE, Baehner RL, Cole FS, Curnutte JT and Orkin SH. (1986) Cloning the gene for an inherited human disorder – chronic granulomatous disease – on the basis of its chromosomal location. *Nature* **322,** 32–38.

Smith CAB. (1963) Testing for heterogeneity of recombination fraction values in human genetics. *Ann. Hum. Genet.* **27,** 175–182.

Spielman RS, McGinnis RE and Ewens WJ. (1993) Transmission test for linkage disequilibrium: the insulin gene region and insulin-dependent diabetes mellitus (IDDM). *Am. J. Hum. Genet.* **52,** 506–516.

Strautnieks SS, Thompson RJ, Hanukoglu A, Dillon MJ, Hanukoglu I, Kuhnle U, Seckl J, Gardiner RM and Chung E. (1996) Localization of pseudohypoaldosteronism genes to chromosome 16p12.2-13.11 and 12p13.1-pter by homozygosity mapping. *Hum. Mol. Genet.* **5** (2), 293–299.

Weber JL. (1990) Informativeness of human (dC-dA)n.(dG-dT)n polymorphisms. *Genomics* **7,** 524–530.

Weissenbach J, Gyapay G, Dib C, Vignal A, Morissette J, Millasseau P, Vaysseix G and Lathrop M. (1992) A second-generation linkage map of the human genome. *Nature* **359,** 794–801.

2

Twin research: nature versus nurture in common diseases

K.O. Kyvik

2.1 Introduction

Many common diseases tend to run in families, and this clustering of disease is often attributed to the genes shared within families. In many cases, especially if the disease studied involves both genetic and environmental etiological factors (often called multifactorial inheritance), this could just as well be caused by the shared environment within the family. The study of twin pairs has for many years played a major role in trying to distinguish between the importance of genetic and environmental factors for diseases as well as normal traits, taking advantage of the natural experiment created by the formation of two different types of twins, identical and fraternal.

2.2 History

Twin studies have a long-standing history, and their introduction into research is usually credited to Sir Francis Galton who, in 1876, wrote:

> "The objection to statistical evidence in proof of the inheritance of peculiar faculties has always been: 'The persons whom you compare may have lived under similar social conditions and have had similar advantages of education, but such prominent conditions are only a small part of those that determine the future of each man's life. It is to trifling accidental circumstances that the

Genetics of Common Diseases: future therapeutic and diagnostic possibilities,
edited by I. Day and S. Humphries.

> bent of his disposition and his success are mainly due, and these you leave wholly out of account – in fact, they do not admit of being tabulated, and therefore your statistics, however plausible at first sight, are really of very little use.' No method of inquiry which I have previously been able to carry out – and I have tried many methods – is wholly free from this objection. I have therefore attacked the problem from the opposite side, seeking for some new method by which it would be possible to weigh in just scales the *effects of Nature and Nurture*, and to ascertain their respective shares in framing the disposition and intellectual ability of men. *The life-history of twins supplies what I want.*" (Galton, 1876)

It took almost 50 more years, though, before the classical twin method of comparing identical or monozygotic (MZ) and fraternal or dizygotic (DZ) twin pairs in order to establish evidence of genetic determination, was put to use by Merriman in a behavioural twin study of intelligence (Merriman, 1924). This study demonstrated a higher correlation of intelligence quotient (IQ), in MZ twin pairs than in DZ twin pairs, even considering the fact that only the researchers' subjective judgement of physical similarity was used for zygosity determination in the small sample of twin pairs.

A sound foundation for the twin methodology was further developed by Siemens (1924) who showed that: (i) twin series of a suitable size, making investigation of normal variability possible, can be gathered (e.g. by means of schools); (ii) Reliable zygosity of twin pairs can be established by combining a number of criteria instead of using just a single criterion; (iii) DZ as well as MZ twins have to be examined since they are, like MZ twins, born at the same time and exposed to the same environment.

2.3 Biology of twinning

MZ twin pairs derive from a single fertilized ovum that divides and develops into two individuals. These two individuals are for most practical purposes genetically identical. Exceptions are mutations taking place after the splitting of the ovum or different patterns of X inactivation in female MZ twins. One-third of these MZ pairs have totally separate amnions and chorions, and placenta may be either separate or fused depending on the distance of implantation in the uterus. Two-thirds of MZ twin pairs have common chorion and placenta. Of this last group, 1–4% have both a common chorion and amnion. The etiology of MZ twinning is still unknown, but suggestions include developmental retardation due to delayed implantation and lack of oxygen (Bulmer, 1970), rupture of the pellucid zone or separation of a group of cells recognized as alien to the zygote after some kind of genetic change (Hall, 1996).

DZ twin pairs result from the ovulation, fertilization and implantation of two eggs, and these twins share their genes to the same extent as ordinary siblings (i.e. 50%). These twin pairs always have separate amnions and chorions, but placenta may be fused if the two zygotes implant close to each other. DZ twinning is known to be related to maternal age, parity and height (MacGillivray *et al.*, 1988) and to genetic effects (Meulemans *et al.*, 1994).

2.4 Demography

In populations of Caucasian origin approximately one in every 100 births is a twin birth. Of these roughly one-third are MZ twin pairs and two-thirds are DZ twin pairs. The MZ twinning rate had been thought to have been stable, but has for unknown reasons been increasing for some years; for example, in Denmark (Kyvik *et al.*, 1995a). The DZ twinning rate has been known to be decreasing since about 1950. This has been partly explained by a maternal age effect. Although the mean age of first pregnancy has increased, very few women nowadays continue to have children into their forties. It might also be ascribed to the widespread use of contraception which limits the number of children and thus the significance of biological fecundity for reproduction.

Ultrasound studies have shown that twin conceptions are much more prevalent than twin births, and that over 50% of twin pregnancies are lost or converted to singleton pregnancies because of the vanishing of one twin.

The new reproductive techniques are known to induce multiple pregnancies, mostly DZ twin pairs, but MZ twinning is also known to result from *in vitro* fertilization.

2.5 Twin studies

2.5.1 The classical twin study

As mentioned above, DZ twin pairs share approximately 50% of their genes, like ordinary siblings, and MZ twin pairs are genetically identical. Both types of pair share intra-pairwise environment to the same degree despite the two-fold difference in genetic similarity. The basic concepts of twin methodology are thus: (i) discordance in MZ pairs can be attributed to environmental effects; and (ii) a greater phenotypic similarity among MZ than DZ twins must be caused by the greater genetic similarity, that is, the trait or disease under study is influenced by genetic factors. This is the basic concept of what is usually called the classical twin study.

There are three very important assumptions underlying the concept of the classical twin study: the first is the assumption of intra-pairwise equal environments in MZ and DZ twin pairs; the second is the assumption that twins are representative of the general population; the third is that we can establish zygosity in a reliable way.

2.6 Assumptions

2.6.1 Equal environments

This assumption is of course vital for the classical twin study. If it is not correct (i.e. if MZ twin pairs share equal environment to a greater degree than DZ twin pairs), then a greater similarity among MZ than DZ twin pairs cannot be ascribed to the twofold greater genetic similarity, but has to be due also to the greater intra-pairwise environmental similarity.

For many important variables the assumption is correct. Both types of twins are of the same age and sex (if only same-sexed pairs are studied), they have shared the same uterus at the same time and thus shared the same exposures from the mother (e.g. her infections, smoking, drinking or exposures related to her work). Furthermore, they grow up at exactly the same time.

The most common criticism is that parents and family tend to treat MZ twin pairs more similarly, but it may just as easily be argued that they tend to accentuate differences in order to be able to distinguish between the twins. This criticism of a more similar treatment of MZ twin pairs is probably most vital for twin studies of behavioural traits, and it has been tested in studies of, for example, cognitive and personality measures in twin pairs where both the parents and the twins themselves were unaware of the real zygosity. These studies have shown that real zygosity is more important for behaviour than perceived zygosity (Plomin *et al.*, 1990).

As the equal environment assumption is valid in behavioural studies, there is no reason to suspect that it is not so in studies of somatic diseases and traits.

2.6.2 Representativeness

The concept of being able to generalize the results from the twin study to the general population is important. Twins tend: to be born 3–4 weeks before term; to be on average 1000 g lighter than singletons, and to suffer a greater risk of birth complications and malformations. Twin studies involving these issues cannot be generalized to singletons.

Before presuming that results from twin studies can be generalized, it is therefore important to check that means and variances of continuous measures in twins are the same as in the general population.

For binary traits, it should be checked that the prevalence is the same in twins as in the general population. Finally, there is a tendency for MZ twin pairs and female twin pairs to be more willing to volunteer than DZ and male twin pairs. For this reason it is important for the researcher to check for an effect of zygosity and, if necessary, also sex in the analysis (e.g. by checking that means and variances in the groups are the same).

In recent years the representativeness of twin studies has been criticized on the basis of the so-called adverse intra-uterine environment twin pairs are exposed to (Phillips, 1993). If this criticism is correct, one would expect twins to have a different pattern of morbidity and mortality than the general population. Studies of mortality and a number of diseases, among them insulin-dependent diabetes mellitus, in Danish twins has found that this is not the case (Kyvik *et al.*, 1995b). Twins have the same prevalence of diseases and the same mortality as the general population (Christensen *et al.*, 1995).

2.6.3 Zygosity determination

In large twin surveys such as questionnaire studies, a diagnosis of zygosity can be based on the twins' answers to a panel of standardized questions (see below) relating to similarity. Based on the answers to these questions, zygosity can be established.

It has earlier been shown that being as like as "two peas in a pod" will classify most MZ twin pairs correctly. All twins answering yes to this question, to at least one of the questions on being mistaken and to having same eye and hair colour can be classified as MZ. Twins answering yes to being as alike as ordinary siblings, to both questions of being mistaken and to having same eye and hair colour can also be classified as MZ. The rest can be classified as DZ, and all inconsistent answers, or disagreeing answers between co-twins, implies classification as being of unknown zygosity (UZ).

The validity of this method is over 90–95% in all ages [summarized in Kyvik *et al.* (1995a)]. In twin pairs where only one co-twin responds, the zygosity is based on this one answer which is also regarded as a reliable method.

2.6.4 Questions of similarity

While 5% misclassification might be acceptable in twin studies involving several hundred pairs of twins, it is certainly not acceptable in studies with few pairs. A method involving the analysing of several blood and enzyme type systems has been used for decades. In the Danish Twin Register, a combination of 11 of these systems has been used.

Twin pairs showing complete concordance for all systems are regarded as MZ pairs, and it has been calculated that the frequency of misclassification is below 1% for DZ twin pairs (Hauge, 1981).

Are/were you and your twins as like as:	☐ Two peas in a pod ☐ Ordinary sibling ☐ Not at all ☐ Do not remember
Are/were you mistaken by family and friends?	☐ Yes ☐ No ☐ Do not remember
Are/were you mistaken by teachers and schoolmates?	☐ Yes ☐ No ☐ Do not remember
Do/did you have the same eye and hair colour?	☐ Yes ☐ No ☐ Do not remember

With the development of the DNA technology, molecular genetic markers are increasingly used also for this purpose and are an excellent tool for zygosity diagnostics, diminishing the risk of misclassification of DZ pairs to negligible.

2.7 The concept of concordance rates

When studying discontinuous traits, such as most common diseases, the degree of concordance between MZ and DZ twins are compared. This measure is by tradition called the concordance rate, although it is in fact a proportion. When talking about concordance rates, many researchers inexperienced in twin studies quite simply mean the proportion or percentage of concordant pairs. The concordance rates are in fact measures of the probability or risk of being diseased, and there are three different types of rate.

In order to understand the concept of concordance rates, it is important first to discuss the methodology of ascertaining twin pairs. There are three different approaches to this.

(i) *Case reports.* Descriptions of MZ, concordant or discordant twin pairs have been abundant in the literature, mostly for the curiosity value. In fact, there is no scientific value in the reports of the concordant pairs, only in the discordant ones (i.e. discordant twin pairs show that a disease or trait is *not* attributable to genetic effects alone).

(ii) *Disease specific twin samples.* This method of sampling twin pairs has been in widespread use in countries with no centralized registration of the population. Sampling of diseased twins has been carried out by means of, for example, advertizing or by contact with hospitals, out-patient clinics, patient organizations; their co-twins are examined to determine whether they are also affected. It is often possible by this approach to gather large samples of diseased twins, but the representativeness of such a cohort is doubtful for several reasons: such twin cohorts usually have an under representation of DZ twin pairs, due to the fact that these pairs often regard themselves as being of less value scientifically than the MZ pairs; furthermore, concordant pairs are over represented as they have a double chance of being ascertained.

(iii) *Population-based twin registries.* In these registries, total twin populations are sampled and screened for traits and diseases. A good registration of the total population is necessary in order to achieve the means of sampling all twins, and this approach has mostly been used in the Scandinavian countries. In Denmark, a population-based twin register has existed since 1954. From 1997 it will cover most Danish twins born between 1870 and 1992 (inclusive) of whom most have been contacted at least once (Kyvik *et al.*, 1996).

Modified versions of population-based twin registers also exist [e.g. all twins born in one specific hospital or all twins serving in an army (Hrubec *et al.*, 1978)].

It follows from the above that ascertainment of twin pairs for a study can never be expected to be complete. With the first two approaches, only twin pairs with at least one affected partner will be sampled and, even in a population-based sample, it is impossible to know for sure that all affected pairs are sampled.

Pairs sampled will be of the following three types:

(i) Concordant pairs, both partners ascertained independently (i.e. both partner probands) = C1.
(ii) Concordant pairs, only one partner ascertained independently (i.e. one proband and one secondary case in each pair) = C2.
(iii) Discordant pairs = D.
(iv) Unaffected pairs (not used for estimating concordance rates) = U.

2.7.1 Types of concordance rate

The first concordance rate estimated is usually the *pairwise concordance rate*, which is the probability that both partners in a pair are affected given that one partner is.

Pairwise concordance rate:

$$P_P = \frac{C1 + C2}{C1 + C2 + D} \quad \text{or} \quad \frac{C1}{C1 + D}$$

This concordance rate is often used for comparing MZ and DZ rates, and it is what most people mean when talking about concordance rates. This rate cannot be used for predicting risk on an individual level; that is, the healthy twin in an affected pair cannot be informed about his/her individual risk. As a result this rate or probability cannot be compared to risk estimates for other individuals (e.g. the general population or ordinary siblings).

In order to be able to do this, it is necessary to estimate a concordance rate where each affected twin individual is counted. If there is complete ascertainment the *casewise concordance rate* would estimate the population probability of a twin partner being diseased given that his/her partner is.

Casewise concordance rate:

$$P_C = \frac{2C1 + 2C2}{2C1 + 2C2 + D} \quad \text{or} \quad \frac{2C1}{2C1 + D}$$

If we cannot be sure of complete ascertainment in the twin sample, and especially if all affected twin partners have not been ascertained independently, the probandwise concordance rate, in which only partners of independently ascertained individuals are counted, should be used. This rate is on estimate of the population casewise rate under incomplete ascertainment and gives the risk that a twin individual is affected if his/her twin partner is. Due to its robustness of incomplete ascertainment, it is comparable from study to study, and can be used to produce estimates for other individuals (e.g. the general population or ordinary siblings).

Probandwise concordance rate:

$$P_{Pr} = \frac{2C1 + C2}{2C1 + 2C2 + D}$$

It follows that the choice of concordance rate depends on: (i) which research question is being asked (i.e. which risk measure is interesting); (ii) what the degree of ascertainment is. Since ascertainment will mostly be incomplete, the *probandwise rate* is the correct one to use.

In cross-sectional studies, diseases with varying age of onset are a problem. This can be overcome by estimating the probandwise concordance by means of survival analysis (e.g. Kaplan Meier or actuarial analysis). This approach is based on an analysis of risk time in partners of probands, meaning that twin pairs with two probands count twice. In pairs with a secondary case and a proband, only the risk time of the secondary case counts. In pairs with a proband and a healthy partner, the healthy partner counts as being 'at risk' and the proband does not count. This way it is possible to get an estimate of the risk in individual twins from birth to the last observed age and this is directly comparable to risk figures for the population or other family member (i.e. first or second degree relatives). It is also possible to do the analysis based on discordance time in each pair and thus estimate the risk of the second twin given that the first one is diseased. However, this risk cannot be compared with other risk measures like the probandwise approach.

2.8 Heritability

Heritability is widely used in quantitative and behavioural genetics and has spread also to twin studies of binary traits. It is based on the concept that the total phenotypic variance is a sum of the genotypic and environmental variances.

$$V_P = V_A + V_D + V_C + V_E$$

where

V_A is the part of the genotypic variance that can be attributed to an additive genetic component.

V_D is the part of the genotypic variance that can be attributed to a non-additive genetic component (i.e. dominance and epistatic effects).

V_C is the part of the environmental variance that can be attributed to effects of a common environment.

V_E is the variance attributed to unique environment. This could also include measurement variance.

Heritability is the proportion of the total variance that is attributable to genotypic variance, either in the broad sense:

$$h^2 = V_G / V_P$$

Where V_G is the total genetic variance, i.e. including both additive and non-additive genetic effects.

Or in the narrow sense:

$$h^2 = V_A / V_P$$

It follows from the above that heritability is a measure of what causes variation in populations. It is not a measure that refers to individuals; that is, it does not tell us how much of a person's score can be attributed to genes. Furthermore, as it depends on both genotypic and environmental variance, it is not constant and cannot be extrapolated from one population to another. Finally, it is valid only if there is no gene–environment correlation or interaction.

Heritability has been extended to cover categorical or binary traits by defining a liability model, in which it is presumed that everyone in the population has a liability to disease. This liability is presumed to be normally distributed, and, on a scale of liability, people have different thresholds for being diseased depending on their relationship to probands with the disease (i.e. the more genes in common, the lower threshold). Heritability is then estimated based on the known patterns of genetic and environmental associations in twins and the observed concordance of the disease, based on the same equations as above. For such a trait, heritability is thus not the proportion of disease caused by genes, but simply the proportion of variance in liability that can be attributed to genetic effects relative to the total phenotypic variance.

An alternative approach to heritability is to use twin pair materials for in-depth studies of concordance according to zygosity and presence of environmental markers and susceptibility genes.

Model fitting procedures (e.g. multiple regression or structural equation modelling) are widely used to estimate heritability and to try to identify the best model of the etiology of the disease or trait under study (Neale and Cardon, 1990).

2.9 Twin studies and common diseases

When studying common disease in twin pairs some of the issues raised above must be kept in mind. The first is that both MZ and DZ twin pairs should be studied as no conclusion can be drawn about the importance of genetic factor based on MZ twin pairs only. Infectious diseases is a good example, in which the concordance rate among MZ twin pairs might be very high (almost 100% for measles), but it turns out to be just as high in DZ twin pairs. Thus genetic factors are not important. The second issue concerns diseases that might be influenced by the twin status in itself; for example, malformations or diseases that can be ascribed to birth complications.

Twin studies have established the significance of genetic factors in a number of common diseases, such as coronary heart disease (Berg, 1983) and psychiatric diseases such as schizophrenia and manic depressive psychosis (Bertelsen *et al.*, 1977; Fischer, 1973). A study of insulin-dependent diabetes mellitus has shown a recurrence risk in MZ twins up to age 40 years of nearly 70% and for DZ twins of approximately 12%, compared to a risk of 6–7% for ordinary first-degree relatives up to the same age in the same population. This shows that genetic effects are certainly important, but also that shared environment may contribute to etiology. Furthermore, this study showed a much higher recurrence risk in MZ twins than in HLA-identical siblings up to the same age, indicating that genetic factors other than HLA are important for this disease (Kyvik *et al.*, 1995b). This last finding is also consistent with identification of other risk loci such as the insulin gene and several other chromosomal sites (Bennett and Todd, 1996). Contrary to this, for non-insulin-dependent diabetes mellitus (NIDDM), a twin study has shown that genetic factors are of less importance for the disease itself than hitherto believed, thus drawing attention to the precursor states of the disease instead [e.g. insulin resistance and β-cell-dysfunction] (Poulsen *et al.*, 1996).

2.10 Other types of twin studies

2.10.1 Twin–family studies

The study of families of twins may be regarded as a usual family study and used in linkage and segregation analysis. The other two approaches to twin–family studies are based on families of MZ twin pairs.

The first and most classical approach takes advantage of the fact that, despite the total genetic similarity, some MZ twin pairs are concordant and others discordant for disease. This fact can be accounted for either by non-genetic factors influencing the manifestation of disease or by etiological heterogeneity (i.e. the disease can be found in an inherited or a non-inherited form). A similar recurrence risk in relatives of discordant and concordant MZ twin pairs is an indication that the discordance is caused by non-genetic factors, while a lower risk in relatives of discordant pairs suggests that discordance is due to sporadic non-inherited cases. This classical approach has been used in studies of psychiatric disorder (Vogel *et al.*, 1997).

The second approach, which has mostly been used in behavioural genetics, is based on the interesting genetic structures occurring when both individuals within an MZ twin pair have children (Plomin *et al.*, 1990). In this case, there is the same genetic relationship between the twin individual and his/her own offspring, as between him/her and those of his/her twin partner. Furthermore, the genetic cousin relationship is similar to the genetic half-sibling relationship, but the offspring are reared in two different families. By comparing the children of female MZ twin pairs with male MZ twin pairs, maternal effects and sex linkage can be detected, since these would tend to make the children of female pairs more identical than the children of male pairs. An important assumption of this type of study is of course that MZ twin pairs do not set up equal home environments – for example, as a consequence of their genetic similarity. It is possible in the analysis to handle the problem, by using path analysis.

2.10.2 Twin–control studies

Twin–control studies fall into several categories (Plomin *et al.*, 1990; Vogel *et al.*, 1997). The one most used to study common diseases until now is the study of environmental exposures in concordant versus discordant MZ twin pairs. This is directly comparable with the classic epidemiological case–control study, but with perfect matching on genetic factors. It can be used also as an experimental design to test aptitude for physical or mental training or to test therapeutic measures.

The twin–control study may be extended with DZ twin pairs, who do share some potentially important environmental factors to the same degree as MZ twin pairs. Such factors might be age and sex as well as maternal age, height, weight and parity. A twin–control study drawing on the Danish twin register and relating the occurrence of antibodies to IDDM, insulin auto-antibodies, islet-cell antibodies and GAD antibodies, has shown the same prevalence of antibodies in MZ and DZ healthy twin partners of diabetic twins despite the much higher recurrence risk of MZ twins. Furthermore, antibodies were more prevalent among these healthy twin partners than among ordinary first-degree relatives. This means not only that the occurrence of antibodies is environmentally determined, but also that it is possibly an environmental factor acting in a period of shared environment [e.g. during pregnancy or very early childhood (unpublished manuscript)] (Petersen *et al.*, 1997).

The twin–control study can also be used for studies of specific genetic markers; for example, by comparing whether the same markers or combinations of markers occur in concordant and discordant MZ twin pairs. Furthermore, DZ twin pairs can be used in allele-sharing studies or, if they are concordant, in affected sib-pair studies.

Twin–control studies generally require reasonably large sample sizes and, if the disease under study is rare, this might be a problem.

2.10.3 Studies on separated twins

The classical twin study and the adoption study, which were introduced in the same year (1924), in many ways complement each other. Both aim at separating the importance of genetic and environmental factors for the etiology of disease and normal traits. Studies on separated MZ twin pairs reared in different environments combine these two types of study and are a powerful tool in separating these effects. However, the number of MZ twin pairs reared apart is rather small, and this approach is therefore confined to the study of quantitative or very common traits. It has so far been limited to behavioural and psychiatric research (Plomin, 1990). Currently, there are ongoing studies of separated twins in USA, Sweden and Finland adding to the number of these twin pair studies (Plomin *et al.*, 1990).

2.11 Conclusion

The study of twins is a powerful tool for the identification of genetic and environmental contributions to common diseases. They take advantage of the experimental situation created by nature, by the fact that MZ twin pairs are twice as similar from a genetic point of view as DZ twin pairs. If genetic factors are important for the development of a particular disease, MZ twin pairs should be more concordant for the disease than DZ twin pairs.

A number of assumptions underlie these studies, especially the classical twin study, and these have given rise to criticism of the twin design. However, none of them is important enough to justify that twin studies should be abandoned.

As stated above the methods of ascertainment of twins and the sample sizes are basic problems to consider in twin research. It is important to use the right estimate of concordance when carrying out twin studies, and furthermore, to be careful when interpreting estimates of heritability. However, their estimates of heritability of common disease traits available from twins are among the best available, and they inform the molecular researcher of the magnitude of genetic component she/he is ultimately seeking to unravel.

References

Bennett ST and Todd JA. (1996) Human type 1 diabetes and the insulin gene, principles of mapping polygenes. *Annu. Rev. Genet.* **30**, 343–370.

Berg K. (1983) Genetics of coronary heart disease. *Prog. Med. Genet. (New Series)*, **5**, 35–90.

Bertelsen A, Harvald B and Hauge M. (1977) A Danish study of manic-depressive disorders. *Br. J. Psychiat.* **139,** 330–351.

Bulmer MG. (1970) *The Biology of Twinning in Man*. Clarendon Press, Oxford.

Christensen K, Vaupel J, Holm NV and Yashin AI. (1995) Twin mortality after age six: fetal origin hypothesis versus twin method. *Br. Med. J.* **310**, 432–436.

Fischer M. (1973) *Genetic and Environmental Factors in Schizophrenia, A Study of Schizophrenic Twins and their Families.* Munksgaard, Copenhagen.

Galton F. (1876) The history of twins, as a criterion of the relative powers of nature and nurture. *J. Roy. Anthr. Inst.* **5**, 391–406.

Hall JG. (1996) Embryologic development and monozygotic twinning. *Acta Genet. Med. Gemellol.* **45**, 53–57.

Hauge M. (1981) The Danish twin register. In: *Prospective Longitudinal Research* (eds SA Mednich, AE Baert and BP Bachmann), pp. 217–222. Oxford Medical Publications, Oxford.

Hrubec Z and Neel JV. (1978) The National Academy of Sciences – National Research Council twin registry, ten years of operation. In: *Twin Research, Proceedings of the Second International Congress of Twin Studies, Part B, Biology and Epidemiology* (ed. WE Nance), pp. 153–172. Alan R. Liss, New York.

Kyvik KO, Green A and Beck-Nielsen H. (1995a) The new Danish twin register. Establishment and analysis of twinning rates. *Int. J. Epidemiol.* **24**, 589–596.

Kyvik KO, Green A and Beck-Nielsen H. (1995b) Concordance rates of insulin-dependent diabetes mellitus, a population-based study of young Danish twins. *Br. Med. J.* **311**, 913–917.

Kyvik KO, Christensen K, Skytthe A, Harvald B and Holm NV. (1996) The Danish twin register. *Dan. Med. Bull.* **43**, 467–470.

MacGillivray I, Samphier M and Little J. (1988) Factors affecting twinning. In: *Twinning and Twins* (eds I MacGillivray, DM Campbell and B Thompson), pp. 67–92. John Wiley & Sons, Chichester.

Merriman C. (1924) The intellectual resemblance of twins. *Psychol. Monog.* **33**, 1–58.

Meulemans WJ, Lewis CM, Boomsma DI, Derom CA, VandenBerghe H, Orlebeke JF, Vlietinck RF and Derom RM. (1996). Genetic modelling of dizygotic twinning in pedigrees of spontaneous dizygotic twins. *Am. J. Med. Genet.* **61**(3), 258–263.

Neale MC and Cardon LR. (1990) *Methodology for Studies of Twins and Families.* Kluwer Academic, Dordrecht.

Petersen JS, Kyvik KO, Bingley PJ, Gale EAM, Dyrberg T and Beck-Nielsen H. (1997). Islet cell autoantibodies in monozygotic and dizygotic Danish twin pairs with IDDM: evidence for environmental exposure early in life leading to islet cell autoimmunity. *Br. Med. J.* (in press).

Phillips DIW. (1993) Twin studies in medical research: can they tell us whether diseases are genetically determined? *Lancet*, **341**, 10080–10089.

Plomin R, DeFries JC andMcClearn GE. (1990) *Behavioral Genetics, A Primer.* WH Freeman, New York.

Poulsen P, Kyvik KO, Vaag A and Beck-Nielsen H. (1996) Concordance rates for non insulin-dependent diabetes mellitus. A population based study of Danish twins. *Diabetologia* **39** (suppl. 1) A15.

Siemens HW. (1924) *Die Zwillingspathologie.* Springer, Berlin.

Vogel F and Motulsky AG. (1997) *Human Genetics, Problems and Approaches,* 3rd Edn. Springer, Berlin.

3

The molecular basis of genetic variation: mutation detection methodologies and limitations

Emmanuel Spanakis and Ian N.M. Day

3.1 Introduction

Genetic variation is naturally occurring variation in either the number or the primary structure (nucleotide sequence) of the DNA molecules (chromosomes) that constitute the genetic material of a species. Some terminally differentiated somatic cells are naturally polyploid. Otherwise, variation in the number of chromosomes is the result of abnormal replication and/or segregation of chromosomes during cell division. Aberrant chromosome numbers are very frequent in transformed cancer cells and in laboratory cell lines. In contrast, very few chromosome-number abnormalities have been described in humans. Trisomy 21 is a common example. The reason seems to be that loss or gain of an entire chromosome is usually lethal at a very early stage of embryonic development, due to the massive number of functions affected at the cell and organ level. Gross chromosome aberations are detected by cytogenetic techniques. Here, we are more concerned about the nature and the detection of variation in the primary structure of the DNA. This may alter by a number of molecular mechanisms, including non-homologous recombination, which are collectively called 'mutations'. Mutations are classified in the diagram in *Figure 3.1*.

It is important to distinguish the terms 'mutation rate' and 'gene frequency'. Mutation rate is the reciprocal of the number of cell divisions

Genetics of Common Diseases: future therapeutic and diagnostic possibilities,
edited by I. Day and S. Humphries.

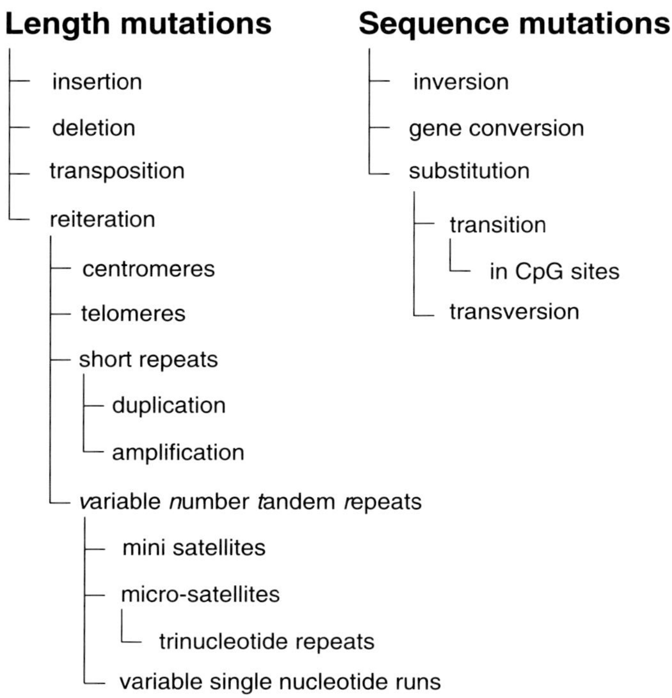

Figure 3.1. A classification of mutations.

needed for a particular mutation to occur at a particular locus (Drake, 1992; Sommer and Ketterling, 1994). Gene, or allele, frequency is the frequency of a particular variant in a population. Mutation rate depends on purely biochemical and biophysical factors, such as the type of mutation, the complexity of the sequence and of the mechanism involved, the amount of energy required and/or the efficiency of repair; it refers to the causes of *de novo* mutations. Gene frequency depends on relative survival and reproduction of carrier and non-carrier subpopulations. In turn, these parameters are dependent on the phenotypic effects, rather than the causes, of a mutation as well as on random factors (Aquadro, 1992).

3.2 Current theories of mutation

3.2.1 Causes of mutations

Stochastic mutation. Mutation is generally thought to be a random process of erroneous DNA replication. Due to interference from chemical and physical micro-environmental factors (mutagenic substances, ultraviolet light, radiation, etc.), the probability of misincorporation or mispairing of nucleotides in newly synthesized DNA molecules is so high that several mutations occur in every cell at each division. Cells have, however, complex DNA-repair enzymatic systems that recognize and replace mispaired nucleotides (Heywood and Burke, 1990; Wevrick and Buchwald, 1993). Uncorrected errors may render a cell unable to survive or divide. Otherwise, such errors become the norm in the genetic material of progeny cells (somatic mutations). A common type of human disease due to somatic mutations is cancer (Umar and Kunkel, 1996). Mutations occurring in germ cells (cells differentiating into gametes) may pass over to an individual's progeny (Sommer, 1995). The fate of a *de novo* mutation in a population depends on the reproductive performance of the carrier individual(s) and on chance. Without mutation there would be no evolution. Unfortunately, this is about the only statement population genetic theorists unanimously accept. How much mutation does evolution require, and how much can a species afford, are still unanswered questions.

Non-random mutation. With the development of mutation detection methods, geneticists can study and measure the relative occurrence of an ever-increasing number of mutations, and the evidence for a non-random distribution of mutations is accumulating (Boulikas, 1992). Some loci remain unchanged (conserved) for long periods of evolutionary time whereas some loci present intense mutability (Burns and Surridge, 1990; Jolly *et al.*, 1996); these are called mutational hot spots. Some genes mutate more frequently than others (Glickman *et al.*, 1994). Intergenic sequences are more variable than intragenic sequences (Kissi et al., 1995) and transcribed sequences are less variable than non-transcribed ones. These inequalities are usually explained by the theory of natural selection (Moxon *et al.*, 1994) but transcription-dependent repair mechanisms have also been suggested and documented (Selby and Sancar, 1994).

Some types of mutation occur more, or less, frequently than expected by chance alone. For example, transition (replacement of a purine by another purine, or of a pyrimidine by another pyrimidine) is more common than transversion (replacement of a purine by a pyrimidine, or *vice versa*) although the prior probability of a transversion is twice as much as the prior probability of a transition (Kumar, 1996). CpG sites are genetically unstable and TpG or CpA mutants are found at exceptionally high frequences in vertebrates. This mutability is associated with methylation, a natural chemical modification of DNA that is also involved in DNA repair and regulation of transcription (Hsieh *et al.*, 1992). 'CpG-islands', however, are methyl-free loci that contain many highy conserved CpG sites (Cross and Bird, 1995).

Mutation rates are probably also influenced by chemical modification of DNA other than methylation. Alkylation, for example, seems to be a pre-mutational event (Sanderson and Shield, 1996). Substances that recognize and modify specific DNA sequences have been discovered or rationally designed (Broggini *et al.*, 1995). Other chemicals, like mitomycin, may not really recognize specific sequences but produce non-random lesions (Basu *et al.*, 1993; Mercado and Tomasz, 1977). Sequence-, site- or nucleotide-specific modifications suggest a non-random stepwise mechanism of mutation under many possiple influences, including a number of recognized modulating genes (mutators) and a range of chemical mutagens.

Non-homologous recombination as well as other mutational phenomena can also be sequence-specific and can be modulated by genetic alterations in the enzymes involved. A number of extra-chromosomal genetic elements, like insertion elements, invertrons, transposons, retrotransposons *et cetera* (Amariglio and Rechavi, 1993; Finnegan, 1989; Sakaguchi, 1990) can mediate non-homologous sequence recombination within a chromosome, as well as between chromosomes, cells or species (McDonald, 1993). Units of tandemly repeated DNA are prone to insertion or deletion by unequal recombination (Epplen, 1988; Okada, 1991). This is probably a major mechanism creating minisatelite polymorphisms (VNTR).

Gene conversion (Schimenti, 1994) represents non-reciprocal transfer of sequences between non-homologous regions due to mispairing followed by mismatch repair. Mispairing of interspersed repeats can also produce major rearrangements such as deletions or duplications of sequences with little homology. The *Alu* repeat units occur at approximate intervals of 4 kb throughout the human genome and are seen as recombination hot spots (Schmid, 1996). *Alu* repeats occur even within genes and can cause intragenic deletions or duplications. Some large genes contain many *Alu* repeats and are particulary predisposed to these types of mutation. Intragenic deletions of the human low-density-lipoprotein-receptor (*LDLR*) gene, for example, often have *Alu* sequences at both endpoints (Hobbs *et al.*, 1990).

A particular mutational mechanism called 'replication slippage' or 'polymerase slippage' is specifically associated with repetitive DNA (Hancock, 1996; Wolff *et al.*, 1988). Mismatched repeat units may be missed out or duplicated during replication. This mechanism involves no recombination and can create small deletions and insertions or expansions within repetitive sequences. The end-points of some deletions e.g. in the cystic fibrosis gene), are marked by very short direct repeats suggesting slipped strand mispairing (Magnani *et al.,* 1996). The same mechanism is thought to be responsible for trinucleotide expansions (Hummerich and Lehrach, 1995), though triplets are practically never deleted from trinucleotide repeats. Another, yet unkown, sequence-specific mechanism may be involved (McLaughlin *et al.,* 1996). Slippage and unequal recombination are probably also the causes of hypervariability in micro-satellites [i.e. small arrays of tandem repeats of a simple sequence (usually of a 1–4 nucleotides long unit)]. Inverted repeats can cause inversions by a mechanism of chromatin folding. When inverted repeats occur in the proximity of a gene, the latter is susceptible to inversions (Bi and Liu, 1996; Gierl and Frey, 1991; Gubbins *et al.,* 1977; Naylor *et al.,* 1995).

According to some theorists, non-random mutation mechanisms involving specific sequences and sophisticated enzymatic systems could be part of a natural genetic 'program' of evolution. The selective value of mutator genes that modulate the general rate of mutation has been confirmed experimentally in bacteria (Cox, 1976). Bacteriophages and plasmids have been considered as highly sophisticated vehicles for transferring genetic material from cell to cell, population to population, species to species. Proviruses could, perhaps, be doing the same in higher animals (Riley and Anilionis, 1978). Recently, the more contentious concept of a 'directed' or 'adaptive' mutation has been presented (Foster and Cairns, 1992; Samson and Cairns, 1977): mutations also occur in the absence of replication; mutations that are advantageous in a particular environment occur at higher rates than the stochastic mutation theory would predict. If a natural molecular mechanism of adaptive mutation could be convincingly documented, this almost Lamarckian hypothesis would be very appealing. But the adaptive mutation theory is, at present, under strong attack because it is experimentally difficult to discriminate factors that influence mutation rates from factors that influence survival and proliferation of mutants (Lenski and Mittler, 1993). Nevertheless, there may well be more in mutation than random error. Novel methodologies allowing direct and fast access to total gene pools may revolutionize mutation theory.

3.2.2 Effects of mutations

Phenotypic manifestation of mutations. We find it hard to classify mutations according to their phenotypic effects in a sensible way. One reason is the complexity of the concept of a phenotype and the number of meanings this has been given. The term phenotype implies something apparent (φαινομαι = being apparent). But what is apparent in biology depends on what one looks at and on the resolution of the employed measurement techniques. At the beginning, only macroscopically visible traits were consider as phenotypic (e.g. colour, size, disease, etc.). Now, a mutation may be considered to be functional (to have a phenotype) if it changes a macroscopic, microscopic, biochemical or molecular trait of the carrier individual in a measurable and significant manner. Undoubtedly, there are mutations of which the effects cannot be, or have not yet been, measured. These mutations are usually referred to as 'silent' (in molecular, cellular or organismic genetics) or 'neutral' (in population/evolutionary genetics). However, history has shown that such terms should be used with caution. Inheritance of some quantitative phenotypic traits, such as body size or skills, has been recognized almost at the same time as that of discrete Mendelian traits. It is only in the past few years, however, that systematic genetic studies of mutations with very small quantitative contributions to non-Mendelian traits have become possible. The same is true for psychological and behavioural phenotypes, of which only a handful of human studies have been published (Bartfai *et al.*, 1991; Loehlin, 1993; Schachter and Stone, 1985) as well as for mutations of which the phenotypic effects are restricted to cellular or molecular structure (Falvo *et al.*, 1995; Milton *et al.*, 1990).

A second reason is that a phenotype is usually not a static property of a mutation but it is dynamically linked to the genetic background on which the mutation is found as well as to environmental factors. Traditionally, mutations are classified according to their relative phenotypic effects at the homozygous and heterozygous state. The old Mendelian classes of mutation – gender-dependent, lethal, dominant, co-dominant, recessive – are now being split into subclasses as new mechanisms of phenotypic expression are being discovered (Wilkie, 1994). Haploinsufficiency, for example, is the case of a dominant loss-of-function mutation in a gene of which two copies are required for normal function (Nobukuni *et al.*, 1996). Dominant-negative (gain-of-function) mutations interfere with the proper function of the normal copy of a gene. Huntington's disease (Albin and Tagle, 1995) and many types of cancer (Damm, 1993; Sluyser, 1994) may be due to dominant-negative mutations. The phenotypic effects of mutations may be only quantitative, even when change in the affected protein(s) is qualitative (stuctural). An example is the human *APOE* gene coding for apolipoprotein E. Two common mutations resulting in amino acid substitutions reduce the biological activity of this apolipoprotein and contribute differentially to the risk of developing Alzheimer's dementia or corronary disease (see Chapter 6). Although different mutations may measurably alter the amount of a gene product to different extents, manifestation of a clinical phenotype may require the amount of that product to cross a threshold; this is the case in CF (Corrado *et al.*, 1995).

Further complications in the phenotypic classification of mutations arise from the current theory of gene expression. Genes interact within a complex web of functions. Therefore, the phenotype of an allele may change depending on whether the mutated gene and its partners are expressed in a particular tissue, at a particular developmental stage, in a particular environment, or not. Mutations may exert their effects at different developmental stages or ages (e.g. early- or late-onset diseases) even when the mutated gene is constitutively expressed. Mutations in the glucokinase gene, playing a key role in the regulation of glycolysis, may be manifested by the development of diabetes by the age of 25 (Bell *et al.*, 1996).

Natural selection of mutations. Since mutations may, and some certainly do, affect survival and reproduction potential of carriers, the frequency of alleles in a population depends on the rate at which a particular mutation is generated (mutation pressure) and on the relative survival and reproduction (fitness) of the various allele carriers. This is the neo-Darwinian theory. It incorporates the notion of stochastic mutation and that of a conservative natural selection that suppresses most genetic variation. The observed frequency of a mutation depends on its phenotypic effects. There is a common genotype (wild type); mutations are generally rare and most of them are detrimental. Mutations that cause death of the cell or individual where they occur are never seen. Some mutations may exert more or less severe detrimental effects only under particular environmental circumstances and/or at late stages of the individual's life; so, they may be observed. Common mutations are believed to exert their detrimental effects at ages after reproduction and may, therefore, be sustained at high frequencies. Conventionally, loci that present variation at frequencies higher than 1% are called polymorphic, and their various forms, alleles. Some mutations, although detrimental or lethal in the homozygote state, may confer a selective advantage to the heterozygotes. This theoretical concept, known as 'hybrid vigour', 'heterosis', 'over-dominance' or, simply, 'heterozygote advantage' (Watterson, 1982; Wolanski, 1994; Ziehe and Gregorius, 1985) has been used to explain the persistance of alleles associated with CF (Anderson *et al.*, 1967; Romeo *et al.*, 1989), phenylketonuria (Woolf *et al.*, 1975) and a few other human disorders (Brenner Ullman *et al.*, 1994; Wang and Schilling, 1995). Beneficial mutations are very rare but do occur and punctuate established genetic equilibria causing evolution (Elena *et al.*, 1996).

According to the neo-Darwinian theory all mutations have selective value, even those with no obvious phenotypic effects. This value is inferred by the variation of an allele's frequency in ecological or evolutionary time (population history). At the extreme, all information carried in the genome of a species including non-expressed and intergenic sequences is of selective value because replicating useless genetic material would be energy- and resource-consuming, hence selectively disadvantageous. Selection against useless biological material is believed to be the cause of the experimentally documented instability of plasmids conferring no selective advantage to their hosts (Nakamura, 1974; Proctor, 1994).

Selectively neutral mutations. Neutralism (Kimura, 1989; Volkenstein, 1987) is the major opponent of the neo-Darwinian school since the development of protein electrophoresis. Its basic argument is that only a small proportion of the genetic material is translated into protein and only a small proportion of the proteins are essential for normal biological function. In fact, only the functional domains of essential proteins are charged with essential activity; and these domains are indeed conserved in evolutionary time. Other domains may change without compromising function. In all, most mutations must be selectively neutral, and natural populations have many such mutations as protein electrophoresis reveals. Their actual frequencies depend on population history [i.e. the genotypic distribution of a subpopulation when this was genetically isolated (founder effect)], the variation in population size and density since isolation (bottleneck effect) and random genetic drift. Such stochastic factors can 'explain' most observed changes in the genotypic distributions of small populations. If large populations are random collections of smaller populations, then neutral mutation is quantitatively the most important cause of genetic diversity after recombination.

Epistemological drawbacks. There is probably some truth in both views. The debate is actually not on the plausibility of the proposed mechanisms but, rather on the relative numbers of neutral and selected mutations. Preferences are based on personal intuition and beliefs rather than on formal proof. A proper scientific hypothesis should instead be accompanied by precise predictions against which it can be tested. By definition, a genetic variant has a selective (dis)advantage over its competitors (negative or positive relative fitness) only if its frequency changes in time (Lenski, 1991). A selective (dis)advantage may, therefore, not be accepted as an 'explanation' of gene frequency changes but merely as a formal description of such phenomena. For this reason, the theory of natural selection has been criticized as being a tautology (Popper, 1959).

A handful of examples of selective persistence in human populations of highly deleterious mutations appear in every textbook. The most common is that of sickle cell anaemia due to a mutation that renders heterozygote carriers resistant to malaria (Flint *et al.*, 1993). Another is that of CF of which the carriers are thought to be resistant to cholera and/or to other causes of death (see heterozygote advantage above). These models do make predictions that are in principle falsifiable but, in practice, very difficult to test. One has to wait and see what will happen to the frequency of these genes should the parasites become pandemic or extinct. It is even more difficult comparatively to evaluate alternative hypotheses as to the biological nature of the causative selective force(s). For example, follow the debate on the selective forces maintaining the common alleles of *CFTR* gene causing CF (Brackenridge, 1978; Hollander, 1982; Pritchard, 1991; Romeo, et al., 1989; Schroeder, 1995; Shier, 1979; Thompson, J.V. *et al.*, 1997).

The epistemological status of neutralism, as a simple negation of selection, is not any better. A theory calling random factors to explain natural phenomena cannot make any directly testable predictions because 'random' means 'unpredictable'. Not surprisingly, there is a selectionist/neutralist debate over practically every trait and every evolutionary phenomenon (Zlotogora, 1994). The two views overlap. On one hand, the relative fitness of a genetic variant may change in space and time depending on uncontrolled (often uncontrollable), hence random, environmental conditions. For example, a mutation conferring resistance or sensitivity to a parasite may remain unperceived for as long as the carrier is not infected. Before the AIDS epidemic, mutations conferring resistance to HIV (Fauci, 1996; Samson *et al.*, 1996) could be considered to be neutral. In high risk populations such mutations could have high selective values. On the other hand, some of the 'random' factors causing fluctuations in population size and genetic drift may well be 'selective' in nature, since most of them, including migration, may ultimately be linked to resource avalability, predation and parasitism, survival and/or reproduction. Models of density- or frequency-dependent selection are common in the literature (Allen, 1988; Asmussen, 1983; Gavrilets and

Hastings, 1995; Gregorius, 1979; Hammerstein, 1996; Parham, 1994; Potts and Slev, 1995; Zhang and Jiang, 1994). The size of (sub)populations is, therefore, important in both theories.

Fitness is, anyhow, by far the most complex parameter biologists have ever attempted to model and, as such, it is subject to enormous statistical error. A trait must, therefore, have an extremely strong (dis)advantage, and the environment must be extremely simple, for a predicted fitness of a genotype to be significantly different from zero and for a mutation to be non-neutral. Otherwise, selectionist and neutralist hypotheses cannot be discriminated. It has been estimated, for example, that a 2.3% higher survival of *CFTR* heterozygotes during cholera outbreaks would explain the persistence of the common deleterious *CFTR* allele in human populations (Strachan and Read, 1996). To test the selection hypothesis that predicts the above figure we need to measure and to monitor survival of the various *CFTR* genotypes in the presence and absence of cholera. If such measurements ever became possible, their statistical error should be well below 2.3%. A recent view is that every phenotypic trait, every genotype and the entire genotypic distribution of a population is the result of the joint action of three 'forces': adaptation, history and chance. A strategy for quantifying the relative contributions of these forces to a phenotypic trait in simple experimental systems has been published (Travisano *et al.*, 1995) but its applicability to higher organisms and complex ecosystems remains to be shown.

In sum, the magnitude of the phenotypic effect of a mutation ranges continuously in space and time from an extinction of the entire carrier (sub)population to nothing at all. Current theories attempting to explain mutation rates and gene frequencies have been based on concepts which are extremely complex, poorly defined and difficult to measure such as 'fitness', 'history' and 'chance'. Given the rapid progress of mutation research, the entire genetic/evolutionary theory of mutation is bound to evolve. Molecular biology, and interspecies comparisons of genes in particular, would provide hard data for testing evolutionary theories, but the available samples are still relatively small and their analysis has just begun (Takahata, 1996). Methods for accurate measurement of mutation rates and mutation frequencies in large population samples will undoubtedly speed up theoretical development.

3.3 Parallel development of theory with methodology

Invention and discovery are intimately linked in genetics as they are in all science. Some recent breakthroughs are listed in *Table 3.1*. Recombinant

Table 3.1. Recent landmarks in molecular genetic-variation research

Development	Developer(s)	Year
DNA electrophoresis on agarose gel	Sambrook J	1973
Restriction, cloning into *E. coli*	Boyer H and Cohen S	1974
DNA sequencing by chain terminators	Sanger F	1975
DNA hybridization analysis	Southern E	1975
Chemical DNA sequencing	Maxam AM and Gilbert W	1977
RFLP mapping	Botstein D *et al.*	1980
GenBank established		1982
Automated sequencing	Carruthers M and Hood L	1983
PCR	Mullis K *et al.*	1985
Human Genome Organization		1987
H. influenzae genome map completed		1989
Human Genome Project launched		1990

DNA technology provided molecular probes for the study of gene structure and expression. Molecular probes allowed access to mutations without phenotypes. From then on, a totally new genetic thinking became possible. Sequencing and gene-expression detection methods contributed to the current concept of the gene. Transgenic animal technology (Capecchi, 1989a, b) allowed, for the first time, experimental modelling of animal geneotypes. Polymerase chain reaction (Lenstra, 1995; Mullis *et al.*, 1992) allowed isolation of specific predesigned nucleic acid sequences and revolutionized molecular biology. Automated sequencing technologies and recent developments in genetic informatics made it possible to map and sequence entire genomes and enabled studies of the complete genomic organization of species. The Human Genome Project is expected to produce a draft sequence of the entire genome by the year 2005 (Hudson *et al.*, 1995). It will certainly provide new insights to many questions related to the genetic structure and function of cells and individuals. Here are some theoretical advances resulting from modern genetic technologies.

3.3.1 Locus organization

The phenotype of an organism has been frequently reduced to the primary structure of its constituent proteins. A mutation may indeed change the sequence of the protein (structural mutation) when it occurs within a coding domain of a gene, or the amount of protein produced, when it occurs at a gene-regulatory locus. However, molecular biology and sequencing in particular are revealing a complex organization of genes and genomes. As a result, the one-gene-one-protein and one-function-one-disease concepts are being abandoned although they may not be false for all genes. We are left with genes and gene systems with a plastic structure and very vaguely defined boundaries. These consist of coding sequences ('codons' or 'exons') interrupted by non-coding sequences ('introns') and regulatory sequences [promoters, enhancers, signal sequences, splicing sites, sequences determining the stability of transcript ribonucleic acid (RNA), etc.]. Genes overlap partially or completely (nested genes), share sequences, and produce more than one transcript and more than one structurally and/or functionally distinct protein. Similarly, a functional protein may contain subunits coded for by more than one gene, sometimes by several unlinked genes (Cook and Tomlinson, 1995; Hood and Ein, 1968). It has also become possible to isolate and map unkown genes and, on the basis of recognizable sequence elements, to infer protein structure, function and regulation, as well as to predict allelic phenotypic variation. Engineered genetic rearrangements may now be introduced into cells or into otherwise intact animals (Capecchi, 1989a, b) in order to test hypotheses about the phenotypic effects of mutations.

3.3.2 Genome structure

Genes are organized spatially along chromosome maps. Coding or non-coding sequences may be found in one to several thousand copies (ribosomal-DNA, satellite-DNA, mini-satellites, micro-satellites). Reiteration may be species-specific (VNTR), or individual/cell-specific (gene amplification). In the latter case reiteration is considered to have arisen by mutation during the organism's life and is often pathogenic (Berchuck, 1995; Brodeur, 1995; Collins, 1995). Gene copies may be expressed to produce functional RNA and protein or be silent pseudogenes (Wilde, 1986). Some highly repetitive sequences do not code for protein but are known to have major and absolutely essential structural/functional roles (telomeres, centromeres). Small sequences may present natural low to moderate reiteration (mini-satellites, microsatellites) but their expansions can be pathogenic (Anthoney *et al.*, 1996; Singh, 1995). The origin of reiterated DNA is thought to be non-homologous recombination or slippage but its functional role and evolutionary significance are still obscure. Repeats are frequently found within genes, even within coding sequences, and may have a range of phenotypic effects, from a complete disruption of protein structure and function to nothing at all (Lopes Cendes *et al.*, 1996). Repetitive DNA is in general highly polymorphic and has recently been very useful in gene mapping because it provides molecular genetic markers spread throughout the genome (Koreth *et al.*, 1996).

Genes with diverse structures and functions, even from different species present sequence homology suggesting common evolutionary origin and sometimes common mode of action (Hanks *et al.*, 1988). Homologous genes are believed to have arisen through duplication(s) of an ancestral form and subsequent diversification of the copies by natural selection (Jones and Kafatos, 1982; MacIntyre, 1994).

3.3.3 Hierarchical organization, gene regulation and pleiotropic genes

A second dimension of gene organization is regulation. The study of gene expression regulation owes its progress to the development of molecular probes, eukaryotic, mammalian and human cell lines and to gene expression detection technology. Central 'regulatory' genes control the expression of structurally and functionally (un)related genes, which may exert secondary and higher order controls on other genes. The array of genes under common control is call a 'regulon' (a term more common in microbiology). Regulation may be direct: a gene product is a DNA-binding protein that controls transcription from various promoters; or indirect: metabolites regulate the activity of gene products and ultimately transcription. There are probably more mechanisms of regulation than there are genes, since a gene and/or its products are usually subject to multiple positive and negative controls. Obviously, a mutation occurring anywhere within a gene regulation web is likely to affect the entire system. This is an explanation of pleiotropy: a single mutation producing multiple phenotypic effects (Besmer, 1991).

Common control may, however, have a purely structural basis. Several neighbouring genes often share common promoters. Any upstream mutation may affect the expression of downstream genes. Thus, the phenotype of a mutation may be irrelevant to the function of the mutated sequence, and a biological function may alter without any of the involved genes presenting any sequence change (Leighton *et al.*, 1995).

3.3.4 Functional interrelations and polygenes

Because gene products collaborate to carry out complex biochemical pathways, genes that are completely unrelated in terms of origin, structure, position and, sometimes, control may be found related under a common task. The genes coding for the enzymes that carry out a multi-step metabolic pathway (Humphries *et al.*, 1995), or ligand-receptor systems (Besmer, 1991), may be taken as examples. Different mutations in either gene along the chain may alter such complex functions in the same way. A metabolite will not be synthesized if any one of the necessary enzymes is missing or has not been activated. A signal will not be transmitted if either the ligand or the receptor is defective. Often, secondary mutations in the same, or in a second, gene can partially or completely restore the phenotypic effects of a first mutation (Luderer Gmach *et al.*, 1996). This is called gene co-evolution (not to be confused with population or species co-evolution) particularly when successive mutations accumulate on the interacting genes during a process of co-adaptation in evolutionary time (Hartlein and Cusack, 1995).

This notion of gene collaboration is not far from the notion of 'polygenes'. In these gene systems, the products contribute to a particular phenotype in a quantitative (additive) manner without necessarily interacting at the biochemical level, sharing control mechanisms, or being organized in any of the above senses (Tanksley, 1993). Many, if not most, of the important quantitative phenotypic traits, including very common diseases, are suspected today to be polygenic (Williams *et al.*, 1990; Leibel *et al.*, 1993; Humphries *et al.*, 1995). Phenotypic traits with environmental as well as genetic components are referred to as 'complex' or 'multifactorial' traits (Ebers *et al.*, 1996; Ghosh and Collins, 1996; Massy and Keane, 1996; Strohman, 1995). The same genotype may produce various phenotypes according to the presence and intensity of one or more environmental factors. Some forms of cancer have been considered as typical complex diseases (Schatzkin *et al.*, 1995) but the reader may also refer to the numerous heritable metabolic disorders that can be corrected with diet (Combe *et al.*, 1993; Koch *et al.*, 1995; Rybak, 1995).

3.3.5 Developmental mutations and epigenetic phenotypes

The concept of a phenotype is further complicated regarding genes that are differentially expressed in space and/or time. Apart from a relatively small number of genes that are expressed in all tissues all the time (constitutive or housekeeping genes) most genes are expressed at various levels only in some tissues while some are exclusively expressed in one type of cells (cellular markers). Furthermore, many genes are regulated during development and ageing. Genes that are associated with development are called developmental genes. Their expression may be switched on/off at a particular developmental stage. Such changes are usually irreversible (commitment, terminal differentiation) and are carried out by largely unknown epigenetic mechanisms (Issa and Baylin, 1996; Monk, 1995; Strohman, 1995). Epigenetic mechanisms are also responsible for specific inactivation of one of the two parental copies of a gene (genomic imprinting), which may also be tissue specific and timed (Latham *et al.*, 1995). A similar phenomenon is the inactivation of one of the two X chromosomes in female mammals (Migeon, 1994). Mutations in developmental genes will obviously be silent during the period of time and in the tissues where expression is suppressed. Mutations interfering with the epigenetic mechanisms themselves may cause pleiotropic developmental disorders (Ronemus *et al.*, 1996).

3.3.6 Epistasis

Mutation in one locus or gene may modify the phenotype of mutations in another locus or another gene. There are many possible epistatic mechanisms derived easily from the above multi-dimensional organization of genes in a genome. Double-mutant epistatic effects have been recognized in CF (Savov *et al.*, 1995; Sereth *et al.*, 1993) whereas *retinitis pigmentosa* provides an example of a digenic epistasis (Kajiwara *et al.*, 1994). Partial penetrence of two genes due to epistasis has been documented in the human LDLR/Lp(a) gene system (Berg, 1990). In population genetics, epistasis is a mechanism whereby alleles of one locus modify the selective values of alleles of another locus through functional interaction. Analogous, but mechanistically simpler 'co-selection' phenomena are 'hitchhiking' and 'linkage-disequilibrium' (Wagener and Cavalli Sforza, 1975).

3.4 Methods of mutation detection

Compared to the Human Genome (mapping and sequencing) Project, the Human Gemome Diversity Project (Cavalli Sforza *et al.*, 1991; Hoang *et al.*, 1996) is lagging far behind, as is any species diversity project. A genetic diversity project aims to identify every genetic disease, every disease-resistant allele, every rare phenotype and every existing single mutation or recombination event. It also aims to determine the global, regional and local frequencies of each variant and, ultimately, to monitor gene frequency changes in time. The task is huge, but it will be very rewarding.

We should then be able to predict genetic disease, not merely the risk of it, and to intervene before disease develops. We will also learn about the genetic structure and history of populations and we will, perhaps, be able to predict future evolution. We will probably also see genes in their process of birth, development, ageing and death.

Sequencing the entire genome of every individual, however, would produce excessive information. The procedure could be speeded up by high throughput methods for screening very large population samples (ultimately the entire global population) with two distinct aims: first, to detect unknown variation; second, to recognize reliably and score already known variants. We also need improved databases, interfaces and algorithms to store and retrieve variation information. Currently available mutation detection methods are summarized in *Table 3.2* and are overviewed below.

3.4.1 Detection of unknown genetic variants

Protein electrophoresis. Linear protein electrophoresis was the first technique to detect molecular variants. In its first applications tissue extracts were loaded on paper or cellulose acetate and ionically charged protein molecules were allowed to migrate in an electric field (Fagerhol and Braend, 1965; Merril and Goldman, 1982; Nyman 1965; Seaman and Pethica, 1964; Taylor and Giometti, 1992). Natural pigments such as haemoglobins were the first macro-molecules to be resolved by molecular weight and charge. The first matrices were later replaced by agar, starch and polyacrylamide gels. Dyes able to stain all proteins non-specifically as well as specific histochemical (enzymological), immunological, fluorescent and isotopic labelling techniques were developed so that practically all types of proteins could be detected on gels. The method had a tremendous theoretical impact. A mass of data for a discrete genetic variation (e.g. allelic isozymes) that is not associated with disease or other macroscopic traits became available. Molecular polymorphisms were shown to be far more frequent in natural populations than neo-Darwinists predicted. New genetic concepts such as that of neutral variation (Kimura, 1968) and that of a measurable genetic distance in phylogeny (Nei, 1976) were born.

Yet, electrophoretic variation represents only a fraction of all the potential variation of proteins and their coding genes. So, further developments of protein electrophoresis (e.g. two-dimensional electrophoresis by molecular weight and isoelectric point, western blot and capillary electrophoresis) were all to provide more sensitive quantitative techniques to measure gene expression and posttranslational protein processing rather than to detect novel types of genetic variation. For the latter task attention was soon diverted to the DNA itself.

Gene cloning and sequencing. The isolation of DNA metabolism enzymes (restriction endonucleases, ligases and polymerases) from micro-organisms opened the way to recombinant DNA technology (Arber and Linn, 1969; Meselson and Yuan, 1968). Genomic DNA could be fragmented, and the fragments could be inserted into self-replicating vectors (plasmids, bacteriophages or, lately, yeast chromosomes). Isolated genomic loci could thus be isolated, produced in large amounts and sequenced (Cohen and Chang, 1974; Hershfield *et al.*, 1974; Maxam and Gilbert, 1977; Nelson, 1995; Sanger *et al.*, 1977; Schaefer, 1995; Seed, 1995; Voss *et al.*, 1995).

Sequencing is the ultimate universal technique for detecting genetic variation. It can detect 100% of the genetic variation, and it can now be applied directly on selected loci amplified by polymerase chain reaction (PCR) (Gyllensten and Erlich, 1988). Despite automation, however, sequencing remains a relatively slow procedure and it is not cost-effective for all those specimens that represent the commonest genotype ('wild type'). For this reason, techniques for discriminating unknown from known variants have always been welcome, and new methods are still being developed.

Restriction analysis (RA). It was soon discovered that DNA derived from different individuals could be differentially fragmented by a restriction endonuclease (Wyman and White, 1980). Using this, an important application of restriction enzymes has been to generate physical and linkage maps of genomes and genes (Botstein *et al.*, 1980; Kiko *et al.*, 1979). Restriction fragment-length polymorphisms are, now, the most widely documented and used genetic markers (Gillet, 1991). RA consists of a short incubation of DNA with a restriction endonuclease and a straightforward electrophoresis of the mixture. Genomic DNA fragments have a continuous range of sizes and form a smear along the electrophoresis track. The fragments of a particular locus were usually identified by Southern blot followed by hybridization with a radioactively labelled homologous probe and autoradiography (Southern, 1975). With the introduction of PCR, the loci of interest may be selectively amplified, digested by restriction and electrophorized (Her and Weinshilboum, 1995). The fragments can be simply detected using non-radioactive dyes (Grimm and Pflugfelder, 1995).

Table 3.2. Mutation detection methods

Technique	First genetic application	Medline entries[a]	Main genetic applications	Theoretical sensitivity	Throughput	Disadvantage
Protein electrophoresis Isozyme electrophoresis Western blotting	<1964 1971 1982	1.5 5.0 7.6	Mutations in translated loci; post-transcriptional modifications	Low	Low	Very few mutations detected
Northern blotting Reverse transcriptase PCR	1979 1989	23.5 2.5	Mutation in transcribed loci; splicing variation; quantification of gene expression	Moderate	Low	Few mutations detected
Restriction analysis Southern blotting	1977 1978	14.5 21.8	Restriction fragment-length mutations; point mutations in restriction sites	Moderate	High	Misses point mutations; cost of enzymes; difficult set-up Electrophoresis and blotting time
PCR Long PCR	1988 1990	31.6 <0.1	Multiple; large reiterations; insertions or deletions	Low, special	High	
Heteroduplex analysis Enzymatic cleavage of mismatches	1986 1986 1989	1.1 <0.1 <0.1	Point mutations, small deletions or insertions; *de novo* scanning for point mutations	High	High	Misses mutant homozygotes Misses mutant homozygotes; cost enzymes

Chemical cleavage of mismatches						Misses mutant homozygotes; use of dangerous mutagens
Denaturant gradient gel electrophoresis	1979	0.5	*De novo* scanning for point mutations (especially diagnostics)	High	Moderate	Very difficult set up; lengthy electrophoresis; special apparatus
Temperature gradient gel electrophoresis	1987	<0.1				
Single strand conformational polymorphism analysis	1989	1.7	*De novo* scanning for point mutations	?	Moderate	Mostly empirical lengthy electrophoresis
Constant denaturant gel electrophoresis	1991	<0.1	Known mutations (?)	Special	High	
Allele-specific hybridization	1983	<0.1	Known mutations	Special	High	
Amplification refractory mutation system	1989	<0.1	Known mutations	Special	High	
Chips	1993	<0.1	All mutations (?); re-sequencing	High	Very high	Very expensive set up; over powerful

[a] Approximate number of entries in Medline up to 1996 (in thousands).

RA provided the means for large-scale studies of genetic variation of non-coding sequences. Among its major achievements was also the discovery of the genetic instability of CpG sites and the wide-spread occurence of methylation (McKeon *et al.,* 1982). These were due to the parallel use of restriction endonucleases that recognized and cleaved sequences containing CpG nucleotides and of isoschizomer methylation-sensitive restriction endonucleases. When combined with high resolution electrophoretic matrices, RA can detect insertions or deletions as small as one base pair long as well as reiterations. It can also detect single nucleotide substitutions occurring within a restriction site. It cannot detect point mutations occurring outside the restriction sites nor double, inversion and deletion events or complex rearrangements that produce no change in fragment size.

Heteroduplex analysis (HA). This relies on techniques that detect imperfect double helices formed when two partially homologous single strands anneal. Originally, HA was developed for physical mapping, as loops formed by lengthy non-homologous parts may be visible by electron microscopy (Godson, 1973; Hu *et al.,* 1975). During electrophoresis, heteroduplexes migrate at substantially lower velocities than homoduplexes of equal size (Gubbins *et al.,* 1977). This property gives HA significant

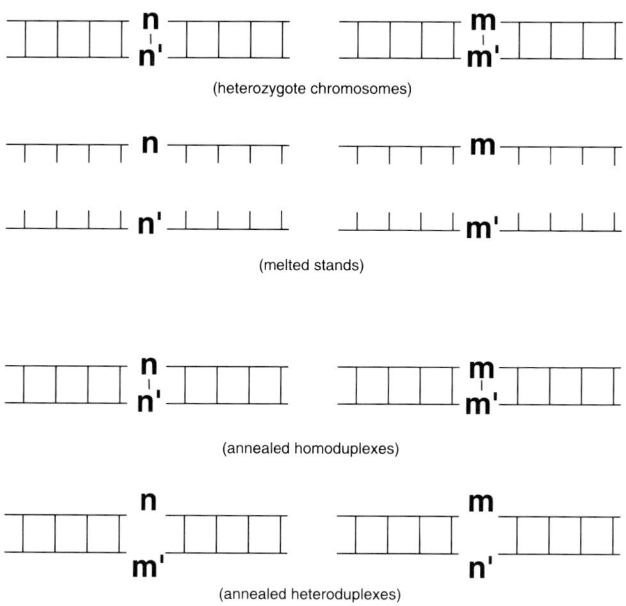

Figure 3.2. DNA heteroduplex formation in heterozygotes.

advantages over RA in large-scale screenings for genetic variation (Glavac and Dean, 1995; Lessa and Applebaum 1993). First, there is no incubation and no costly enzymes. The amplified DNA is heated to melt and cooled slowly to anneal. Two partially homologous double helices will reanneal in four possible ways (*Figure 3.2*). Thus, heterozygotes may be recognized by the presence of additional slowly migrating heteroduplex bands in a simple electrophoresis. Second, HA may detect all mutations within a locus including single nucleotide substitutions. The disadvantage of HA is for the detection of mutant homozygotes. Unless the alleles differ in size by at least one base pair (the limit depending on the resolution of the gel) allelic homoduplexes cannot be resolved.

Cleavage of mismatches. More recent derivatives of HA are techniques that employ nucleases (Myers *et al.*, 1985; Shenoy *et al.*, 1986) or chemical substances (Cotton *et al.*, 1988; Smooker and Cotton, 1993) to cleave heteroduplexes at the site of mismatch. Common examples of such chemical substances are hydroxylamine which modifies mismatched deoxycytosines, and osmium tetroxide which modifies mismatched thymidines. Treatment with the modifying agent is followed by a piperidine treatment that cleaves the DNA at the site of mismatch. Electrophoresis through polyacrylamide gels may reveal the size of the fragments and the exact location of the mismatch(es) can be deduced. The efficiency of cleavage and, therefore, the utility of such techniques was found to be dependent on the type of mismatch and its sequence context (Yatagai *et al.*, 1992). The toxicity of the employed chemicals is, obviously, an additional disadvantage. Nevertheless, the observation of nucleotide- and consensus-sequence-specific modifications by such chemicals add more weight to the theory of non-random mutagenesis; the techniques of chemical cleavage of mismatch in general may have a lot more to reveal in that direction.

Denaturing gradient gel electrophoresis (DGGE). This methodology is based on the theory of DNA-strand dissociation (melting) and aims at an electrophoretic separation of allelic homoduplexes as well as heteroduplex molecules (Fischer and Lerman 1980; Lerman and Silverstein, 1987; Sheffield *et al.*, 1989). The melting point of a DNA molecule is dependent on the nucleotide sequence. More accurately, the probability of dissociation of each base pair along the sequence of a locus (melting profile) depends on the adjacent base pairs. Any change in sequence will theoretically alter the melting profile of a locus. The electrophoretic mobility of partially dissociated double strands decreases as a function of the proportion of the molecule that has dissociated. The reason is that the effective length of a DNA molecule increases because of partial dissociation of the strands. Complete dissociation results in single strands which, assuming no secondary structure, are as long as the mother double helix. To avoid complete strand dissociation, which would defeat the principle of DGGE, the locus to be analysed is 'clamped'. DNA clamps are 5'-, or 3'-end-specific primers bearing a GC-rich tail (typically 40 nucleotides long). This GC-rich domain has an extremely high melting point and ensures that the clamped amplimer will remain in a partially duplex state through a wide range of temperatures or denaturant concentrations before it is completely dissociated.

DGGE is performed at a constant temperature around 60°C. The effective range of denaturant (usually urea and formamade) concentrations where a double-stranded clamped amplimer gradually melts is determined by a tranverse DGGE. The clamped amplimer is loaded along a denaturant gradient perpendicular to the electric field. In lanes with very low denaturant concentrations the amplimer is entirely double stranded and migrates at its maximum velocity. In lanes with very high denaturant concentrations the amplimer is melted and migrates at a minimum velocity. Within a relatively short range of intermediate denaturant concentrations the amplimer migrates at a continuously decreasing velocity, as the density of the gradient increases. Within that short range of denaturant densities, amplimers with different melting profiles are expected to be retarded differentially.

The above pilot transverse DGGE is only to determine the effective range of denaturant concentrations that would resolve allelic amplimers. Following that, a parallel DGGE is set. Clamped amplimers from various genomic DNA specimens are allowed to migrate along a gradient of effective denaturant concentrations. Driven by the electric field the amplimers migrate through increasingly higher densities of denaturant. Because every variant has a different melting profile, allelic amplimers are expected to dissociate and to slow down at different denaturant concentrations (i.e. at diffenent distances from the origin of the gel).

The elegance of this methodology is that it leaves nothing to chance and that it promises very nearly 100% accurate detection of genetic variants. However, since it was first described in 1979 (Fischer and Lerman, 1979) the method has remained relatively unpopular (compare number of published applications in *Table 3.2*). The reasons are, we believe, multiple. It requires a great deal of accurate theoretical and empirical analysis prior to a routine application. In fact, there is no universal protocol for DGGE.

The technique has to be readapted for each locus, and this adaptation is a all research project in itself. It includes computer-assisted amplimer design and, therefore, it requires more than elementary computing skills as well as specialized training and experience. Although a computer programme ('melt87') that calculates melting profiles is freely available from the developers (Lerman and Silverstein, 1987), this has no graphical output and requires further software support. DGGE also requires a special apparatus with high accuracy and precision of temperature control. Most importantly, band resolution is dependent on the slope of the gradient rather than the distance run (*Figure 3.3a, b*). This means that a particular gradient may be adequate for resolving certain alleles only and that, in practice, detection is a lot less than 100% sensitive. Lastly, in its present form, DGGE is a low-throughput technique, since it requires overnight runs in a vertical gel format, where only some 50 samples can be loaded per gel and apparatus. Higher throughput formats are commercially available but their cost is still prohibitive for most laboratories.

Temperature gradient gel electrophoresis (TGGE). The principle of this technique is the same as that of DGGE. A uniform denaturant gel is used instead, on which a spatial temperature gradient is applied to the gel by means of a metal contactor plaque heated at one edge and cooled at the other (Henco *et al.*, 1994; Hollmen and Kulonen, 1966; Po *et al.*, 1987). The temperature gradients may be applied perpendicular to the electrophoretic field for pilot evaluations, or parallel to the tracks for routine runs. The theory and practice of amplimer design is identical to that for DGGE. A special apparatus is required, and we are not aware of any high-throughput commercial versions. TGGE has been even less popular than DGGE, probably due to the difficulty of maintaining an accurate spatial temperature gradient.

Single strand conformational polymorphism analysis (SSCPA). Unlike DGGE or TGGE, SSCPA has no prior theoretical support but has had a much greater appeal. It exploits the empirical observation that single DNA strands migrating under non-denaturant conditions at room temperature will adopt sequence dependent secondary structures (Orita *et al.*, 1989). The phenomenon is analogous to RNA folding (Nielsen *et al.*, 1995). Radio-labelled amplimers are denatured by heating, and then loaded on a non-denaturing vertical gel. After electrophoresis the gel is dried and autoradiographed. The relative mobility of differentially folding variants is unpredictable but robust. Both mutant homozygotes and heterozygotes are readily detected: the former produce one band; the latter, two. SSCPA has been applied widely for the detection of polymorphisms and pathogenic mutations. Its sensitivity has been claimed to be close to 100% in some instances (Fan *et al.*, 1993; Hayashi, 1992; Hayashi and Yandell, 1993). However, there is no theoretical reason why all mutations should produce mutually exclusive conformational polymorphisms. The literature is likely to show a positive bias. Systematic evaluation indicates sensitivity between 20 and 70% according to gel constituents, length of fragment and base change (Liu and Sommer, 1994).

The major disadvantage of SSCPA is the lack of a theoretical framework that would predict which mutations can be detected, why, and what electrophoretic patterns should be expected. Without such theory we have difficulty to foresee further rational development of this methodology. Other important disadvatages are low throughput and the use of isotopes, although non-isotopic versions have been described (Dockhorn Dworniczak *et al.*, 1991).

3.4.2 Diagnostic detection of known mutations

Under this title we consider high throughput methods designed to detect particular common alleles (suspected to be) associated with disease or disease risk.

Constant denaturant gel electrophoresis (CDGE). This is a simple version of DGGE. The denaturant concentration, in which amplimers of the commonest variant and of an important allele of a locus melt differentially, is determined theoretically and empirically. The amplimers are then electrophorized at a constant temperature in gels containing the effective denaturant concentration (Hovig *et al.*, 1991). Since the variants migrate at different velocities from the outset, resolution is dependent on run time (*Figure 3.3c*). Although this method may detect circumstantially several variants in the same gel during relatively short runs, a 100% detection efficiency is not theoretically expected. CDGE has been applied to the detection of common mutations particularly in the field of cancer (Hovig *et al.*, 1992; Ridanpaa *et al.*, 1995).

Methods employing allele-specific oligonucleotides (ASO). When the sequences of the alleles to be detected are known, it is posible to synthesize multiple short oligonucleotides, each complementary to one allele. Genotyping can then be performed by hybridization of genomic DNA extracts with the oligonucleotides on a solid support (filter) followed by high strigency washes (ASO hybridization). Depending on the relative numbers of specimens to be typed and alleles to be detected, one may opt to fix the genomic DNA on the filter by an ordinary dot blot and label the oligonucleotides (Studencki *et al.*, 1985) or fix the oligonucleotides on the filter and label the DNA to be typed (Sajantila *et al.*, 1991).

A high throughput technology based on the latter methodological principle is being developed (Favor, 1995; Fodor *et al.*, 1993; Jacobs and Fodor, 1994; Mazzola and Fodor, 1995; Pease *et al.*, 1994). It employs light-directed chemical synthesis of oligonucleotides in a dense microscopic array on a glass 'chip'. Ninety-six thousand six hundred overlapping oligonucleotides can represent several kilobases of a known sequence region to be interrogated for genetic variation in PCR products from individual samples. Fluorescent fragmented RNA strands prepared post-PCR are hybridized to the chip, which is then scanned and analysed for fluorescent signal. One worker can complete 50 mitochondrial genome sequences (16 000 base pairs) per day. Oligonucleotide chips have not yet been independently evaluated in a real dignostic environment, but in a pilot trial, with a 3.45 kb BRCA1 exon, sensitivity for mutation detection was 93%.

Alternativelly, ASO may be used for allele-specific amplification of a variable locus. This method is also known as 'amplification refractory mutation system' (ARMS) (Newton *et al.*, 1989). The locus is enzymatically amplified only if the particular allele-specific primer is present in the PCR mixture. This method has been developed on the basis of the observation that a perfect match of the 3'-end nucleotide of a primer is required for efficient priming of the PCR under certain conditions. ARMS can, therefore, detect known point mutations by a simple PCR followed by an agarose gel electrophoresis.

Forced restriction sites. Site-directed mutagenesis methods (Flavell *et al.*, 1974; Ho *et al.*, 1989; Marth, 1996; Mendel *et al.*, 1995; Sang *et al.*, 1996) were originally developed for purposes other than mutation detection. Artificial mutation methodology contributes to multiple fields of mutation research and can be applied to enhance the potentials of mutation detection techniques. For example, restriction sites may be artificially introduced into, or abolished from, an amplimer as desired, using appropriately designed oligonucleotide primers. Such designs may assist restriction typing of polymorphic loci, thus extending the research-oriented uses of restriction analysis to diagnostic applications.

3.5 Limitations of current methods for mutation detection

We note considerable limitations of the existing methodologies which have to be overcome if any genetic diversity project were to proceed with a speed comparable to that of a genome project. The first bottleneck is in sample acquisition and preparation. If thousands of individuals are to be typed, DNA-extraction protocols should be shortened, without compromising the quality and the amount of the extract, and should be automated. If the subjects are humans, then the starting tissue should be such that everybody without exception would be happy to offer it. Current protocols are mostly designed for blood, limiting the potential samples to patients, blood donors and some brave volunteers. A good alternative is mouth mucosa. This tissue can be sampled by a simple mouth wash (Liu *et al.*, 1995). No licenced personnel are needed. Subjects can sample themselves. Specimen vials can be sent out and returned by post. So far, however, DNA extraction is still lengthy and has not yet been automated.

Noise levels are sometimes too high in automatic sequencing, and when this occurs, nucleotides cannot be recognized; probably most of the time the problem lies with the quality of the sequencing template. General and robust PCR guidelines based on sequence details and a general theory of PCR would cut down the time spent in PCR optimization and would improve the quality of amplimers. Beside the present difficulties of optimization there are also intrinsic problems in current PCR technology. The relatively low fidelity of the polymerases used today (even that of the improved enzymes) creates part of the sequencing noise. It also causes trouble in heteroduplex-based methods. Because of replication infidelity, non-homologous 'clones' are created during amplification and form a range of artificial heteroduplexes during slow annealing. Another internal problem of PCR is in the length of fragments that can currently be amplified. Although considerable progress has been made (Taylor and Logan, 1995), long-PCR technology requires further development. Thus, with few exceptions (Watanabe *et al.*, 1996a, b), large tandem repeats cannot yet be easily amplified.

The existing automatic sequencers have not been designed to 'call' heterozygotes. Since heterozygote amplimers contain two genuine sequences, biochemical hardware and software improvements will be necessary, unambiguously to discriminate genuine variation from noise. Development of sequencing technology towards multiplex sequencing is now being considered (Cherry *et al.*, 1994; Griffin and Griffin, 1993).

Apart from sequencing, none of the current mutation detection techniques is fully efficient. The limitations, as mentioned above, are theoretical and/or practical. Techniques that can theoretically detect all types of mutations at both the homozygote and heterozygote state are complex and difficult to execute; so far in practice, their performance is not fully satisfactory. Mutation detection techniques that would recognize at least some mutations obviating sequencing have not been reported.

A major limitation is in the throughput of current methods. A human genetic variation project sounds frighteningly large with today's means, but so did the Human Genome Project 20 years ago. Chip technology promises extremely high throughput, but we can already note its potential limitations. It needs specialized instrumentation and initial chip synthesis. It is significantly insensitive in detecting, and difficult to pre-set for, all possible insertion or deletion variations. It is expected to have difficulties with low-complexity sequences. It is not obvious how it can be efficient for common polymorphism typing given that even high grade PCR multiplexing will not readily access the power of the chip (e.g. 10s to 100s of known-sequence polymorphisms accessed per chip). Pre-chip biochemical steps are quite complex. The chip is a committed integral system. Mutations near polymorphisms are expected to be problematic. It is essential to know the full sequence, and accurately so, before a chip can be constructed. Most importantly, the start-up cost for any gene sequence is approximately $10 000, which is mainly the cost of making the photolithographic overlays for chip synthesis. While this is not a problem for established genes of diagnostic interest, it renders 'once-off' research questions on each of the estimated 100 000 genes, many of which would need several chips at current chip oligo density, prohibitively expensive.

3.6 Our approaches

We are interested in the comprehensive analysis of large population samples (e.g. 1000–100 000 individuals) for impact of common and rare variation in traits of cardiovascular disease. For high throughput, liquid phase analysis is often regarded as preferable to electrophoresis-based analyses. However electrophoresis readily enables access to parameters of size, charge and shape, advantages in many categories of DNA analysis. We have invented an electrophoresis system, Microplate-Array Diagonal Gel Electrophoresis (MADGE), which is fully compatible with industrial standard 96-well microplates used for many liquid phase procedures including PCR. MADGE offers extremely convenient set-up, compactness, and access to any matrix including polyacrylamide (Day and Humphries, 1994). This gives access to rapid PCR checking, sizing *et cetera,* and hence to a wide range of single nucleotide polymorphism typing (Bolla *et al.,* 1995; Day *et al.,* 1995a,b,c; O'Dell *et al.,* 1995; Talmud *et al.,* 1996). PAGE-MADGE opens the ready use of 'ultra-short' PCR. One illustration is a universal means of analysing CpG sites in which we place both PCR primers adjacent to a CpG site. Both primers have a 3'-T base and always force, thus, a *Taq*I restriction site (TCGA); mutations ablate this site (Day *et al.,* 1997; O'Dell *et al.,* 1996).

In addition to size-dependent genotyping, analyses dependent upon DNA melting have been implemented on MADGE by using real-time–variable-temperature electrophoresis. The first development was the profiling mismatch and perfect-match oligonucleotide dissociation (off-target strands) by gel electrophoresis (PODGE). PODGE is an extension of standard ASO technology which is limited to a single temperature 'snap-shot' of perfect match. The second development is termed PMD (Day *et al.,* 1997), which is effectively a reconfiguration of DGGE, converting the denaturing gradient from chemical to thermal and the independent variable from space to time. This permits spatial sample arrays (e.g. MADGE) and also confers the convenience and flexibility of programmability (*Figure 3.3d*). Multiphasic programmed gradients will permit resolution of different heteroduplex and homoduplex moeities. PMD makes it feasible for one individual to scan one million bases per day. Comparison of PMD with SSCPA for *LDLR* gene exon 3 in 1000 FH probands showed approximately two-fold greater sensitivity to base changes using PMD (Day *et al.,* 1997). To facilitate PMD (or DGGE) set-up, we have streamlined the process from (i) web-sequence to (ii) theoretical melting profiles to (iii) transverse-DGGE (using a simple high throughput modification of our horizontal gel system; H-DGGE) to (iv) PMD (or DGGE) trials to (v) PMD, with order of magnitude improvement at steps i–ii and ii–iii (Spanakis and Day, in preparation).

Taken together, this suite of methods enhances our capability to call both common and unknown rare sequence variants in thousands of samples simultaneously, by 10–100-fold. In addition, we have work in progress on rapid highly parallel mouthwash DNA preps, on SSCPA–MADGE and on the hardware necessary to support SSCPA–MADGE.

In contrast with the chip, MADGE, CpG-PCR, PODGE, H-DGGE and PMD open much greater parallelism of sample analysis for population studies, both of common and rare single nucleotide variants. One person can readily run 2000 tracks per day on MADGE, which is as much as we have so far needed. One expressed sequence tag/radiation hybrid mapping laboratory has pushed this to 2000 tracks per hour (M. James

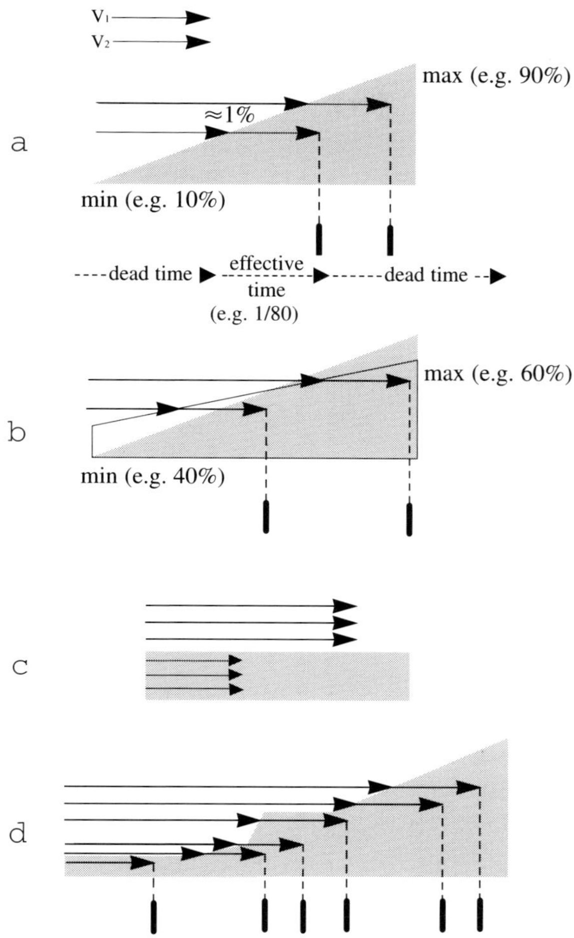

Figure 3.3. (a) Allele amplimers melt and slow down at different points during electrophoresis along a denaturant gradient: v_1 = high velocity of double strands; v_2 = low velocity of partially melted molecules. (b) Resolution of denaturing gradient gel electrophoresis depends solely on the slope of the gradient. Typically, single-nucleotide substitutions change the effective concentration of denaturant by about 1%. Most of the time, the variants migrate with the same velocity, v_1 or v_2. (c) In uniform gels containing an effective concentration of denaturant, allelic amplimers achieve different velocities from the outset; resolution increases with distance run. (d) In programmable melting display (PMD) the chemical gradient in space is replaced with a temperature gradient in time; the slope of the gradient is programmable, allowing significantly higher throughput by more efficient use of gels and shorter electrophoreses.

and P.J.R. Day, personal communication), which is a track-usage rate more than twice as great as the entire Sanger Centre runs per day as sequencing tracks on 83 Applied BioSystems (ABI) automatic sequencers. PMD enables 500 bases scanned per track (perhaps more) and PMD–MADGE has the capability to scan 1–10 million bases per day per person through greater sample parallelism. MADGE is very cheap to set up. PMD is expected to be at least as sensitive as DGGE, which is generally favoured by diagnostic labs over SSCPA on account of its high sensitivity for most PCR products (perhaps limited with some GC-rich loci). MADGE and add-on systems are adaptable to the introduction of new techniques such as, among others, enzymatic mismatch cleavage, new gel-based detection methods, and new PCR methods. Double-mutant PCR products are discernible using DGGE and are, thus, also expected to be in PMD. Only the PCR primer sequences need be known at the point of scanning. PMD–MADGE (or SSCPA–MADGE) entail minimal start-up costs (new PCR primers only) for a new gene.

Doubtless though, several other systems will develop worldwide which will have the capability to contribute to the examination of gene variation on a population scale.

3.7 Conclusions

Despite the impressive advances in modern genetics, the relative contributions of stochastic and deterministic mechanisms of mutation and evolution remain unresolved. There are many more mutations and polymorphisms in natural populations than previously predicted. Much, perhaps most, of this variation may appear to be silent or neutral, but a great proportion of it is potentially functional and selectively important. Registration of all occurring variants and precise mesurement of their frequencies is required not only for better diagnosis, prognosis and management of genetic deseases but also for further development of genetic, epidemiological and evolutionary theory. After a series of technological revolutions including electrophoresis, DNA-restriction, cloning, sequencing, PCR and sequence informatics, we can readily detect most rearrangements as well as single nucleotide substitutions. Technological research now focuses on throughput and cost-effectiveness of mutation detection so that we can scan larger population samples.

Acknowledgements

Emmanuel Spanakis is supported by MRC Grant G9605150MB; Ian Day is supported by a Lister Institute Research Fellowship.

References

Albin RL and Tagle DA. (1995) Genetics and molecular biology of Huntington's disease. *Trends Neurosci.* **18,** 11–14.

Allen JA. (1988) Frequency-dependent selection by predators. *Philos. Trans. R. Soc. Lond. B. Biol. Sci.* **319,** 485–503.

Amariglio N and Rechavi G. (1993) Insertional mutagenesis by transposable elements in the mammalian genome. *Environ. Mol. Mutagen.* **21,** 212–218.

Anderson CM, Allan J and Johansen PG. (1967) Comments on the possible existence and nature of a heterozygote advantage in cystic fibrosis. *Bibl. Paediatr.* **86,** 381–387.

Anthoney DA, McIlwrath AJ, Gallagher WM, Edlin AR and Brown R. (1996) Microsatellite instability, apoptosis, and loss of p53 function in drug–resistant tumor cells. *Cancer Res.* **56,** 1374–1381.

Aquadro CF. (1992) Why is the genome variable? Insights from *Drosophila. Trends Genet.* **8,** 355–362.

Arber W and Linn S. (1969) DNA modification and restriction. *Annu. Rev. Biochem.* **38,** 467–500.

Asmussen MA. (1983) Density-dependent selection incorporating intraspecific competition. II. A diploid model. *Genetics* **103,** 335–350.

Bartfai A, Pedersen NL, Asarnow RF and Schalling D. (1991) Genetic factors for the span of apprehension test: a study of normal twins. *Psychiatry Res.* **38,** 115–124.

Basu AK, Hanrahan CJ, Malia SA, Kumar S, Bizanek R and Tomasz M. (1993) Effect of site-specifically located mitomycin C-DNA monoadducts on *in vitro* DNA synthesis by DNA polymerases. *Biochemistry* **32,** 4708–4718.

Bell GI, Pilkis SJ, Weber IT and Polonsky KS. (1996) Glucokinase mutations, insulin secretion, and diabetes mellitus. *Annu. Rev. Physiol.* **58,** 171–186.

Berchuck A. (1995) Biomarkers in the ovary. *J. Cell. Biochem. Suppl.* **23,** 223–226.

Berg K. (1990) Risk factor variability and coronary heart disease. *Acta Genet. Med. Gemellol. Roma* **39,** 15–24.

Besmer P. (1991) The kit ligand encoded at the murine Steel locus: a pleiotropic growth and differentiation factor. *Curr. Opin. Cell Biol.* **3,** 939–946.

Bi X and Liu LF. (1996) DNA rearrangement mediated by inverted repeats. *Proc. Natl Acad. Sci. USA* **93,** 819–823.

Bolla MK, Haddad L, Humphries SE, Winder AF and Day INM. (1995) High-throughput method for determination of apolipoprotein E genotypes with use of restriction digestion analysis by microplate array diagonal gel electrophoresis. *Clin. Chem.* **41,** 1599–1604.

Botstein D, White RL, Skolnick M and Davis RW. (1980) Construction of a genetic linkage map in man using restriction fragment length polymorphisms. *Am. J. Hum. Genet.* **32,** 314–331.

Boulikas T. (1992) Evolutionary consequences of nonrandom damage and repair of chromatin domains. *J. Mol. Evol.* **35,** 156–180.

Brackenridge CJ. (1978) The seasonal variation of births of offspring from couples heterozygous for cystic fibrosis. *Ann. Hum. Genet.* **42,** 197–201.

Brenner Ullman A, Melzer Ofir H, Daniels M and Shohat M. (1994) Possible protection against asthma in heterozygotes for familial Mediterranean fever. *Am. J. Med. Genet.* **53,** 172–175.

Brodeur GM. (1995) Genetics of embryonal tumours of childhood: retinoblastoma, Wilms' tumour and neuroblastoma. *Cancer Surv.* **25,** 67–99.

Broggini M, Coley HM, Mongelli N, Pesenti E, Wyatt MD, Hartley JA and D'Incalci M. (1995) DNA sequence-specific adenine alkylation by the novel antitumor drug tallimustine (FCE 24517), a benzoyl nitrogen mustard derivative of distamycin. *Nucleic Acids Res.* **23,** 81–87.

Burns RG and Surridge C. (1990) Analysis of beta-tubulin sequences reveals highly conserved, coordinated amino acid substitutions. Evidence that these 'hot spots' are directly involved in the conformational change required for dynamic instability. *FEBS Lett.* **271,** 1–8.

Capecchi MR. (1989a) The new mouse genetics: altering the genome by gene targeting. *Trends Genet.* **5,** 70–76.

Capecchi MR. (1989b) Altering the genome by homologous recombination. *Science* **244,** 1288–1292.

Cavalli Sforza LL, Wilson AC, Cantor CR, Cook Deegan RM and King MC. (1991) Call for a worldwide survey of human genetic diversity: a vanishing opportunity for the Human Genome Project. *Genomics* **11,** 490–491.

Cherry JL, Young H, Di Sera LJ, Ferguson FM, Kimball AW, Dunn DM, Gesteland RF and Weiss RB. (1994) Enzyme-linked fluorescent detection for automated multiplex DNA sequencing. *Genomics* **20,** 68–74.

Cohen SN and Chang AC. (1974) A method for selective cloning of eukaryotic DNA fragments in *Escherichia coli* by repeated transformation. *Mol. Gen. Genet.* **134,** 133–141.

Collins VP. (1995) Gene amplification in human gliomas. *Glia* **15,** 289–296.

Combe C, Deforges Lasseur C, Caix J, Pommereau A, Marot D and Aparicio M. (1993) Compliance and effects of nutritional treatment on progression and metabolic disorders of chronic renal failure. *Nephrol. Dial. Transplant* **8,** 412–418.

Cook GP and Tomlinson IM. (1995) The human immunoglobulin VH repertoire. *Immunol. Today* **16,** 237–242.

Corrado G, Palmisano P, Cavaliere M, Capuano M, Frandina G and Antonelli M. (1995) Cystic fibrosis: genetics and clinical applications. *Riv. Eur. Sci. Med. Farmacol.* **17,** 67–76.

Cotton RG, Rodrigues NR and Campbell RD. (1988) Reactivity of cytosine and thymine in single-base-pair mismatches with hydroxylamine and osmium tetroxide and its application to the study of mutations. *Proc. Natl. Acad. Sci. USA* **85,** 4397–4401.

Cox EC. (1976) Bacterial mutator genes and the control of spontaneous mutation. *Ann. Rev. Genet.* **10,** 135–156.

Cross SH and Bird AP. (1995) CpG islands and genes. *Curr. Opin. Genet. Dev.* **5,** 309–314.

Damm K. (1993) ErbA: tumor suppressor turned oncogene? *FASEB J.* **7,** 904–909.

Day INM and Humphries SE. (1994) Electrophoresis for genotyping: microtiter array diagonal gel electrophoresis on horizontal polyacrylamide gels, hydrolink, or agarose. *Anal. Biochem.* **222,** 389–395.

Day INM, Humphries SE, Richards S, Norton D and Reid M. (1995a) High-throughput genotyping using horizontal polyacrylamide gels with wells arranged for microplate array diagonal gel electrophoresis (MADGE). *Biotechniques* **19,** 830–835.

Day INM, Whittall R, Gudnason V and Humphries SE. (1995b) Dried template DNA, dried PCR oligonucleotides and mailing in 96-well plates: LDL receptor gene mutation screening. *Biotechniques* **18,** 981–984.

Day INM, O'Dell S, Cash ID, Humphries S and Weavind G. (1995c) Electrophoresis for genotyping: temporal thermal gradient gel electrophoresis for profiling of oligonucleotide dissociation. *Nucleic Acids Res.* **23**, 2404–2412.

Day INM, Haddad L, O'Dell SD, Day LB, Whittall R and Humphries S. (1997) Identification of a common low density lipoprotein receptor mutation (R329X) in the South of England: complete linkage disequilibrium with an allele of microsatellite D19S394. *J. Med. Genet.* **34**, 111–116.

Dockhorn Dworniczak B, Dworniczak B, Brommelkamp L, Bulles J, Horst J and Bocker WW. (1991) Non-isotopic detection of single strand conformation polymorphism (PCR-SSCP): a rapid and sensitive technique in diagnosis of phenylketonuria. *Nucleic Acids Res.* **19,** 2500.

Drake JW. (1992) Mutation rates. *Bioessays* **14,** 137–140.

Ebers GC, Kukay K, Bulman DE *et al.* (1996) A full genome search in multiple sclerosis. *Nat. Genet.* **13,** 472–476.

Elena SF, Cooper VS and Lenski RE. (1996) Punctuated evolution caused by selection of rare beneficial mutations. *Science* **272,** 1802–1804.

Epplen JT. (1988) On simple repeated GATCA sequences in animal genomes: a critical reappraisal. *J. Hered.* **79,** 409–417.

Fagerhol MK and Braend M. (1965) Serum prealbumin: polymorphism in man. *Science* **149,** 986–987.

Falvo JV, Thanos D and Maniatis T. (1995) Reversal of intrinsic DNA bends in the IFN beta gene enhancer by transcription factors and the architectural protein HMG I(Y). *Cell* **83,** 1101–1111.

Fan E, Levin DB, Glickman BW and Logan DM. (1993) Limitations in the use of SSCP analysis. *Mutat. Res.* **288,** 85–92.

Fauci AS. (1996) Resistance to HIV-1 infection: it's in the genes. *Nat. Med.* **2,** 966–967.

Favor J. (1995) Mutagenesis and human genetic disease: dominant mutation frequencies and a characterization of mutational events in mice and humans. *Environ. Mol. Mutagen. (supplement)* 25/26, 81–87.

Finnegan DJ. (1989) Eukaryotic transposable elements and genome evolution. *Trends Genet.* **5,** 103–107.

Fischer SG and Lerman LS. (1979) Two-dimensional electrophoretic separation of restriction enzyme fragments of DNA. *Methods Enzymol.* **68,** 1982.

Fischer SG and Lerman LS. (1980) Separation of random fragments of DNA according to properties of their sequences. *Proc. Natl Acad. Sci. USA* **77,** 4420–4424.

Flavell RA, Sabo DL, Bandle EF and Weissmann C. (1974) Site–directed mutagenesis: generation of an extracistronic mutation in bacteriophage Q beta RNA. *J. Mol. Biol.* **89,** 255–272.

Flint J, Harding RM, Clegg JB and Boyce AJ. (1993) Why are some genetic diseases common? Distinguishing selection from other processes by molecular analysis of globin gene variants. *Hum. Genet.* **91,** 91–117.

Fodor SP, Rava RP, Huang XC, Pease AC, Holmes CP and Adams CL. (1993) Multiplexed biochemical assays with biological chips. *Nature* **364,** 555–556.

Foster PL and Cairns J. (1992) Mechanisms of directed mutation. *Genetics* **131,** 783–789.

Gavrilets S and Hastings A. (1995) Intermittency and transient chaos from simple frequency-dependent selection. *Proc. R. Soc. Lond. B. Biol. Sci.* **261,** 233–238.

Ghosh S and Collins FS. (1996) The geneticist's approach to complex disease. *Ann. Rev. Med.* **47,** 333–353.

Gierl A and Frey M. (1991) Eukaryotic transposable elements with short terminal inverted repeats. *Curr. Opin. Genet. Dev.* **1,** 494–497.

Gillet EM. (1991) Genetic analysis of nuclear DNA restriction fragment patterns. *Genome* **34,** 693–703.

Glavac D and Dean M. (1995) Applications of heteroduplex analysis for mutation detection in disease genes. *Hum. Mutat.* **6,** 281–287.

Glickman BW, Saddi VA and Curry J. (1994) International Commission for Protection Against Environmental Mutagens and Carcinogens working paper no. 2. Spontaneous mutations in mammalian cells. *Mutat. Res.* **304,** 19–32.

Godson GN. (1973) DNA heteroduplex analysis of the relation between bacteriophages phiX174 and S.13. *J. Mol. Biol.* **77,** 467–477.

Gregorius HR. (1979) Deterministic single-locus density-dependent selection. *J. Math. Biol.* **8,** 375–391.

Griffin HG and Griffin AM. (1993) DNA sequencing: recent innovations and future trends. *Appl. Biochem. Biotechnol.* **38,** 147–159.

Grimm S and Pflugfelder GO. (1995) Nonradioactive method for restriction mapping of lambda phage DNA. *Biotechniques* **18,** 400–401.

Gubbins EJ, Newlon CS, Kann MD and Donelson JE. (1977) Sequence organization and expression of a yeast plasmid DNA. *Gene* **1,** 185–207.

Gyllensten UB and Erlich HA. (1988) Generation of single-stranded DNA by the polymerase chain reaction and its application to direct sequencing of the HLA-DQA locus. *Proc. Natl Acad. Sci. USA* **85,** 7652–7656.

Hammerstein P. (1996) Darwinian adaptation, population genetics and the streetcar theory of evolution. *J. Math. Biol.* **34,** 511–532.

Hancock JM. (1996) Simple sequences and the expanding genome. *Bioessays* **18,** 421–425.

Hanks SK, Quinn AM and Hunter T. (1988) The protein kinase family: conserved features and deduced phylogeny of the catalytic domains. *Science* **241,** 42–52.

Hartlein M and Cusack S. (1995) Structure, function and evolution of seryl-tRNA synthetases: implications for the evolution of aminoacyl-tRNA synthetases and the genetic code. *J. Mol. Evol.* **40,** 519–530.

Hayashi K. (1992) PCR-SSCP: a method for detection of mutations. *Genet. Anal. Tech. Appl.* **9,** 73–79.

Hayashi K and Yandell DW. (1993) How sensitive is PCR-SSCP? *Hum. Mutat.* **2,** 338–346.

Henco K, Harders J, Wiese U and Riesner D. (1994) Temperature gradient gel electrophoresis (TGGE) for the detection of polymorphic DNA and RNA. *Methods Mol. Biol.* **31,** 211–228.

Her C and Weinshilboum RM. (1995) Rapid restriction mapping by use of long PCR. *Biotechniques* **19,** 530–532.

Hershfield V, Boyer HW, Yanofsky C, Lovett MA and Helinski DR. (1974)

Plasmid ColEl as a molecular vehicle for cloning and amplification of DNA. *Proc. Natl Acad. Sci. USA* **71,** 3455–3459.

Heywood LA and Burke JF. (1990) Mismatch repair in mammalian cells. *Bioessays* **12,** 473–477.

Ho SN, Hunt HD, Horton RM, Pullen JK and Pease LR. (1989) Site-directed mutagenesis by overlap extension using the polymerase chain reaction. *Gene* **77,** 51–59.

Hoang L, Byck S, Prevost L and Scriver CR. (1996) PAH Mutation Analysis Consortium Database: a database for disease-producing and other allelic variation at the human PAH locus. *Nucleic Acids Res.* **24,** 127–131.

Hobbs HH, Russell DW, Brown MS and Goldstein JL. (1990) The LDL receptor locus in familial hypercholesterolemia: mutational analysis of a membrane protein. *Ann. Rev. Genet.* **24,** 133–170.

Hollander DH. (1982) Etiogenesis of the European cystic fibrosis polymorphism: heterozygote advantage against venereal syphilis? *Med. Hypotheses* **8,** 191–197.

Hollmen T and Kulonen E. (1966) Combination of temperature gradient with gel electrophoresis and its applications to analysis of collagen-gelatin transition. *J. Chromatogr.* **21,** 454–459.

Hood L and Ein D. (1968) Immunologlobulin lambda chain structure: two genes, one polypeptide chain. *Nature* **220,** 764–767.

Hovig E, Smith Sorensen B, Brogger A and Borrensen AL. (1991) Constant denaturant gel electrophoresis, a modification of denaturing gradient gel electrophoresis, in mutation detection. *Mutat. Res.* **262,** 63–71.

Hovig E, Smith Sorensen B, Uitterlinden AG and Borresen AL. (1992) Detection of DNA variation in cancer. *Pharmacogenetics* **2,** 317–328.

Hsieh CL, Gauss G and Lieber MR. (1992) Replication, transcription, CpG methylation and DNA topology in V(D)J recombination. *Curr. Top. Microbiol. Immunol.* **182,** 125–135.

Hu S, Ohtsubo E and Davidson N. (1975) Electron microscopic heteroduplex studies of sequence relations among plasmids of *Escherichia coli:* structure of F13 and related F-primes. *J. Bacteriol.* **122,** 749–763.

Hudson TJ, Stein LD, Gerety SS, Ma J, Castle AB, Silva J, *et al.* (1995) An STS-based map of the human genome. *Science* **270,** 1945–1954.

Hummerich H and Lehrach H. (1995) Trinucleotide repeat expansion and human disease. *Electrophoresis* **16,** 1698–1704.

Humphries SE, Peacock RE and Talmud PJ. (1995) The genetic determinants of plasma cholesterol and response to diet. *Baillières Clin. Endocrinol. Metab.* **9,** 797–823.

Issa JP and Baylin SB. (1996) Epigenetics and human disease. *Nat. Med.* **2,** 281–282.

Jacobs JW and Fodor SP. (1994) Combinatorial chemistry – applications of light-directed chemical synthesis. *Trends Biotechnol.* **12,** 19–26.

Jolly CJ, Wagner SD, Rada C, Klix N, Milstein C and Neuberger MS. (1996) The targeting of somatic hypermutation. *Semin. Immunol.* **8,** 159–168.

Jones CW and Kafatos FC. (1982) Accepted mutations in a gene family: evolutionary diversification of duplicated DNA. *J. Mol. Evol.* **19,** 87–103.

Kajiwara K, Berson EL and Dryja TP. (1994) Digenic retinitis pigmentosa due to mutations at the unlinked peripherin/RDS and ROM1 loci. *Science* **264,** 1604–1608.

Kiko H, Niggemann E and Ruger W. (1979) Physical mapping of the restriction fragments obtained from bacteriophage T4 dC-DNA with the restriction endonucleases SmaI, KpnI and BglII. *Mol. Gen. Genet.* **172,** 303–312.

Kimura M. (1968) Evolutionary rate at the molecular level. *Nature* **217,** 624–626.

Kimura M. (1989) The neutral theory of molecular evolution and the world view of the neutralists. *Genome* **31,** 24–31.

Kissi B, Tordo N and Bourhy H. (1995) Genetic polymorphism in the rabies virus nucleoprotein gene. *Virology* **209,** 526–537.

Koch R, Acosta PB and Williams JC. (1995) Nutritional therapy for pregnant women with a metabolic disorder. *Clin. Perinatol.* **22,** 1–14.

Koreth J, O'Leary JJ and O'D McGee J. (1996) Microsatellites and PCR genomic analysis. *J. Pathol.* **178,** 239–248.

Kumar S. (1996) Patterns of nucleotide substitution in mitochondrial protein coding genes of vertebrates. *Genetics* **143,** 537–548.

Latham KE, McGrath J and Solter D. (1995) Mechanistic and developmental aspects of genetic imprinting in mammals. *Int. Rev. Cytol.* **160,** 53–98.

Leibel RL, Bahary N and Friedman JM. (1993) Strategies for the molecular genetic analysis of obesity in humans. *Crit. Rev. Food Sci. Nutr.* **33,** 351–358.

Leighton PA, Ingram RS, Eggenschwiler J, Efstratiadis A and Tilghman SM. (1995) Disruption of imprinting caused by deletion of the H19 gene region in mice. *Nature* **375,** 34–39.

Lenski RE. (1991) Quantifying fitness and gene stability in microorganisms. *Biotechnology* **15,** 173–192.

Lenski RE and Mittler JE. (1993) The directed mutation controversy and neo-Darwinism. *Science* **259,** 188–194.

Lenstra JA. (1995) The applications of the polymerase chain reaction in the life sciences. *Cell. Mol. Biol. Noisy le grand* **41,** 603–614.

Lerman LS and Silverstein K. (1987) Computational simulation of DNA melting and its application to denaturing gradient gel electrophoresis. *Methods Enzymol.* **155,** 482–501.

Lessa EP and Applebaum G. (1993) Screening techniques for detecting allelic variation in DNA sequences. *Mol. Ecol.* **2,** 119–129.

Liu Q and Sommer SS. (1994) Parameters affecting the sensitivities of dideoxy fingerprinting and SSCP. *PCR Meth. Appl.* **4,** 97–108.

Liu YH, Bai J, Zhu Y, Liang X, Siemieniak D, Venta PJ and Lubman DM. (1995) Rapid screening of genetic polymorphisms using buccal cell DNA with detection by matrix-assisted laser desorption/ionization mass spectrometry. *Rapid Commun. Mass. Spectrom.* **9,** 735–743.

Loehlin JC. (1993) Nature, nurture, and conservatism in the Australian twin study. *Behav. Genet.* **23,** 287–290.

Lopes Cendes I, Maciel P, Kish S *et al.* (1996) Somatic mosaicism in the central nervous system in spinocerebellar ataxia type 1 and Machado-Joseph disease. *Ann. Neurol.* **40,** 199–206.

Luderer Gmach M, Liebig HD, Sommergruber W, Voss T, Fessl F, Skern T and Kuechler E. (1996) Human rhinovirus 2A proteinase mutant and its second-site revertants. *Biochem. J.* **318,** 213–218.

MacIntyre RJ. (1994) Molecular evolution: codes, clocks, genes and genomes. *Bioessays* **16,** 699–703.

Magnani C, Cremonesi L, Giunta A, Magnaghi P, Taramelli R and Ferrari M. (1996) Short direct repeats at the breakpoints of a novel large deletion in the CFTR gene suggest a likely slipped mispairing mechanism. *Hum. Genet.* **98,** 102–108.

Marth JD. (1996) Recent advances in gene mutagenesis by site-directed recombination. *J. Clin. Invest.* **97,** 1999–2002.

Massy ZA and Keane WF. (1996) Pathogenesis of atherosclerosis. *Semin. Nephrol.* **16,** 12–20.

Maxam AM and Gilbert W. (1977) A new method for sequencing DNA. *Proc. Natl Acad. Sci. USA* **74,** 560–564.

Mazzola LT and Fodor SP. (1995) Imaging biomolecule arrays by atomic force microscopy. *Biophys. J.* **68,** 1653–1660.

McDonald JF. (1993) Evolution and consequences of transposable elements. *Curr. Opin. Genet. Dev.* **3,** 855–864.

McKeon C, Ohkubo H, Pastan I and de Crombrugghe B. (1982) Unusual methylation pattern of the alpha 2 (l) collagen gene. *Cell* **29,** 203–210.

McLaughlin BA, Spencer C and Eberwine J. (1996) CAG trinucleotide RNA repeats interact with RNA-binding proteins. *Am. J. Hum. Genet.* **59,** 561–569.

Mendel D, Cornish VW and Schultz PG. (1995) Site-directed mutagenesis with an expanded genetic code. *Ann. Rev. Biophys. Biomol. Struct.* **24,** 435–462.

Mercado CM and Tomasz M. (1977) Circular dichroism of mitomycin-DNA complexes. Evidence for a conformational change in DNA. *Biochemistry* **16,** 2040–2046.

Merril CR and Goldman D. (1982) Quantitative two-dimensional protein electrophoresis for studies of inborn errors of metabolism. *Clin. Chem.* **28,** 1015–1020.

Meselson M and Yuan R. (1968) DNA restriction enzyme from *E. coli. Nature* **217,** 1110–1114.

Migeon BR. (1994) X-chromosome inactivation: molecular mechanisms and genetic consequences. *Trends Genet.* **10,** 230–235.

Milton DL, Casper ML and Gesteland RF. (1990) Saturation mutagenesis of a DNA region of bend. Base steps other than ApA influence the bend. *J. Mol. Biol.* **213,** 135–140.

Monk M. (1995) Epigenetic programming of differential gene expression in development and evolution. *Dev. Genet.* **17,** 188–197.

Moxon ER, Rainey PB, Nowak MA and Lenski RE. (1994) Adaptive evolution of highly mutable loci in pathogenic bacteria. *Curr. Biol.* **4,** 24–33.

Mullis K, Faloona F, Scharf S, Saiki R, Horn G and Erlich H. (1992) Specific enzymatic amplification of DNA *in vitro:* the polymerase chain reaction. *Biotechnology* **24,** 17–27.

Myers RM, Larin Z and Maniatis T. (1985) Detection of single base substitutions by ribonuclease cleavage at mismatches in RNA:DNA duplexes. *Science* **230,** 1242–1246.

Nakamura H. (1974) Plasmid-instability in acrA mutants of *Escherichia coli* K12. *J. Gen. Microbiol.* **84,** 85–93.

Naylor JA, Buck D, Green P, Williamson H, Bentley D and Giannelli F. (1995) Investigation of the factor VIII intron 22 repeated region (int22h) and the associated inversion junctions. *Hum. Mol. Genet.* **4,** 1217–1224.

Nei M. (1976) Mathematical models of speciation and genetic distance. In: *Population Genetics and Ecology* (eds S Karlin and E Nevo) pp. 723–765. Academic Press, New York.

Nelson DL. (1995) Positional cloning reaches maturity. *Curr. Opin. Genet. Dev.* **5,** 298–303.

Newton CR, Graham A, Heptinstall LE, Powell SJ, Summers C, Kalsheker N, Smith JC and Markham AF. (1989) Analysis of any point mutation in DNA. The amplification refractory mutation system (ARMS). *Nucleic Acids Res.* **17,** 2503–2516.

Nielsen DA, Novoradovsky A and Goldman D. (1995) SSCP primer design based on single-strand DNA structure predicted by a DNA folding program. *Nucleic Acids Res.* **23,** 2287–2291.

Nobukuni Y, Watanabe A, Takeda K, Skarka H and Tachibana M. (1996) Analyses of loss-of-function mutations of the MITF gene suggest that haploinsufficiency is a cause of Waardenburg syndrome type 2A. *Am. J. Hum. Genet.* **59,** 76–83.

Nyman L. (1965) Species specific proteins in freshwater fishes and their suitability for a 'protein taxonomy'. *Hereditas* **53,** 115–126.

O'Dell SD, Humphries SE and Day INM. (1995) Rapid methods for population-scale analysis for gene polymorphisms: the ACE gene as an example. *Br. Heart J.* **73,** 368–371.

O'Dell SD, Humphries SE, and Day INM. (1996) PCR induction of a TaqI restriction site at any CpG dinucleotide using two mismatched primers (CpG-PCR). *Genome Res.* **6,** 558–568.

Okada N. (1991) SINEs. *Curr. Opin. Genet. Dev.* **1,** 498–504.

Orita M, Iwahana H, Kanazawa H, Hayashi K and Sekiya T. (1989) Detection of polymorphisms of human DNA by gel electrophoresis as single-strand conformation polymorphisms. *Proc. Natl Acad. Sci. USA* **86,** 2766–2770.

Parham P. (1994) The rise and fall of great class I genes. *Semin. Immunol.* **6,** 373–382.

Pease AC, Solas D, Sullivan EJ, Cronin MT, Holmes CP and Fodor SP. (1994) Light-generated oligonucleotide arrays for rapid DNA sequence analysis. *Proc. Natl Acad. Sci. USA* **91,** 5022–5026.

Po T, Steger G, Rosenbaum V, Kaper J and Riesner D. (1987) Double-stranded cucumovirus associated RNA 5: experimental analysis of necrogenic and non–necrogenic variants by temperature-gradient gel electrophoresis. *Nucleic Acids Res.* **15,** 5069–5083.

Popper K. (1959) *The Logic of Scientific Discovery*. Huntchinson, London.

Potts WK and Slev PR. (1995) Pathogen-based models favoring MHC genetic diversity. *Immunol. Rev.* **143,** 181–197.

Pritchard DJ. (1991) Cystic fibrosis allele frequency, sex ratio anomalies and fertility: a new theory for the dissemination of mutant alleles. *Hum. Genet.* **87,** 671–676.

Proctor GN. (1994) Mathematics of microbial plasmid instability and subsequent differential growth of plasmid-free and plasmid-containing cells, relevant to the analysis of experimental colony number data. *Plasmid* **32,** 101–130.

Ridanpaa M, Burvall K, Zhang LH, Husgafvel Pursiainen K and Onfelt A. (1995) Comparison of DGGE and CDGE in detection of single base changes in the hamster hprt and human N-ras genes. *Mutat. Res.* **334,** 357–364.

Riley M and Anilionis A. (1978) Evolution of the bacterial genome. *Ann. Rev. Microbiol.* **32,** 519–560.

Romeo G, Devoto M and Galietta LJ. (1989) Why is the cystic fibrosis gene so frequent? *Hum. Genet.* **84,** 1–5.

Ronemus MJ, Galbiati M, Ticknor C, Chen J and Dellaporta SL. (1996) Demethylation-induced developmental pleiotropy in Arabidopsis. *Science* **273,** 654–657.

Rybak LP. (1995) Metabolic disorders of the vestibular system. *Otolaryngol. Head Neck Surg.* **112,** 128–132.

Sajantila A, Strom M, Budowle B, Tienari PJ, Ehnholm C and Peltonen L. (1991) The distribution of the HLA-DQ alpha alleles and genotypes in the Finnish population as determined by the use of DNA amplification and allele specific oligonucleotides. *Int. J. Legal. Med.* **104,** 181–184.

Sakaguchi K. (1990) Invertrons, a class of structurally and functionally related genetic elements that includes linear DNA plasmids, transposable elements, and genomes of adeno-type viruses. *Microbiol. Rev.* **54,** 66–74.

Samson L and Cairns J. (1977) A new pathway for DNA repair in *Escherichia coli. Nature* **267,** 281–283.

Samson M, Libert F, Doranz BJ *et al.* (1996) Resistance to HIV-1 infection in caucasian individuals bearing mutant alleles of the CCR-5 chemokine receptor gene. *Nature* **382,** 722–725.

Sanderson BJ and Shield AJ. (1996) Mutagenic damage to mammalian cells by therapeutic alkylating agents. *Mutat. Res.* **355,** 41–57.

Sang N, Condorelli G, De Luca A, MacLachlan TK and Giordano A. (1996) Generation of site-directed mutagenesis by extralong, high-fidelity polymerase chain reaction. *Anal. Biochem.* **233,** 142–144.

Sanger F, Nicklen S and Coulson AR. (1977) DNA sequencing with chain-terminating inhibitors. *Proc. Natl Acad. Sci. USA* **74,** 5463–5467.

Savov A, Angelicheva D, Balassopoulou A, Jordanova A, Noussia Arvanitakis S and Kalaydjieva L. (1995) Double mutant alleles: are they rare? *Hum. Mol. Genet.* **4,** 1169–1171.

Schachter FF and Stone RK. (1985) Pediatricians' and psychologists' implicit personality theory: significance of sibling differences. *J. Dev. Behav. Pediatr.* **6,** 295–297.

Schaefer BC. (1995) Revolutions in rapid amplification of cDNA ends: new strategies for polymerase chain reaction cloning of full-length cDNA ends. *Anal. Biochem.* **227,** 255–273.

Schatzkin A, Goldstein A and Freedman LS. (1995) What does it mean to be a cancer gene carrier? Problems in establishing causality from the molecular genetics of cancer. *J. Natl Cancer Inst.* **87,** 1126–1130.

Schimenti JC. (1994) Gene conversion and the evolution of gene families in mammals. *Soc. Gen. Physiol. Ser.* **49,** 85–91.

Schmid CW. (1996) Alu: structure, origin, evolution, significance and function of one-tenth of human DNA. *Prog. Nucleic Acid Res. Mol. Biol.* **53,** 283–319.

Schroeder SA, Gaughan DM and Swift M. (1995) Protection against bronchial asthma by CFTR delta F508 mutation: a heterozygote advantage in cystic fibrosis. *Nat. Med.* **1,** 703–705.

Seaman GV and Pethica BA. (1964) A comparison of the electrophoretic characteristics of the human normal and sickle erythrocyte. *Biochem. J.* **90,** 573–578.

Seed B. (1995) Developments in expression cloning. *Curr. Opin. Biotechnol.* **6,** 567–573.

Selby CP and Sancar A. (1994) Mechanisms of transcription-repair coupling and mutation frequency decline. *Microbiol. Rev.* **58,** 317–329.

Sereth H, Shoshani T, Bashan N and Kerem BS. (1993) Extended haplotype analysis of cystic fibrosis mutations and its implications for the selective advantage hypothesis. *Hum. Genet.* **92,** 289–295.

Sheffield VC, Cox DR, Lerman LS and Myers RM. (1989) Attachment of a 40-base-pair G + C-rich sequence (GC-clamp) to genomic DNA fragments by the polymerase chain reaction results in improved detection of single-base changes. *Proc. Natl Acad. Sci. USA* **86,** 232–236.

Shenoy S, Daigle K, Ehrlich KC, Gehrke CW and Ehrlich M. (1986) Hydrolysis by restriction endonucleases at their DNA recognition sequences substituted with mismatched base pairs. *Nucleic Acids Res.* **14,** 4407–4420.

Shier WT. (1979) Increased resistance to influenza as a possible source of heterozygote advantage in cystic fibrosis. *Med. Hypotheses* **5,** 661–667.

Singh L. (1995) Biological significance of minisatellites. *Electrophoresis* **16,** 1586–1595.

Sluyser M. (1994) Hormone resistance in cancer: the role of abnormal steroid receptors. *Crit. Rev. Oncog.* **5,** 539–554.

Smooker PM and Cotton RG. (1993) The use of chemical reagents in the detection of DNA mutations. *Mutat. Res.* **288,** 65–77.

Sommer SS. (1995) Recent human germ–line mutation: inferences from patients with hemophilia B. *Trends Genet.* **11,** 141–147.

Sommer SS and Ketterling RP. (1994) How precisely can data from transgenic mouse mutation–detection systems be extrapolated to humans?: lesions from the human factor IX gene. *Mutat. Res.* **307,** 517–531.

Southern EM. (1975) Detection of specific sequences among DNA fragments separated by gel electrophoresis. *J. Mol. Biol.* **98,** 503–517.

Strachan T and Read AP. (1996) *Human Molecular Genetics*. BIOS Scientific Publishers Ltd, Oxford.

Strohman RC. (1995) Linear genetics, non-linear epigenetics: complementary approaches to understanding complex diseases. *Integr. Physiol. Behav. Sci.* **30,** 273–282.

Studencki AB, Conner BJ, Impraim CC, Teplitz RL and Wallace RB. (1985) Discrimination among the human beta A, beta S, and beta C-globin genes using allele-specific oligonucleotide hybridization probes. *Am. J. Hum. Genet.* **37,** 42–51.

Takahata N. (1996) Neutral theory of molecular evolution. *Curr. Opin. Genet. Dev.* **6,** 767–772.

Talmud PJ, Tamplin OJ, Heath K, Gaffney D, Day INM and Humphries SE. (1996) Rapid testing for three mutations causing familial defective apolipoprotein B100 in 562 patients with familial hypercholesterolaemia. *Atheroscl.* **125,** 135–137.

Tanksley SD. (1993) Mapping polygenes. *Ann. Rev. Genet.* **27,** 205–233.

Taylor GR and Logan WP. (1995) The polymerase chain reaction: new variations on an old theme. *Curr. Opin. Biotechnol.* **6,** 24–29.

Taylor J and Giometti CS. (1992) Use of principal components analysis for mutation detection with two-dimensional electrophoresis protein separations. *Electrophoresis* **13,** 162–168.

Thompson EA and Neel JV. (1997) Allelic Disequilibrium and allele frequency distribution as a function of social and demographic history. *Am. J. Hum. Genet.* **60,** 197–204.

Travisano M, Mongold JA, Bennett AF and Lenski RE. (1995) Experimental tests of the roles of adaptation, chance, and history in evolution. *Science* **267,** 87–90.

Umar A and Kunkel TA. (1996) DNA-replication fidelity, mismatch repair and genome instability in cancer cells. *Eur. J. Biochem.* **238,** 297–307.

Volkenstein MV. (1987) Punctualism, non-adaptationism, neutralism and evolution. *Biosystems* **20,** 289–304.

Voss H, Schwager C, Wiemann S, Zimmermann J, Stegemann J, Erfle H, Voie AM, Drzonek H and Ansorge W. (1995) Efficient low redundancy large-scale DNA sequencing at EMBL. *J. Biotechnol.* **41,** 121–129.

Wagener DK and Cavalli Sforza LL. (1975) Ethnic variation in genetic disease: possible roles of hitchhiking and epistasis. *Am. J. Hum. Genet.* **27,** 348–364.

Wang CH and Schilling RF. (1995) Myocardial infarction and thalassemia trait: an example of heterozygote advantage. *Am. J. Hematol.* **49,** 73–75.

Watanabe M, Abe K, Aoki M, Kameya T, Itoyama Y, Shoji M, Ikeda M, Iizuka T and Hirai S. (1996a) A reproducible assay of polymerase chain reaction to detect trinucleotide repeat expansion of Huntington's disease and senile chorea. *Neurol. Res.* **18,** 16–18.

Watanabe M, Abe K, Aoki M *et al.* (1996b) Analysis of CAG trinucleotide expansion associated with Machado-Joseph disease. *J. Neurol. Sci.* **136,** 101–107.

Watterson GA. (1982) Testing selection at a single locus. *Biometrics* **38,** 323–331.

Wevrick R and Buchwald M. (1993) Mammalian DNA-repair genes. *Curr. Opin. Genet. Dev.* **3,** 470–474.

Wilde CD. (1986) Pseudogenes. *CRC Crit. Rev. Biochem.* **19,** 323–352.

Wilkie AO. (1994) The molecular basis of genetic dominance. *J. Med. Genet.* **31,** 89–98.

Williams RR, Hunt SC, Hasstedt SJ, Hopkins PN, Wu LL, Berry TD, Stults BM, Barlow GK and Kuida H. (1990) Genetics of hypertension: what we know and don't know. *Clin. Exp. Hypertens. A.* **12,** 865–876.

Wolanski N. (1994) Assortative mating in somatic traits and its consequences. *Stud. Hum. Ecol.* **11,** 73–111.

Wolff RK, Nakamura Y and White R. (1988) Molecular characterization of a spontaneously generated new allele at a VNTR locus: no exchange of flanking DNA sequence. *Genomics* **3,** 347–351.

Woolf LI, McBean MS, Woolf FM and Cahalane SF. (1975) Phenylketonuria as a balanced polymorphism: the nature of the heterozygote advantage. *Ann. Hum. Genet.* **38,** 461–469.

Wyman AR and White R. (1980) A highly polymorphic locus in human DNA. *Proc. Natl Acad. Sci USA* **77,** 6754–6758.

Yatagai F, Hachiya S, Hama Y, Gordon AJ and Hanaoka F. (1992) Resolution and characterization of polymorphic DNA by SSCP and chemical cleavage methodologies. *J. Radiat. Res. Tokyo (Supplement)* **33,** 95–108.

Zhang DY and Jiang XH. (1994) Is heterozygous advantage necessary for polymorphism in one-locus two-allele systems? *J. Theor. Biol.* **166,** 245–250.
Ziehe M and Gregorius HR. (1985) The significance of over- and underdominance for the maintenance of genetic polymorphisms. I. Underdominance and stability. *J. Theor. Biol.* **117,** 493–504.
Zlotogora J. (1994) High frequencies of human genetic diseases: founder effect with genetic drift or selection? *Am. J. Med. Genet.* **49,** 10–13.

4

Finding susceptibility genes for schizophrenia

M.J. Owen and M.C. O'Donovan

4.1 Introduction

There is overwhelming evidence that schizophrenia runs in families. Gottesman (1991) has reviewed Western European studies comprehensively and has shown that the lifetime risk of schizophrenia in first-degree relatives of affected probands is around 10%, compared with approximately 1% in the general population. Convincing evidence that this familial clustering reflects mainly genetic, rather than shared environmental, factors comes from a large number of twin and adoption studies. Thus the concordance rates of schizophrenia are consistently higher in MZ twins than in DZ twins, with proband-wise MZ concordance rates averaging about 40% versus DZ rates of about 14% (Gottesman and Shields, 1982). In addition, adoption studies of various designs have clearly shown that the biological relatives of affected individuals have significantly higher rates of schizophrenia and related disorders than their adopted relatives (McGuffin *et al.*, 1994).

These findings certainly imply that genes are major etiological factors, but it is equally clear from the high rates of discordance in MZ twins that what is inherited is not the certainty of developing schizophrenia but rather a predisposition or liability to develop the disorder. Unfortunately, there is no reliable way to detect or measure this disease susceptibility in terms of altered psychological or physiological

Genetics of Common Diseases: future therapeutic and diagnostic possibilities,
edited by I. Day and S. Humphries.

processes, and therefore, at present, the exact nature of the genetic 'trait' is unknown.

As well as being unable to define exactly what is transmitted, we are also unable to say how it is transmitted. Certainly, a classical Mendelian or sex-linked single gene cannot account for the patterns of familial transmission of schizophrenia which, like other common disorders, shows an apparently complex mode of inheritance. Such disorders are usually attributable to the combined action of several genes of modest effect (oligogenic inheritance) or to many genes of small effect (polygenes) together with environmental factors (Risch, 1990). However, the etiology of schizophrenia is likely to be heterogeneous, and it is therefore difficult to exclude the possibility that some cases are due to the action of a single gene of major effect acting either alone or against a background of multiple small genetic and environmental factors. Furthermore, a number of mechanisms have recently been identified that can cause the inheritance of phenotypes caused by a single gene to depart from simple Mendelian patterns. It has been hypothesized that at least one of these, the dynamic mutation, might play a role in the etiology of schizophrenia.

4.2 Molecular genetic approaches to schizophrenia

Considerable effort has been devoted to adapting and refining the methods of molecular genetics (which have been so successful in the identification of genes responsible for Mendelian disorders) to allow their application to common diseases with complex patterns of inheritance (Lander and Schork, 1994; Owen and Craddock, 1996; Owen and McGuffin, 1993). These methods – linkage analysis in families with multiple affected members, allele-sharing studies in affected relative pairs and association studies – will be discussed in turn below.

4.2.1 Linkage studies in 'multiplex families'

The approach that was adopted first by most researchers was to overlook the complex pattern of transmission of schizophrenia and to carry out linkage studies in large, multiply affected families. The strategy of searching for genes for complex disorders by studying apparent Mendelian subforms is based upon two main assumptions: first that etiological heterogeneity exists; and, second, that such high density families, or at least a proportion of them, are segregating genes of major effect. Similar approaches have been successful in other complex diseases including Alzheimer's disease (Goate *et al.*, 1991; Rogaev *et al.*, 1995;

Sherrington *et al.*, 1995), non-insulin-dependent diabetes (Vionnet *et al.*, 1992) and breast cancer (Anon, 1994). However, although families containing multiple schizophrenic patients in several generations do exist, there are no clear phenotypic features (e.g. early age at onset), that allow such cases to be distinguished from schizophrenia as a whole. Also, even in these highly atypical families, there are frequent irregularities of transmission, such as skipped generations and phenotypic heterogeneity. It is perhaps therefore not surprising that the early optimism engendered by a report of a dominant gene with incomplete penetrance mapping to chromosome 5q (Sherrington *et al.*, 1988) has not been substantiated by replication studies (McGuffin *et al.*, 1990). Furthermore, despite the fact that several groups have conducted large-scale genome searches using large numbers of highly informative markers, there is no good evidence that disease in any of these so called 'multiplex' families is caused by a mutation in a single major gene mapping anywhere in the genome.

4.2.2 Allele-sharing studies in affected relative pairs

Owing to the failure of classic linkage approaches using large pedigrees, most groups studying the molecular genetics of complex traits have modified their strategy and are applying methods based upon smaller families, or upon smaller subunits within multiplex families. The detail of these methods vary but most are based upon identity-by-descent methods (see chapter by Povey in this volume) and involve the analysis of affected siblings, although other pairs of relatives can also be used. Evidence of linkage is derived when affected pairs inherit a chromosomal fragment more often than not would be expected simply by chance.

There are several advantages to these approaches. Firstly, knowledge of the mode of transmission of the disorder is not required. Secondly, affected sib-pair analysis can detect genes of fairly modest effect (Risch, 1990). Thirdly, relative to classical analyses, they are relatively resilient to moderate degrees of diagnostic error which is potentially a particular problem in psychosis since there is no way to validate the diagnosis. Furthermore, families in which only a couple of members are affected are probably more typical of diseases like schizophrenia than large multiplex families and therefore are more readily obtained and the results may be more generalizable. The main disadvantage of this approach is that it is much less powerful than classical linkage analysis when it comes to detecting genes of major effect that segregate in a Mendelian fashion.

4.2.3 Linkage disequilibrium approaches

Providing that population stratification can be excluded, the causes of allelic association between a marker and a disease are due either to the DNA variant itself being of pathogenic significance or to tight linkage between the marker and the disease resulting in association being maintained over many generations. In practice, the latter tends to occur only if the distance between the two is 1 cM or less. Allelic association studies therefore offer the opportunity to map genes for complex diseases more precisely than linkage studies. At present, this opportunity is restricted by the effort required to carry out systematic genome searches with many thousands of markers, but this is certainly theoretically possible (Risch and Merikangas, 1996) and may soon be technically feasible.

However, to date, allelic association studies have, by necessity, been restricted either to attempts to refine the map positions of susceptibility genes that have been localized by linkage studies or to studies of candidate genes. The application of linkage disequilibrium approaches to disease mapping is particularly important for complex traits because linkage is unlikely to provide a small enough candidate region for a detailed analysis of all sequences contained within. Conversely, in the case of candidate gene studies, there are clear attractions to studying DNA variations that are likely to affect protein structure or expression (VAPSEs) rather than relying on maintenance of linkage disequilibrium between non-functional and often extragenic polymorphisms (Sobell *et al.*, 1992). Even so, it should be remembered that associations between disease and VAPSEs may still reflect linkage disequilibrium rather than an involvement *per se* in disease pathogenesis, and, in practice, it may be hard to discriminate between the two.

The existence of population stratification needs to be excluded in allelic association studies. This occurs when there has been a recent admixture of two or more populations. Marker–disease association can then result simply from both the disease and a marker allele occurring at a higher frequency in one group forming the new population admixture than in another. Along with chance, such stratification effects probably account for most of the non-replicated associations that have previously reported to date. One solution to stratification effects is to ensure that studies are carried out on homogenous populations with good ethnic matching between patients and controls. However, the source of stratification effects may not be known in any given population, and therefore association carried out on family material consisting of affected individuals and both their parents may be preferable, because in such parent–offspring trios a comparison of the frequency of transmitted and untransmitted alleles provides a perfect internal control.

4.2.4 Positional cloning studies in schizophrenia

The application of modern molecular genetic strategies adapted to identifying disease genes for complex traits has resulted in the emergence of some consistent findings and, at last, there are now reasons for optimism. The first finding of sustained interest came from a conventional genome scan at Johns Hopkins University (JHU) in 39 multiply affected families (Pulver *et al.*, 1994a). Although there was no convincing evidence for linkage based upon classical criteria (LOD score > +3), a *post hoc* analysis involving several different genetic models suggested the presence of a susceptibility locus at 22q12–q13. The same group co-ordinated a four-centre study aimed at replicating these findings, but the overall results were negative (Pulver *et al.*, 1994b). However, a number of groups have also reported positive data from this region (Coon *et al.*, 1994; Vallada *et al.*, 1995) and subsequently a UK group co-ordinated an international meta-analysis of data from 11 centres using the more robust affected sib-pairs approach which suggested (p = .001) a susceptibility locus linked to D22S278 (Schizophrenia Collaborative Linkage Group, 1996). Two groups subsequently attempted to pinpoint a chromosome 22 susceptibility locus with greater accuracy using a family-based association study design [the transmission disequilibrium test (TDT) (Spielman *et al.*, 1993)] and reported the most significant results at two closely linked markers, D22S278 (Moises *et al.*, 1995a) and D22S283 (Vallada *et al.*, 1995). However, the weak association detected in the former study was not statistically significant after correction for multiple testing. Also in the latter study (Vallada *et al.*, 1995), transmission distortion was observed in pedigrees that were already known to show allele-sharing from the sib-pair analysis, and the findings may therefore have resulted from linkage rather than linkage disequilibrium. Uncertainty therefore remains concerning the most likely map position of the putative schizophrenia locus on chromosome 22q12–13; indeed, in a subsequent case–control study we were unable to detect linkage disequilibrium between either marker and schizophrenia (NM Williams *et al.*, 1996).

A second encouraging linkage finding emerged from a study reported by Straub *et al.* (1995) from the Medical College of Viginia (MCV) who published analyses of markers from chromosome 6p24–p22 in 265 Irish families multiply affected by schizophrenia. When heterogeneity was allowed for, the MCV group obtained a maximum LOD score of +3.5 at D6S296 using a broad definition of the phenotype and a co-dominant genetic model. Although the results from this dataset met classical significance levels for linkage, they should be regarded as inconclusive evidence for linkage in complex traits (Lander and Kruglyak, 1996).

Supportive, but on their own not highly significant, LOD scores were subsequently reported by three independent groups (Antonarakis *et al.*, 1995; Moises *et al.*, 1995b; Schwab *et al.*, 1995). Others, however, have failed to detect evidence for linkage in this region (Gurling *et al.*, 1995; Mowry *et al.*, 1995). We have studied nine microsatellite markers spanning 40 cM of this region in a sample of 103 affected sibling pairs. Allele-sharing was examined using likelihood-based sib-pair analysis. No evidence for linkage was obtained, and the highest LOD score was only 0.192 for D6S309 (Daniels *et al.*, 1997).

Recent data from a large multicentre collaborative analysis of 6p markers in schizophrenia (Schizophrenia Collaborative Linkage Group for Chromosome 3, 6 and 8, 1996) suggests that the failure of ourselves and others to replicate the MCV findings could be due to lack of power resulting from the small effect size of the locus on 6p. The collaborative analysis contained 448 pedigrees yielding 367 sibships with two or more affected and included the positive data from Schwab *et al.* (1995) and Antonarakis *et al.* (1995) as well as those from many of the centres contributing to the collaborative study reported by Moises *et al.* (1995b). The numbers of fully informative sib-pairs ranged from 200 to 332 for each of the five markers from 6p and data were analysed using both non-parametric and parametric methods. The maximum evidence for linkage was obtained between D6S470 and D6S259 using multipoint sib-pair analysis with a LOD score of 2.19 in the new sample, and 2.68 when the data from MCV were included. The evidence for linkage did not increase with a broad diagnostic model similar to the one used in the MCV study. This corresponds to a very small genetic effect at the putative 6p locus with a relative risk to sibs (λs) of only 1.25.

Evidence for a third susceptibility locus for schizophrenia mapping to 8p22–21 has been reported from the same study of 39 families that implicated chromosome 22 (Pulver *et al.*, 1995). An affecteds-only analysis under a model of heterogeneity yielded maximum LOD scores of 2.35 for a dominant model and 2.20 for a recessive model. Further support for this finding has been subsequently obtained in a collaborative study reported by Moises *et al.* (1995b) and from the large multicentre collaborative study that also studied 6p. In the latter, the most positive results were a heterogeneity LOD score of 2.22 in the new sample and 3.06 when the JHU families were included, in both instances at D8S261 with a recessive model.

Finally, we have reported preliminary evidence for linkage of schizophrenia to chromosome 13q14–q32 in 13 multiply affected families (Lin *et al.*, 1995). A maximum heterogeneity LOD score of 1.61 was found at D13S144 using a dominant model, and a maximum multipoint LOD score of 2 assuming $\alpha = 0.40$ was obtained at D13S128. More recently the JHU group have reported positive LOD scores under dominant and recessive models, and significant non-parametric analyses for markers at 13q32 (Antonarakis *et al.*, 1996).

To summarize the above, there now appears to be converging evidence from several centres that positive results relating to chromosomes 6, 8, 13 and 22 are unlikely to be explicable by chance. However, in each case the evidence points to these being susceptibility loci, each of comparatively small effect, which alone are neither necessary nor sufficient to cause schizophrenia. It seems then that schizophrenia is more analogous to Type 1 diabetes mellitus (Davies *et al.*, 1994) than either Alzheimer's disease or breast cancer as being a common familial disorder where Mendelian subforms are either rare or non-existent but where the condition overall is substantially heritable and contributed to by multiple loci. The implications of this for further studies are clear. These should focus on methods capable of detecting genes of comparatively small effect, which are applicable regardless of the mode of transmission. Consequently, there has been a shift away from attempting to study the co-segregation of markers and schizophrenia within multiply affected families towards the collection specifically of affected sib-pairs and of samples for allelic association studies. We are currently undertaking a two-stage genome scan in 200 pairs of affected siblings with schizophrenia. In Stage 1, 100 affected sibling-pairs have been screened with 199 markers at an average spacing of 20 cM. Seven regions of interest have been identified and are currently being screened at 5 cM intervals in the whole sample.

4.3 Candidate gene association studies

Abnormalities of dopaminergic neurotransmission have long been implicated in schizophrenia. It is now known that there are two families of dopamine receptors, classed as D1-like and D2-like. Both receptor families are coupled to G-proteins but each has opposite effects on cyclic adenosine monophosphate AMP (CAMP) activity and differing pharmacological specificity (Sokoloff *et al.*, 1992). Typical anti-psychotic drugs have a high affinity for D2-like receptors, and much recent attention has focused on the members of the D2 family (known as D3 and D4) that are largely expressed in limbic areas of the brain. Studies on rat brain have shown that expression of the D3 receptor gene is up-regulated both by the administration of typical (e.g. haloperidol) and a typical neuroleptics (e.g. clozapine) (Buckland *et al.*, 1992). By contrast up-regulation of D1 receptors and the D2 receptor itself occurs only after the administration of typical neuroleptics (Buckland *et al.*, 1993).

The human D3 receptor maps to chromosome 3q13.3 and has a polymorphic site in the first exon that gives rise to a glycine-to-serine substitution in the N-terminal extracellular domain. This produces a *Bal*I restriction site (Lannfelt *et al.*, 1992). Two groups, our own in Cardiff, UK, and another in Rouffach and Paris in France, were independently able to demonstrate increased homozygosity at the *Bal*I D3 receptor polymorphism in schizophrenics compared with controls (Crocq *et al.*, 1992). We were subsequently able to replicate this finding (Mant *et al.*, 1994) and two other studies have provided partial support. Jonssen *et al.* (1993) found increased levels of homozygosity in schizophrenics who were good responders to neuroleptics while Nimgaonkar *et al.* (1993) found a strong effect in schizophrenics with a family history of the disorder.

The functional significance of the glycine-to-serine change in exon 1 of the dopamine D3 receptor is still unknown. Possible explanations of the association with schizophrenia are either that the polymorphism itself confers the susceptibility or that it is in linkage disequilibrium with a different polymorphism in or near the D3 receptor gene. We (Asherson *et al.*, 1996) have screened all six exons that make up the coding region of the gene using SSCP analysis. No other mutations were found that altered protein structure in a total of 36 schizophrenics and the same number of controls. Although no evidence of linkage was found in 24 multiply affected families, an excess of homozygotes was again observed among schizophrenics in a case–control comparison. When added to our earlier sample (Mant *et al.*, 1994), the excess both of homozygotes overall, and the 1–1 (Gly-Gly) genotype, was significant, with (as in some previous studies) most of the effect being contributed by males rather than females.

In total there have now been over 20 studies of the D3 receptor and schizophrenia and, although many of these have shown small, non-significant or absent homozygosity effects in schizophrenia, a meta-analysis demonstrates a small excess of homozygotes of the 1–1 genotype associated with an odds ratio of 1.3 and significant heterogeneity. To overcome unknown stratification effects, have performed the TDT test upon family data from nine European centres and showed a significant distortion in the direction of homozygote transmission. Again schizophrenia appeared most specifically associated with the 1–1 genotype with an odds ratio of approximately 2.

Because of their putative role as a principal site of therapeutic action of atypical neuroleptics such as clozapine, risperidone and sertindole, recent attention has also focused on the possible involvement of 5HT2 receptors in schizophrenia. A comparatively small-scale Japanese study suggested an association between the T102C polymorphism in the 5HT2a receptor gene and schizophrenia (Inayama *et al.*, 1994,) and in the UK an association was reported between this marker and response to clozapine (Arranz *et al.*, 1995). Our own group have been responsible for setting up a large European collaboration involving seven centres, the European Multicentre Association Study of Schizophrenia (EMASS) and have examined the T102C polymorphism in 571 schizophrenic patients and 639 ethnically matched controls. This showed an association between schizophrenia and allele 2 that was highly significant but with a small odds ratio of only 1.3. Nevertheless, because allele 2 is very common in European populations, this polymorphism (or one tightly linked) may make a substantial contribution to morbidity, with an attributable fraction of 0.35 (J. Williams *et al.*, 1996).

It should be noted that the T102C polymorphism occurs at a degenerate base within the coding sequence; that is, it does not alter the predicted amino acid sequence of the 5HT2a protein. This raises the strong suspicion that this particular polymorphism does not directly affect susceptibility to schizophrenia but is in linkage disequilibrium with some other functionally significant variant within the 5HT2a receptor gene or, less likely, another gene in close proximity. Erdmann *et al.* (1996) have carried out a mutation screen of the whole coding region and have reported two coding mutations neither of which appears to show an allelic association with schizophrenia. Nevertheless there may be other variants of pathogenic significance in the promoter or other regulatory regions and a search for these is in progress. Also, it remains possible that the T102C polymorphism might influence susceptibility to schizophrenia by quantitative rather than qualitative effects upon the protein, and possible mechanisms for this include post-transcriptional effects upon the efficiency of translation or the stability of the messenger (RNA) (Arranz *et al.*, 1995).

4.3.1 Dynamic mutations

Recently, evidence from two sources has suggested that a novel class of mutation, the expanded trinucleotide repeat, may be involved in the pathogenesis of schizophrenia. The involvement of this kind of mutation in the transmission of a disease is suggested by the phenomenon of 'anticipation', which is the term used to describe a pattern of decreasing age at onset or increasing severity of a disease through successive generations of a family. Although there may be as yet unknown mechanisms that can lead to anticipation, the only known cause of the phenomenon is the pathogenic trinucleotide repeat. As the name implies, trinucleotide repeats are repetitive DNA sequences consisting of three bases, for example $(CAG)_n$ or $(CCG)_n$, The difference between 'wild-type' and pathogenic sequences is only the number of repeat units in the sequence. Thus, for example, the possession of 37 or more CAG repeats in the huntingtin gene causes Huntington's disease whereas polymorphisms in repeat size less than 34 are, as far as is known, without obvious phenotypic consequence. Furthermore, there is a striking correlation between the number of repeat units that are present in a disease gene and the severity of the phenotype and an inverse correlation with age at onset.

Expanded trinucleotide repeats display a tendency to increase in size between generations, a characteristic that has led to their alternative title of 'dynamic mutations'. It is this phenomenon, combined with the correlation between repeat size and severity of phenotype, that is responsible for anticipation. These are most dramatically seen in myotonic dystrophy which is caused by a CTG expansion in the 3' untranslated region of the myotonin gene (Brook *et al.*, 1992), but also occur in Huntington's disease and other expanded CAG repeat disorders (O'Donovan and Owen, 1996).

The discovery of trinucleotide repeats as the mechanism behind anticipation has rekindled interest in observations, made at the beginning of the 20th century, that anticipation occurs in psychosis, and several studies using modern diagnostic criteria (e.g. Asherson *et al.*, 1994; Bassett *et al.* 1994) have confirmed that, in families where schizophrenia occurs in multiple generations, age of onset decreases in successive generations. However, it should be acknowledged that while these findings are consistent with transmission through a trinucleotide repeat mechanism, they can just as convincingly be attributed to systematic sampling biases (Asherson *et al*, 1994).

However, the trinucleotide repeat hypothesis of schizophrenia has recently been supported by several molecular genetic studies. Using the method of repeat expansion detection (RED) (Schalling *et al.*, 1993), we and others have observed that patients with schizophrenia have larger CAG/CTG repeats in their genome than unaffected controls (Morris *et al.*, 1995; O'Donovan *et al.*, 1995). These findings have been subsequently confirmed in a European multicentre study (O'Donovan *et al.*, 1996a). Surprisingly, although these RED studies were inspired by observation of possible anticipation in schizophrenia, the data from the largest sample suggests that there is no correlation between repeat size and age at onset

(O'Donovan *et al.*, 1996a) or any other classification of the phenotype (Cardno *et al.*, 1996). This suggests that CAG or CTG repeat size is not of central importance in determining the age at onset of schizophrenia or disease severity. Instead, the data suggest that like the other mutations we have discussed above, expanded CAG/CTG confer only susceptibility to the disorder (O'Donovan *et al.*, 1996a).

The caveats concerning association studies are also pertinent to the RED studies. Thus, stratification has not been entirely excluded as a possible cause of the finding. One possible confounder for which there is some evidence is that expanded trinucleotide repeats may actually be more commonly involved in diseases than we expect (O'Donovan *et al.*, 1996b). If this is confirmed, stratification may have occurred by selecting unusually healthy controls in one of the studies (O'Donovan *et al.*, 1996a). However, fortunately, the association remains even after individuals in schizophrenic sample were subjected to the same health exclusion criteria as the controls.

Another possibility that cannot be excluded by the RED data is that the expansions are *secondary* to another mutation (e.g. for example DNA repair enzymes), and it is this mutation that actually confers susceptibility to schizophrenia. Alternatively, expansions may be secondary to an environmental variable that is associated with schizophrenia such as drug treatment. While these caveats cannot be rejected, they seem rather unlikely as we have examined more than 150 CAG/CTG loci and found no evidence for secondary non-specific expansions (e.g. Bowen *et al.*, 1996). Nevertheless, until the specific locus or loci are identified, the trinucleotide repeat hypothesis of schizophrenia is likely to remain controversial.

4.4 Conclusions

Molecular genetic studies of schizophrenia have finally entered a phase where the gains are greater than the losses. Major obstacles remain; for example, the route from suggestive linkage to confirmed linkage may not be achievable with realistic sample sizes for genes of the size of effect indicated so far. Instead, suggestive linkage findings obtained with large samples of families containing affected relative pairs will have to be pursued by linkage disequilibrium mapping in the knowledge that some of the areas will ultimately emerge as false positives. A second challenge will be to develop methodologies that will allow a large number of candidate genes mapping to a particular region to be screened for mutations, and to allow the genotyping of the large sample sizes demanded. However, these difficulties should not prevail and although some of the findings discussed in this chapter may ultimately be rejected we can now expect steady advances in the molecular genetics of schizophrenia.

Acknowledgements

Research in our laboratories is supported by the Medical Research Council, the Wellcome Trust and the Welsh Scheme for the Development of Health and Social Research.

References

Anon. (1994) Breaking down BRCA 1 (Editorial). *Nature Genet.* **8,** 310.

Antonarakis SE, Blouin J-L, Pulver AE *et al.* (1995) Schizophrenia susceptibility and chromosome 6p24–22. *Nature Genetics* **11,** 235–236.

Antonarakis SE, Blouin J-L and Curran M. (1996) Linkage and sib-pair analysis reveal a potential schizophrenia susceptibility gene on chromosome 13q32. *Am. J. Hum. Genet.* **59 (**4), A210.

Arranz MJ, Collier D, Sodhi M, Ball D, Roberts G, Price J, Sham P and Kerwin R. (1995) Association between clozapine response and allelic variation in the 5HT2A receptor gene. *Lancet* **346,** 281–282.

Asherson P, Walsh C, Williams J, Sargeant M, Taylor C, Clement A, Gill M, Owen M and McGuffin P. (1994) Imprinting and Anticipation. Are they relevant to genetic studies of schizophrenia? *Br. J. Psychiat.* **164,** 619–624.

Asherson P, Mant R, Holmans P *et al.* (1996) Linkage, association and mutation analysis at the dopamine D3 receptor gene in schizophrenia. *Mol. Psychiat.* **1,** 125–132.

Bassett AS and Honer WG. (1994) Evidence for anticipation in schizophrenia. *Am. J. of Hum. Genet.* **54,** 864–870.

Bowen T, Guy C, Speight G *et al.* (1996) Expansion of 50 CAG/CTG repeats excluded in schizophrenia by application of a highly efficient approach using RED and a PCR screening set. *Am. J. Hum. Genet.* **59,** 912–917.

Brook JD, McCurrach ME, Harley HG *et al.* (1992) Molecular basis of myotonic dystrophy: expansion of a trinucleotide (CTG) repeat at the 3′ end of a transcript encoding a protein kinase family member. *Cell* **68,** 799–808.

Buckland PR, O'Donovan MC and McGuffin P. (1992) Changes in dopamine D1, D2 and D3 receptor mRNA levels in rat brain following antipsychotic treatment. *Psychopharmacology* **106,** 479–483.

Buckland PR, O'Donovan MC and McGuffin P. (1993) Clozapine and sulpiride upregulate dopamine D3 receptor mRNA levels. *Neuropharmacology* **32,** 901–907.

Cardno AG, Murphy KC, Jones LA *et al.* (1996) Expanded CAG/CTG repeats in schizophrenia: a study of clinical correlates. *Br. J. Psychiat.* **169,** 766–771.

Coon H, Holik J, Hoff M, Reimherr F, Wender P, Mylesworsley M, Waldo M,

Freedman R and Byerley W. (1994) Analysis of chromosome 22 markers in nine schizophrenia pedigrees. *Am. J. Med. Genet.* **54,** 72–79.

Crocq MA, Mant R, Asherson P *et al.* (1992) Association between schizophrenia and homozygosity at the dopamine D3 receptor gene. *J. Med. Genet.* **29,** 858–860.

Daniels JK, Spurlock G, Williams NM *et al.* (1997) A linkage study of chromosome 6p in sib-pairs with schizophrenia. *Am. J. Med. Genet. (Neuropsychiat. Genet.),* (in press).

Davies JL, Kawaguchi Y, Bennett ST *et al.* (1994) A genome-wide search for human type 1 diabetes susceptibility genes. *Nature* **371,** 132–137.

Erdmann J, Shimron-Abarbanell D and Rietschel M. (1996) Systematic screening for mutations in the human serotonin-2A (5-HT2A) receptor gene: identification of two naturally occurring receptor variants and association analysis in schizophrenia. *Hum. Genet.* **97,** 614–619.

Goate A, Chartier-Harlin M-C, Mullan M *et al.* (1991) Segregation of a missense mutation in the amyloid precursor protein gene with familial Alzheimer's disease. *Nature* **349,** 704–707.

Gottesman II. (1991) *Schizophrenia Genesis.* WH Freeman, New York.

Gottesman II and Shields J. (1982) *Schizophrenia, the Epigenetic Puzzle.* Cambridge University Press, Cambridge.

Gurling H, Kalsi G, Chen ACH, Green M, Butler R, Read T, Murphy P, Curtis D and Sharma T. (1995) Schizophrenia susceptibility and chromosome 6024–22. *Nature Genetics* **11,** 234–235.

Inayama Y, Yoneda H, Ishida T *et al.* (1994) An association between schizophrenia and a serotonia receptor DNA marker (5HTR2). *Neuropsychopharmacology* **10,** 56s.

Jonssen E, Lannfelt L, Sokoloff P *et al.* (1993) Lack of association between schizophrenia and alleles of the D3 receptor gene. *Acta Psychiat. Scand.* **87,** 345–349.

Lander E and Kruglyak L. (1996) Genetic dissection of complex traits: guidelines for interpreting and reporting linkage results. *Nature Genetics* **11,** 241–247.

Lander ES and Schork NJ. (1994) Genetic dissection of complex traits. *Science* **265,** 2037–2048.

Lannfelt L, Sokoloff P, Martres M, Pilon C, Giros B, Jonsson E, Sedvall G and Schwartz JC. (1992) Amino-acid substitution in the dopamine D3 receptor as a useful polymorphism for investigating psychiatric disorders. *Psychiat. Genet.* **2,** 249–256.

Lin M-W, Curtis D, Williams N *et al.* (1995) Suggestive evidence for linkage of schizophrenia to markers on chromosome 13q14.1–q22. *Psychiat. Genet.* **5,** 117–126.

Mant R, Williams J, Asherson P, Parfitt E, McGuffin P and Owen MJ. (1994) Relationship between homozygosity at the dopamine D3 receptor gene and schizophrenia. *Am. J. Med. Genet.* **54,** 21–26.

McGuffin P, Sargeant M, Hetti G, Tidmarsh S, Whatley S and Marchbanks RM. (1990) Exclusion of a schizophrenia susceptibility gene from the chromosome 5q11–q13 region. New data and a reanalysis of previous reports. *Am. J. Hum.* Genet. **47,** 524–535.

McGuffin P, Asherson P, Owen M and Farmer A. (1994) The strength of the genetic effect – is there room for an environmental influence in the aetiology of schizophrenia? *Br. J. Psychiat.* **164,** 593–599.

Moises HW, Yang L, Li T, Havsteen B, Fimmers R, Baur MP, Liu XH and Gottesman II. (1995a) Potential linkage disequilibrium between schizophrenia and locus D22S278 on the long arm of chromosome 22. *Am. J. Med. Genet. (Neuropsychiat. Genet.)* **60,** 465–467.

Moises HW, Yang L, Kristbjarnarson H *et al.* (1995b) An international two-stage genome wide search for schizophrenia susceptibility genes. *Nature Genetics* **11,** 321–324.

Morris AG, Gaitonde E, McKenna PJ, Mollon JD and Hunt DM. (1995) CAG repeat expansions and schizophrenia: association with disease in females and with early age-at-onset. *Hum. Mol. Genet.* **4,** 1957–1961.

Mowry BJ, Nancarrow DJ, Lennon DP *et al.* (1995) Schizophrenia susceptibility and chromosome 6p24–22. *Nature Genetics* **11,** 233–234.

Nimgaonkar VL, Zhang XR, Caldwell JG, Ganguli R and Chakravarti A. (1993) Association study of schizophrenia with dopamine D3 receptor gene polymorphisms: probable effects of family history of schizophrenia. *Am. J. Med. Genet.* **48,** 214–217.

O'Donovan MC and Owen MJ. (1996) Dynamic mutations and psychiatric genetics. *Psychol. Med.* **26,** 1–6.

O'Donovan MC, Guy C, Craddock N, Murphy KC, Cardno AG, Jones LA, Owen MJ and McGuffin P. (1995) Schizophrenia and bipolar disorder are associated with expanded CAG/CTG repeats. *Nature Genetics* **10,** 380–381.

O'Donovan M, Guy C, Craddock N *et al.* (1996a) Confirmation of an association between expanded CAG/CTG repeats in both schizophrenia and bipolar disorder. *Psychol. Med.* **26,** 1145–1153.

O'Donovan MC, Craddock N, Guy C, McGuffin P and Owen M. (1996b) Involvement of expanded trinucleotide repeats in common diseases (Letter). *Lancet* **348,** 1739–1740.

Owen MJ and Craddock N. (1996) Modern molecular genetic approaches to complex traits: implications for psychiatric disorders. *Mol. Psychiat.* **1,** 21–26.

Owen MJ and McGuffin P. (1993) Association and linkage: complementary strategies for complex disorders. *J. Med. Genet.* **30,** 638–639.

Pulver AE, Karayiorgou M, Wolyniec P *et al.* (1994a) Sequential strategy to identify a susceptibility gene for schizophrenia: report of a potential linkage on chromosome 22q12–q13.1. Part 1. *Am. J. Med. Genet.* **54,** 36–43.

Pulver AE, Karyiorgou M, Lasseter VK *et al.* (1994b) Follow up of a report of potential linkage for schizophrenia on chromosome 22q12–q13.1. Part 2. *Am. J. Med. Genet.* **54,** 44–50.

Pulver AE, Lasseter VK, Kasch L *et al.* (1995) Schizophrenia: a genome scan targets chromosome 3p and 8p as potential sites of susceptibility genes. *Am. J. Med. Genet.* **60,** 252–260.

Risch N. (1990) Linkage strategies for genetically complex traits. III. The effect of marker polymorphism analysis on affected relative pairs. *Am. J. Hum. Genet.* **33,** 630–649.

Risch N and Merikangas K. (1996) The future of genetic studies of complex human diseases. *Science* **273,** 1516–1517.

Rogaev El, Sherrington R and Rogaeva EA. (1995) Familial Alzheimer's disease in kindreds with missense mutations in a gene on chromosome 1 related to the Alzheimer's disease type 3 gene. *Nature* **376,** 775–778.

Schalling M, Hudson TJ, Buetow KH *et al.* (1993) Direct detection of novel expanded trinucleotide repeats in human genome. *Nature Genetics* **4,** 135–139.

Schizophrenia Collaborative Linkage Group (1996) A combined analysis of D22S278 marker alleles in affected sib–pairs: Support for a susceptibility locus for schizophrenia at chromosome 22q12. *Am. J. Med. Genet. (Neuropsychiat. Genet.)* **67,** 40–45.

Schwab SG, Albus M, Hallmayer J *et al.* (1995) Evaluation of a susceptibility gene for schizophrenia on chromosome 6p by multipoint affected sib-pair linkage analysis. *Nature Genetics* **11,** 325–327.

Sherrington R, Brynjolfsson J, Petursson H, Potter M, Dudleston K, Barraclough B, Wasmuth J, Dobbs M and Gurlin H. (1988) Localisation of a susceptibility locus for schizophrenia on chromosome 5. *Nature* **336,** 164–167.

Sherrington R, Rogaev El, Liang Y *et al.* (1995) Cloning of a gene bearing missense mutations in early-onset familial Alzheimer's disease. *Nature* **375,** 754–767.

Sobell JL, Heston LL and Sommer SS. (1992) Delineation of genetic predisposition to multifactorial disease: a general approach on the threshold of feasibility. *Genomics* **12,** 1–6.

Sokoloff P, Lannfelt L, Martres MP, Giros B, Bouthenet ML, Schwartz JC and Leckerman JF. (1992) The D3 dopamine receptor gene as a candidate gene for genetic linkage studies. In: *A Genetic Research in Psychiatry* (eds J Mendlewicz and JH Hippus), pp. 135–145. Springer Verlag, Berlin.

Spielman RS, McGinnis RE and Ewenst WJ. (1993) Transmission test for linkage disequilibrium: the insulin gene region and insulin-dependent diabetes mellitus (IDDM). *Am. J. Hum. Genet.* **52,** 506–516.

Straub RE, Maclean CJ, O'Neill FA *et al.* (1995) A potential vulnerability locus for schizophrenia on chromosome 6p24–22: evidence for genetic heterogeneity. *Nature Genetics* **11,** 287–293.

Vallada HP, Gill M, Sham P *et al.* (1995) Linkage studies on chromosome 22 in familial schizophrenia. *Am. J. Med. Genet. (Neuropsychiat. Genet.)* **60,** 139–146.

Vionnet N, Stoffel M, Takeda J *et al.* (1992) Nonsense mutation in the glucokinase gene causes early onset non-insulin dependent diabetes. *Nature* **356,** 721–722.

Williams J, Spurlock G, McGuffin P *et al.* (1996). Association between schizophrenia and the T102C polymorphism of 5-hydroxytryptamine type 2a receptor gene. *Lancet* **347,** 1294–1296.

Williams NM, Jones LA, Murphy KC, Cardno AG, Asherson P, Williams J, McGuffin P and Owen MJ. (1996) No evidence for allelic association between schizophrenia and markers D22S278 and D22S283. *Am. J. Med. Genet. (Neuropsychiat. Genet.)* **66,** 1–8.

5

Approaches to determining the genetic basis of non-insulin-dependent diabetes mellitus

Mark McCarthy

5.1 Non-insulin-dependent diabetes mellitus (NIDDM)

5.1.1 The global importance of NIDDM

Non-insulin-dependent (type 2) diabetes mellitus (NIDDM) represents one of the major subtypes of diabetes as defined by the World Health Organisation (World Health Organisation Study Group, 1985) and accounts for the majority of glucose intolerance encountered worldwide. Over 100 000 000 people worldwide have NIDDM (McCarty and Zimmet, 1994). Susceptibility to NIDDM and its complications is not restricted to the industrialized world: the prevalence is rising most rapidly in 'underdeveloped' countries, as increasingly urbanized populations adopt more 'Westernized' lifestyles.

The costs of NIDDM are enormous both for the individual living with diabetes and for health care systems attempting to mitigate the consequences of this condition. Morbidity and mortality result from the specific 'microvascular' complications of diabetes – nephropathy, neuropathy and retinopathy – and from the increased susceptibility to macrovascular disease, manifest as accelerated progression to coronary artery, cerebrovascular and peripheral vascular disease. Within industrialized countries it has been estimated that treatment of NIDDM and its complications consume between 5 and

Genetics of Common Diseases: future therapeutic and diagnostic possibilities,
edited by I. Day and S. Humphries.

10% of health care resources (Huse *et al.*, 1989; King's Fund Policy Institute, 1996).

The limitations of current treatments for NIDDM are obvious to the clinician, the scientist and the patient. Most therapies are empirical and/or of limited effectiveness: although they ameliorate the most distressing and dangerous features of NIDDM, they do not address the basic defects underlying its development. Therapeutic developments are most likely to follow improved understanding of the earliest events in the progression from health to overt NIDDM: this remains the major impetus behind attempts to define the genetic basis of this condition.

5.1.2 Evidence for a genetic basis for NIDDM

There is little doubt that individual susceptibility to NIDDM is governed in the main through genetic factors. The long-observed familial aggregation of late-onset diabetes has been shown by segregation analysis to be the result of genes rather than shared family environment (Cook *et al.*, 1994; Elston *et al.*, 1974; Hanson *et al.*, 1995; McCarthy *et al.*, 1994). Twin studies have supported the role of genes through demonstration of increased concordance for NIDDM between MZ as compared to DZ twins (Barnett *et al.*, 1981a,b; Kaprio *et al.*, 1992; Newman *et al.*, 1987). In these studies, the levels of concordance observed have varied widely, reflecting in part ascertainment bias and differing diagnostic schemes and leading to inconsistency in the actual estimates of heritability obtained. This inconsistency becomes less surprizing once one appreciates that heritability is simply that proportion of the variance in the susceptibility to disease due to genetic factors; as such, estimates of heritability will be dependent *inter alia* on the degree of variation in relevant environmental factors (e.g. food supply).

Perhaps the strongest evidence comes from genetic admixture experiments (Brosseau *et al.*, 1979; Gardner *et al.*, 1984; Knowler *et al.*, 1990; Serjeantson *et al.*, 1983). Nauru islanders have amongst the highest prevalence of NIDDM in the world. However, those islanders shown (HLA typing) to have significant European ancestry are much less likely to develop diabetes, even though culturally indistinguishable from their full-blooded neighbours. This strongly suggests protective European genetic influences.

Nothing quite proves a genetic basis for disease as much as finding the genes responsible! Defects in a number of genes [e.g. insulin, insulin receptor, transfer ribonucleic acid $tRNA^{Leu(UUR)}$ gene] have been shown to lead to glucose intolerance with features consistent with NIDDM (Gabbay, 1980; Kadowaki *et al.*, 1994; McCarthy and Hitman, 1993; Taylor,

1992). In addition, diabetes (of NIDDM subtype) is a feature of many inherited conditions (Rimoin and Rotter, 1982); the genes for several of these have been cloned.

It would be foolish to suggest that genetic factors entirely determine individual susceptibility to NIDDM. A variety of co-factors which modulate individual risk has been clearly established, including obesity, lifestyle and exercise (Manson *et al.*, 1991; Manson *et al.*, 1992). Although obesity and exercise capacity may be partly genetically determined themselves (Bouchard *et al.*, 1990; Nyholm *et al.*, 1994), cultural and environmental differences inevitably contribute to individual risk of NIDDM. Most recently, Barker *et al.* have presented compelling evidence that features such as low birth weight are associated with later diabetes (Barker *et al.*, 1993a,b). The mechanism for this association is as yet unclear, though poor intrauterine and infant nutrition has been proposed as underlying both.

5.1.3 Problems with the genetic study of NIDDM

NIDDM is an archetypal complex trait. Like so many of the diseases underlying the majority of ill-health in human populations (such as asthma, hypertension, schizophrenia, susceptibility to infectious agents) NIDDM aggregates within families, but does not segregate in classical Mendelian fashion. Such patterns of segregation indicate that susceptibility is likely to be under the control of several loci, interacting with each other and with environmental co-factors. As a consequence, correlations between the clinical disease phenotype and the individual underlying genotypes are weak.

These facts have general implications for the methodologies which might be employed to identify the genes responsible for a complex trait. Certain additional complications are presented by the study of NIDDM in particular, as listed below:

(i) *Diagnostic difficulties.* The diagnosis of diabetes rests on dichotomization of a glucose distribution which is continuous within populations. The most widely employed diagnostic criteria (World Health Organisation Study Group, 1985) are based on glucose levels, which reflect the propensity to develop specific diabetic complications. It is by no means inevitable that such thresholds have any particular significance with regard to genetic or environmental parameters determining the development of hyperglycaemia: these may influence glucose levels throughout the range of glucose tolerance. Furthermore, the diagnostic procedure (the oral glucose tolerance test) is poorly reproducible and inaccurate (Swai *et al.*, 1991). Even amongst diabetic subjects, NIDDM is generally a diagnosis of exclusion, made after other causes (e.g. IDDM, pancreatitis, mitochondrial diabetes, Cushing's syndrome) are ruled out. The boundaries between the types of diabetes are blurred: many subjects thought on clinical grounds to have NIDDM, show features more typical of slow-onset IDDM [latent autoimmune diabetes of adulthood (LADA)] (Tuomi *et al.*, 1993).

(ii) *Obscure pathophysiology.* All manner of detailed physiological studies of subjects with NIDDM have been undertaken. Sadly, few consistent findings have been forthcoming, and a rather sterile argument has resulted as to whether defects in insulin action or secretion are paramount (Taylor *et al.*, 1994; Turner *et al.*, 1989). In fact, when one looks at groups of subjects with NIDDM, there are manifest abnormalities in both these processes; it, therefore, seems unlikely that physiological analyses will identify the causal (or other) relation between them. Physiological and metabolic studies have often emphasized differences rather than similarities between NIDDM subjects, parallelling the heterogeneity in morphology and response to treatment well known to any physician treating diabetic patients. It would seem fair to conclude that the clinical description of NIDDM encompasses individuals who have reached a state of glucose intolerance through a multiplicity of diverse pathological processes. Physiological studies (which necessarily study groups of subjects) have failed to dissect this heterogeneity. The task of defining the basic defects has not been helped by the likelihood that the array of defective processes (and genes) shows considerable ethnic differences.

(iii) *Complex inheritance.* Although the various lines of evidence adduced above indicate that genes are involved in susceptibility to NIDDM, we know little about the number or nature of these genes (see below)

(iv) *Difficult demographics.* Most successful cloning studies in single gene disorders have used large families segregating the disease as the basic substrate for linkage studies to locate the susceptibility locus. However, the collection of multigenerational families segregating NIDDM is severely hampered by its late onset and increased mortality. Furthermore, the high frequency of subclinical disease mandates testing of apparently unaffected relatives; even then, the age-related penetrance limits the information which a diagnosis of 'normal' glucose tolerance conveys.

(v) *Problematical quantitative trait analyses.* Researchers in many complex traits have attempted to deal with the weak correlation between genotype and distal clinical phenotype by studying quantitative intermediate traits. The not unreasonable assumption is that the relationship between genotype and intermediate phenotype will be stronger (and thus more easily detected) than that with the clinical disease phenotype. There are two major problems frustrating this approach in NIDDM. Firstly, the most appropriate intermediate traits (β-cell mass, β-cell function, insulin sensitivity, etc.) are difficult and expensive to measure, and reproducibility is often poor. More importantly, once subjects become hyperglycaemic, many of these intermediate traits are influenced by the hyperglycaemia itself and by its treatment (Sing *et al.*, 1992). Consequently, trait measurements in diabetic individuals represent both bottom-up (genetic) and top-down (diabetes-related) effects making interpretation and analysis extremely difficult.

5.1.4 Important unresolved questions about the genetics of NIDDM

If we are to move effectively to identify the genes underlying NIDDM, and use this information to unravel the physiology of this disease, it would be helpful to have some idea of what we are looking for. Unfortunately, we have little to guide us on our way. Although we can easily calculate a measure of familiality for NIDDM (e.g. λ_{sib}, the sibling recurrence rate, is around 4 for European populations) (Rich, 1990; Risch, 1990a) which provides some upper limit to the overall effect of genes, it is not clear how many loci contribute to this aggregation.

Crucially, we still have little idea whether there are major genes for NIDDM waiting for us to find. Susceptibility may turn out to be governed by the concerted action of a slew of genes each exerting relatively modest effects. Techniques such as segregation analysis, which have provided answers in Mendelian disorders, are of limited use in complex diseases (McCarthy *et al.*, 1994; Ott, 1990). If we knew the penetrance values and frequencies of the most important susceptibility alleles, this might direct us to ascertain the pedigrees likely to be most informative for their detection. If we knew which pathophysiological processes were defective in NIDDM, it would influence our choice of candidate genes, and perhaps dictate which intermediate trait parameters we measured. Unfortunately, in NIDDM, as with other complex traits, these questions will probably only be answered when the genes themselves are found, allowing us to estimate directly their parameters, their effects and interactions.

Our ignorance acts both as a spur and an impediment to our progress. When the genetic architecture of a disease is unknown, we cannot predict which manoeuvres will lead us most efficiently to the crucial loci. How should one proceed therefore? All studies of the genetics of NIDDM (implicitly or otherwise) adopt assumptions regarding the likely genetic architecture of the disease in the population being examined: generally, these have limited scientific support. However, having made such assumptions, the researcher's task is clear: to choose appropriate research tools and execute the experiments with as much accuracy and diligence as possible.

5.2 Overview of approaches to dissect NIDDM

5.2.1 Forward, reverse and 'direct' genetics

Much has been made of the distinction between classical 'forward' genetics (which starts from an understanding of the biology of a disease and leads to identification of the causative mutation) and the more contemporary 'reverse' genetic approach, typically applied to complex trait analysis. The basic strategy enshrined in the reverse genetics approach follows backwards from identification of mutations found to be linked or associated with disease to a reconstruction of the biology of the disease (Collins, 1992, 1995). Although this has been a useful paradigm, it has probably outlived its usefulness as new technologies and sources of information have developed.

Figure 5.1 summarizes current approaches to the study of NIDDM genetics and, in the process, resurrects the pivotal role of the 'candidate gene'. This term is used in the widest sense, and can be applied to any gene which (on the basis of prior information) has a plausible claim to involvement in the determining variation in the disease phenotype. Prior information supporting the candidacy can come from a variety of sources; increasingly, it will emerge from the synthesis of several lines of enquiry.

Once a candidate gene has been proposed, evaluation proceeds first by exhaustive attempts to identify genetic variation in the locus and its environs, and, subsequently, by accumulating evidence for (or against) a correlation between the candidate polymorphisms uncovered and phenotypic variation relevant to the disease of interest. Significant genotype–phenotype correlations are a strong indicator of the biological relevance of the candidate polymorphism. Structure–function studies can then bridge the gap between changes at the DNA level and relevant biological systems, leading, one hopes, to improvements in patient care.

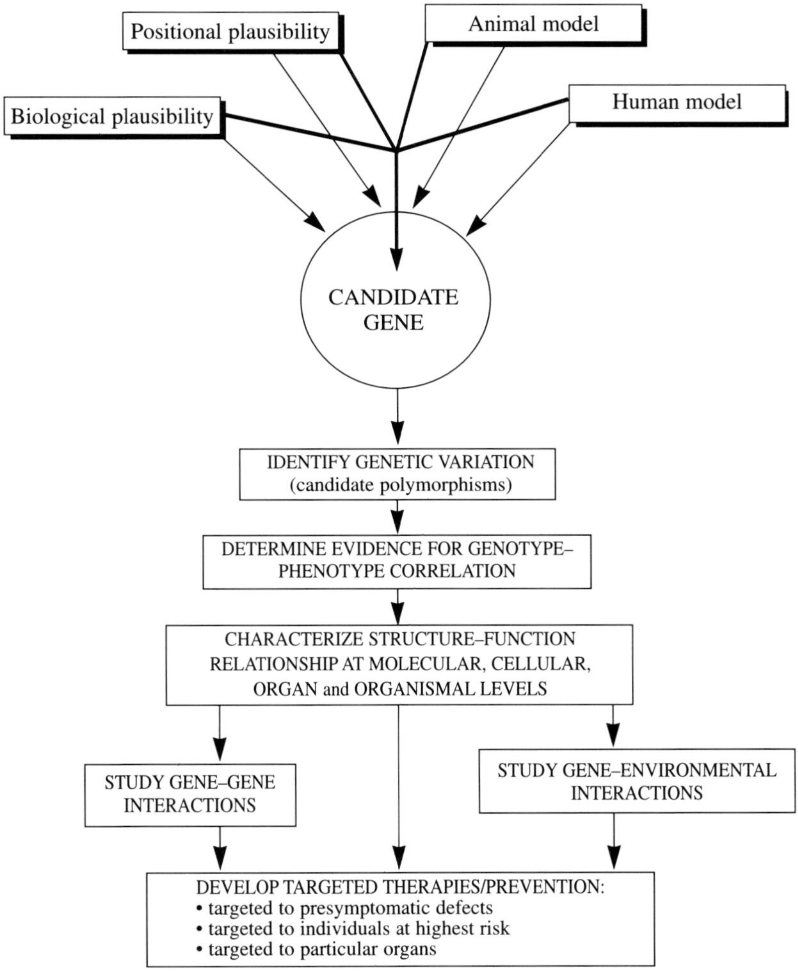

Figure 5.1. Approaches to the study of NIDDM.

In the remainder of this chapter, the elements of this pathway are dissected.

5.3 Choosing a candidate gene

5.3.1 Candidates arising from biological plausibility

Ignorance regarding the pathophysiological basis of NIDDM renders determination of biological plausibility an imprecise art. Subjects with hyperglycaemia display a wide variety of abnormalities in a number of systems, and it is by no means clear which of these defects are primary (in the sense of being early and therefore potentially inherited) and which lie downstream. Dysfunction in any one of a vast number of loci could, in theory at least, lead to some of the metabolic features of NIDDM – any such gene has biological plausibility as a candidate. (*Table 5.1* lists some candidate systems and illustrative candidate genes.) Burgeoning knowledge of the cellular and metabolic processes involved (e.g. in the control of insulin action, insulin secretion, energy balance) is rapidly expanding the list (Kahn, 1994; Porte, 1991).

This approach, based on biological candidacy, has been the predominant one employed to date in the study of NIDDM; many of the genes in *Table 5.1* have already been studied. Given that the major genes underlying NIDDM remain unknown (assuming that they exist), this strategy may be considered a failure. However, several loci making minor contributions have been identified by this method, (Almind *et al.*, 1993; Gabbay, 1980; Hitman *et al.*, 1995; Taylor, 1992; Walston *et al.*, 1995) and

Table 5.1. Candidate pathways with a plausible role in development of NIDDM. A few representative candidate loci are given for each

Candidate pathway	Candidate loci
Insulin action	Insulin receptor Insulin receptor substrate-1 (*IRS1*) Glucose transporter 4 (*GLUT4*)
β-cell function	Glucokinase Sulphonylurea receptor (*SUR*) Glucose transporter 2 (*GLUT2*) Insulin (*INS*)
β-cell development	Regenerating gene (*REG*) Hepatocyte nuclear factor 1-α (*HNF1-α*)
Incretins	Glucagon-like peptide-1 (*GLP1*) GLP-1 receptor (*GLPR*)
Hypothalamic regulation	Neuropeptide Y (*NPY*) Gastric inhibitory polypeptide (*GIP*)
Thermogenesis	β_3 adrenergic receptor (*B3AR*) Uncoupling protein (*UCP*)
Adipocyte differentiation, regulation	Peroxisome proliferator-activated receptor γ (*PPARG*) Adipocyte fatty acid binding protein (*AP2*)

the failure of scientists to choose the correct candidate gene thus far does not provide a logical basis for rejecting the strategy entirely.

However, the limitations are obvious. Many critical genes involved in regulating the processes in *Table 5.1* are not characterized. There are doubtless other pathways that could influence glucose homeostasis than those listed. For example, we know virtually nothing about loci controlling development of the tissues involved in glucose homeostasis.

5.3.2 Candidates from animal models

The advantages in studying animal models of complex human traits are well rehearsed (Lander, 1989), and several rodent models for NIDDM have been developed, including the Goto-Kakizaki (GK) rat (Galli *et al.*, 1996; Gauguier *et al.*, 1996), the Zucker diabetic fatty rat (Tokuyama *et al.*, 1995), the sand rat [*Psammomys obesus* (Barnett *et al.*, 1994)], the Nagoya–Shibata–Yasuda (NSY) mouse (Ueda *et al.*, 1995) and a non-obese, non-diabetic/New Zealand obese (NON/NZO) mouse outcross (Leiter and Herberg, 1996). These models display considerable phenotypic similarities to human NIDDM, including evidence, in most cases, for defects in both insulin secretion and action. Clearly, identification of the loci underlying susceptibility to diabetes in these animals will provide useful lessons applicable to, and candidates for, the study of human NIDDM. It is by no means inevitable that the human homologues of the loci underlying these animal models will play a major role in NIDDM susceptibility in man; however, dissection of the animal models may define metabolic, developmental and regulatory pathways that suggest further candidate genes [parallelling the explosion in understanding arising from the study of the *ob* mouse (Guerre-Millo *et al.*, 1996; Zhang *et al.*, 1994)].

Two groups (Galli *et al.*, 1996; Gauguier *et al.*, 1996) recently reported the results of genome-wide scans identifying the major susceptibility regions for hyperglycaemia in the GK rat, their findings showing a gratifying degree of agreement (Permutt and Ghosh, 1996). In the GK rat, NIDDM is clearly polygenic: surprizingly, fasting and postprandial glucose levels seem to be under distinct genetic control (Galli *et al.*, 1996). These results will spur efforts: (i) to identify the pathogenic loci within the rat and then screen the human homologues for mutations, and (ii) to determine evidence for linkage between NIDDM and the syntenic regions in man.

5.3.3 Candidates from human experiments of nature

If experiments of nature in generating animal models of disease are proving useful, the study of similar 'human models' may be even more rewarding.

Tattersall and Fajans (1975) first described families segregating what was termed 'maturity onset diabetes of the young' (MODY). Affected individuals within these pedigrees had early-onset diabetes (often less than 25 years) but were not insulin-dependent. Indeed, there appeared to be phenotypic similarities with NIDDM, with a β-cell defect as the predominant metabolic finding. Segregation within these pedigrees strongly suggested autosomal dominant genes of high penetrance indicating that these families might allow diabetes genes to be revealed using fairly standard single-gene parametric analysis approaches (Bell *et al.*, 1991).

The sense that there was significant phenotypic heterogeneity within MODY (Fajans, 1990; Tattersall and Mansell, 1991) was confirmed when linkage studies identified at least three MODY genes – *MODY1, 2* and *3* – located on chromosomes 20, 7 and 12 respectively (Bell *et al.*, 1991; Froguel *et al.*, 1992; Hattersley *et al.*, 1992; Menzel *et al.*, 1995; Vaxillaire *et al.*, 1995). Glucokinase was identified as the locus underlying *MODY2* in 1992 (Froguel *et al.*, 1992; Hattersley *et al.*, 1992), and an appreciation of the critical role played by this locus in informing the β-cell about the ambient glucose concentration has followed. In late 1996, Bell and his colleagues identified hepatocyte nuclear factor 1-α (*HNF1-α*) as the locus for *MODY3* (Yamagata *et al.*, 1996a) and its intracellular regulator *HNF-4*α as the locus for *MODY1* (Yamagata *et al.*, 1996b).

What have these studies taught us? Perhaps the most important question remains unanswered – do these loci cause only MODY or are they important determinants of everyday late-onset NIDDM? Glucokinase mutations have been found only in occasional subjects with apparent NIDDM (Hattersley and Turner, 1993; Saker *et al.*, 1993; Shimada *et al.*, 1993); however, a wide variety of case–control studies have suggested associations between microsatellites in the region and NIDDM (Chiu *et al.*, 1992; McCarthy *et al.*, 1993b). Two groups have also suggested that a promoter variant at –30 influences susceptibility to impaired glucose tolerance (Stone *et al.*, 1996) and gestational diabetes (Zaidi *et al.*, 1997). The possibility that other *cis*-acting mutations upstream of the gene influence susceptibility to NIDDM (and explain the positive associations) has not been directly addressed. Assessment of the *MODY1* and *MODY3* loci will be facilitated now that the genes have been cloned. Intriguingly, a number of groups have seen evidence for linkage to the *MODY1* region in genome-wide scans (Bowden *et al.*, 1996; Ji *et al.*, 1996). Further, a group led by Lander and Groop, who performed a genome-wide scan in a small collection of pedigrees from the Botnia region of Finland, found linkage between the *MODY3* region and NIDDM amongst those families with a predominant β-cell defect, even though the average age of diagnosis of diabetes exceeded 55 years (Mahtani *et al.*, 1996).

Whether or not the MODY genes play a role in NIDDM, much has been and will be learnt from these findings. Identifying genes which, when defective, result in an NIDDM-like phenotype enhances our understanding of the β-cell and the ways in which insulin secretory defects contribute to the diabetic phenotype (and thereby goes some way perhaps towards understanding the complex interplay between insulin secretion and insulin sensitivity). MODY also provides some important general lessons for genetic analysis of NIDDM. The possibilities and limits to dissection of a complex phenotype are illustrated. Despite intensive metabolic and physiological studies of NIDDM, it is notable that the three main chips off the NIDDM block (LADA, MODY and mitochondrial diabetes) have been defined, not in terms of intermediate trait physiology, but by their clinical features and segregation patterns. This suggests that attempts to use physiological criteria to identify further subsets within NIDDM are unlikely to be successful. Furthermore, early physiological assessments of subjects with MODY gave inconsistent results, leading to contemporary suggestions of 'genetic and physiological heterogeneity' (Fajans, 1990; Tattersall and Mansell, 1991). Now that the genetic basis has been characterized, and it is possible to study individuals of known genotype, sophisticated metabolic assessment becomes possible and the differences between the cellular and metabolic consequences of glucokinase and HNF mutations appreciated (Byrne *et al.*, 1994; Herman *et al.*, 1994; Sturis *et al.*, 1994). Two other lessons derive from the prolonged efforts to clone the *MODY1* and *MODY3* genes. Firstly, positional cloning remains a time-consuming business, even for simple Mendelian diseases. Secondly, biological plausibility is an unreliable guide to likely loci (since *HNF* would not have been at the top of anyone's list).

Much the same can be said for the lessons learnt from studies of the mitochondrial genome. It was noted that subjects with MELAS often had diabetes, and subsequently that the mitochondrial encephalopathy with lactic acidosis and stroke-like episodes (MELAS) mutation in the $tRNA^{Leu(UUR)}$ gene was the cause of diabetes in about 1–2% of NIDDM (Kadowaki *et al.*, 1994). The importance of the mitochondrial energy-generating system to β-cell function has been emphasized by this observation. Diabetes (of NIDDM subtype) is a feature of a large number of inherited conditions (Rimoin and Rotter, 1982). As the genes and pathways underlying these conditions are uncovered, further candidates will doubtless emerge (Carvajal *et al.*, 1996).

5.3.4 Candidates arising out of genomic position

Uncertainties about the pathophysiological basis of NIDDM, and worries that animal and human models of diabetes are not guaranteed to identify the most important loci governing individual susceptibility to NIDDM, have led to the application of linkage-based methods which make no such assumptions. The genome-wide search strategy seeks to identify genomic regions which show increases in haplotype sharing (i.e. linkage) amongst affected relatives: such regions are likely to harbour susceptibility loci. The approach has seen considerable success in single-gene disorders (Collins, 1992, 1995) and is being increasingly applied to complex traits (e.g. Davies *et al.*, 1994; Ebers *et al.*, 1996; Hashimoto *et al.*, 1994; Sawcer *et al.*, 1996). The essential requirements for this approach are:

(i) *Large family resources.* Precise power calculations are difficult when, as is usually the case with complex traits, the magnitude of the individual locus effects is unknown. For loci of large effect (e.g. *HLA* in IDDM), conclusive evidence for sharing has been obtained with only 100 small nuclear families (Davies *et al.*, 1994). Given reasonable assumptions for NIDDM, major loci conferring susceptibility in outbred populations will have individual λ_{sib} values of around 1.3–2.0; as such, the number of families needed to provide reasonable power for reliable detection will be at least 500, and more likely in the thousands (Risch, 1990b).

(ii) *An array of highly polymorphic markers* arranged throughout the genome. Simple tandem repeats fulfil this requirement and panels of 300–400 markers (i.e. approx 10 cM spacing) are available for this purpose (Dib *et al.*, 1996). Tri- and tetranucleotides are often easier to type using fluorescent technology, but throughput is reduced by the fact that fewer markers can be multiplexed for electrophoresis (Sheffield *et al.*, 1995).

(iii) *Availability of high throughput genetic analysis.* A typical genome-wide scan for NIDDM will require somewhere between 200 000 and 1 000 000 completed genotypes. Such throughput is possible only with highly automated systems for PCR amplification, efficient multiplexing for electrophoresis and rigorous standards for data interpretation and manipulation (Reed *et al.*, 1994).

(iv) *Suitable analytical software.* Recently developed programs such as GENEHUNTER meet many of the requirements: fast, efficient, non-parametric, multipoint analysis of small- and medium-sized pedigrees (Kruglyak and Lander, 1995b; Kruglyak *et al.*, 1996). Armed with this methodological, analytical and technological armamentarium, several large genome-wide scans are underway in a variety of large family resources, collected *inter alia* amongst Pima Indian, French, Finnish, Mexican-American, Danish and British populations. In recent months, the first of these scans have been completed and initial results reported. Even though these initial scans have used rather small clinical resources, the results have been encouraging for the genome-wide strategy in NIDDM (Hanis *et al.*, 1996; Mahtani *et al.*, 1996). Although the theoretical justification for genome-wide scans based on the detection of linkage disequilibrium (rather than linkage) has been advanced (Risch and Merikangas, 1996), the practical application of such endeavours remains some way off.

5.3.5 Issues relating to the planning of genome-wide scans for NIDDM genes

Although the basic idea behind a genome-wide linkage scan is simple enough, interpreting the results is not so clear cut (Bell and Lathrop, 1996; Risch and Botstein, 1996). To understand why, it is vital to appreciate the kinds of assumption that underlie such studies. In essence, the problem with these genome-wide scans is that they are underpowered; even large experiments involving thousands of families and millions of genotypes are not guaranteed to detect unequivocally all the major susceptibility genes (Risch and Merikangas, 1996). Given the likely genetic heterogeneity within NIDDM, it makes sense therefore to try to improve the power of a study by employing manoeuvres which minimize that heterogeneity. Of course, since the precise genetic architecture underlying NIDDM is unknown, attempts to do this must rely on thinly supported assumptions; whether the decisions that individual research groups have taken have been justified will become clear only in retrospect. The main decisions that a researcher embarking on such a collection has to make are:

(i) *Which population should be studied.* Several groups have focused on populations with a history of isolation and probable founder effects (e.g. Finns, Pimas) in the hope that a restricted number of susceptibility loci will be segregating within the population: as a result, these loci should be easier to detect. There is the additional advantage that such populations will facilitate the use of linkage disequilibrium to fine-map loci (see below). Although there is clear evidence for founder effects for some rare single-gene disorders within such populations [e.g. diastrophic dysplasia, autoimmune polyglandular disease, type 1 (APECED)] (Hästbacka *et al.*, 1993), it remains uncertain whether this effect will significantly ease the detection of complex trait loci where susceptibility alleles will be much more common (and hence more likely to slip through a 'bottleneck'). To the extent that this approach proves successful, these same population features will render any locus found more likely to be exerting an inflated population-specific effect. There is a clear need for scans conducted in a variety of ethnic groups, both 'isolated' and 'outbred', to allow both the specifics and the generalities of NIDDM-susceptibility to be defined.

(ii) *What diagnostic criteria should be used.* The limitations associated with measuring intermediate traits in diabetic subjects, and the difficulties imposed by the continuous gradation from normality to diabetes, have been discussed already. Very few groups have sufficient longitudinal data to escape these limitations. However, the Phoenix group have been studying the Pima Indians intensively for some 30 years and have accumulated a wealth of longitudinal physiological (intermediate trait) data (Lillioja *et al.*, 1992). These data diminish some of the problems with the reproducibility of trait measurement and diagnosis, and provide intermediate trait measures from both diabetic subjects (including those taken before the development of hyperglycaemia) and unaffected individuals. The results of the genome-wide scan for NIDDM and relevant traits in the Pimas should be published shortly. Mahtani and colleagues (1996) recently reported analysis of a genome-wide scan in Finnish pedigrees which utilized fasting insulin levels amongst diabetic family members as the basis for subdividing pedigrees: linkage to chromosome 12 was only evident when the families were so divided. A further approach is to identify families segregating traits which share phenotypic similarities with NIDDM, but in whom hyperglycaemia is infrequent – for example, women with polycystic ovarian syndrome (Franks, 1995). The hope is that finding the genes underlying intermediate trait abnormalities in these families will have relevance for NIDDM (Waterworth *et al.*, 1996). Faced with these difficulties (and with the expense associated with obtaining intermediate phenotype data), many groups have simply chosen to be pragmatic and have taken a clinical diagnosis of NIDDM (supported by clinical and biochemical criteria) as the basis for assigning affection status. Although some diagnostic inaccuracy is inevitable under such a scheme, ascertainment should be much quicker (Turner and Levy, 1996). Most of the linkage information for quantitative trait analyses resides in those plucked from the extremes of the relevant trait distributions: one can infer (even without knowing exactly which traits are involved) that subjects diagnosed with NIDDM are likely to represent individuals with those characteristics.

(iii) *Which families should be collected.* The collection of large multigenerational families has been the paradigm for single-gene disorders and for single-gene forms of NIDDM such as MODY, but there are both practical and methodological worries about this approach in the analysis of typical NIDDM. Firstly, large multigenerational pedigrees are rare given the demographic features of NIDDM, and when such a family is encountered, it may well not be segregating 'typical' NIDDM at all. Secondly, given the high prevalence of NIDDM, and the presumed high frequency of susceptibility alleles, intrafamilial heterogeneity is assured (and thereby most of the advantage of the 'large family/single-gene paradigm' is lost). Finally, simply because families are chosen to meet some preconceived notion of inheritance (e.g. one might select medium-sized families that appear to be segregating a highly penetrant autosomal dominant gene), there is no guarantee that these families are any less polygenic or heterogeneous than unselected pedigrees.

For these theoretical (as well as practical) reasons, the main focus of researchers has been on small- and medium-sized nuclear pedigrees: given the late onset of NIDDM, sib-pairs have been the most readily attained resource. Even then, there remain decisions as to which sib-pairs should be collected; not all sib-pairs are equally informative. For example, sib-pairs that come from densely affected sibships (for example, a sibship with four out of five affected sibs) might be expected, in principle, to be relatively limited in informativeness, given the high probability that multiple genes are segregating within the sibship (and therefore that identity by state – for affection status – does not imply identity-by-descent – for disease alleles). In fact, it has been shown that when the genetic architecture underlying a disease is unknown (as with NIDDM), it is not possible to make sensible decisions to restrict recruitment to families with certain configurations of affection status amongst close relatives, and that, as such, phenotypic information gathered on these first degree relatives adds little to the sib-pair study (McCarthy *et al.*, 1996).

(iv) *Genotyping vs. phenotyping effort.* Many of the decisions above require the researcher to make a choice between two extreme strategies.

Under one, s/he collects intensively phenotyped pedigrees from a particular rigidly defined ethnic group, and, in addition, restricts analysis to certain pedigrees in which the segregation pattern of disease meets some preconceived idea of the genetic architecture of the disease. Recruitment is likely to be costly and slow, and the gain in power per family will be dependent on the accuracy of the assumptions underlying this ascertainment strategy. Indeed, if the assumptions are wrong, the power per family may be reduced compared to alternative ascertainment strategies. Imagine the consequences of assuming that the major locus to be found is dominant, and therefore restricting recruitment to families with one parent affected, if the major locus is actually recessive; in this case such families will prove extremely poor substrates for its detection. At the other extreme, the researcher may choose to make as few assumptions as possible and to impose only those recruitment criteria that have some scientific support in terms of limiting genetic heterogeneity under most circumstances. (It seems sensible to avoid families where both parents are known to be diabetic, those which appear to have MODY or mitochondrial diabetes, and those with clinical features or a family history suggesting IDDM or LADA.) Such an ascertainment scheme will involve less investment in phenotyping and will yield large numbers of families in short order, but will generally entail increased genotyping requirements. Which of the two strategies will lead the researcher to the major loci most quickly, cheaply and effectively is impossible to predict when the genetic architecture of the disease is not known. Certainly, as genotyping costs fall and throughput improves, the latter approach becomes more practical. This 'liberal' recruitment strategy has been the approach taken by investigators involved in the British Diabetic Association (BDA)-Warren2 NIDDM repository for example, which has collected over 700 British and Irish pedigrees segregating NIDDM in the last 8 months and is currently starting a genome-wide scan which will eventually include 1000 pedigrees.

5.3.6 Interpreting the results of a genome-wide scan

Following implementation of a genome-wide search on a suitable collection of families segregating the trait of interest, the linkage results need to be analysed and interpreted. The standard LOD score threshold of +3 which has gained currency in the analysis of Mendelian disorders (where a single gene is known to be segregating) (Ott, 1991), will not do for complex traits, for which the genetic architecture is not known.

(Indeed, one must consider the possibility that none of the genes involved has an individual effect large enough to be seen in the sample.) As Lander *et al.* have pointed out in a number of publications (Lander and Kruglyak, 1995; Lander and Schork, 1994), any assessment of the significance of the findings obtained following a genome-wide scan must take account of the fact that multiple semi-independent tests have been performed (as the genome is traversed) and that the linkage statistic will (with a calculable probability) show positive deflections (in the direction of linkage) even when no linkage exists. For example, even if a trait has no genetic basis, a genome-wide scan conducted in a large sample of sib-pairs would be expected to yield one peak with a LOD score of greater than or equal to +2.6 and to produce one exceeding +4.0 in every 20 scans. Thresholds consistent with significant evidence for linkage have therefore been proposed which accord with a probability of less than 0.05 that the peak could have arisen by chance. These thresholds assume that the degree of multiple testing is restricted to the genome-wide analysis; any testing of multiple inheritance models, multiple diagnostic schemes, multiple intermediate traits or multiple statistical approaches within the same analysis calls for further inflation of the LOD score thresholds indicating linkage (Curtis, 1996; Lander and Kruglyak, 1995).

Why are these stringent criteria required? The experience of psychiatric genetics (and notably the work on manic depression) (Risch and Botstein, 1996) emphasizes the confusion that can arise when marginal results are promulgated as representing linkage. These criteria help to provide some yardstick against which the results of genome-wide scans can be judged.

While the rationale for these guidelines seems incontrovertible, their implementation is complicated by the fact that most of the genes in complex traits are not powerful enough to reach the criteria for significant linkage within any practical data set. For example in the work on IDDM, only HLA achieved a LOD score over 4.0: all other peaks have been much smaller (Davies *et al.*, 1994; Hashimoto *et al.*, 1994). Interpretation of the significance of these lesser peaks has to take into account the possibility (indeed the probability) that many will be spurious.

How can one prioritize peaks for further analysis? All else being equal, it makes sense to investigate the largest peaks first; these have the greatest chance of being genuine. One will certainly want to see whether peaks overlie any candidate genes (however defined). Re-analysis of the genome-wide scan after conditioning on genotypes at the largest peak may be a way of 'boosting' other true peaks (Cox *et al.*, 1996; Davies *et al.*, 1994) although there are formidable problems associated with significance testing under such analyses (Frankel and Schork, 1996).

Evidence for linkage disequilibrium within the region (see below) provides independent evidence for linkage and confirms the biological relevance of the region. Typing additional markers within a region showing increased allele sharing will not greatly help to distinguish between allele sharing which reflects a true susceptibility locus from allele sharing which reflects stochastic variation in the linkage statistic (provided the initial analysis was a true multipoint analysis conducted with reasonably polymorphic markers).

The single most reassuring finding for a researcher would be replication of the peak in a second data set analysed by another group, and accordingly great weight has been placed on the presence or absence of such confirmation. However, it needs to be appreciated that replication testing in complex traits has severe limitations. Firstly, even spurious peaks can appear to be replicated, particularly if inadequate attention is paid to multiple testing considerations (Lander and Kruglyak, 1995). Secondly, and even more disturbingly, failure to replicate does not necessarily disprove a positive linkage: it may be extremely difficult to obtain replication of even a true positive result. Amongst the reasons for failure to replicate are:

(i) *Ethnic differences between data sets.* The major loci determining susceptibility in one population group may not be so important in other populations.

(ii) *Diagnostic differences.* Different datasets may use different diagnostic endpoints or intermediate traits. A given locus may have a closer correlation with one phenotype than another.

(iii) *Ascertainment differences.* The types of families ascertained and the collection strategies used may differ between data sets. A data set based on large multigenerational families consistent with autosomal dominant segregation may detect rare genes of large effect not seen in an analysis of a large sib-pair collection (Matthews *et al.*, 1996). Even subtle differences in the explicit or implicit strategies for collecting sib-pairs (e.g. the proportion of affected parents) can have substantial effects on the ability to detect a given locus (McCarthy *et al.*, 1996).

(iv) *Stochastic factors.* A 'regression to the mean' effect operates such that the most detectable locus in an initial data set will tend to have a smaller effect in a subsequent replication study (Bell and Lathrop, 1996; Lander and Kruglyak, 1995; Suarez *et al.*, 1994). This arises because, on average, that first locus will have been 'helped' to detection by stochastic (random) allele sharing. If this seems somewhat counter-intuitive, consider a hypothetical trait governed by five identical non-linked genes of equal effect and assume that only one of these five breaches the threshold for significant linkage in the first genome-wide screen. Random effects will have 'determined' which of the five was detected and would be likely to boost one of the other loci in a replication study using a different data set. To guarantee reasonable power to achieve replication, the second data set needs to be considerably larger than the first (even if identical ascertainment schemes are applied to the two data sets). These causes of non-replication present serious (but not insuperable) obstacles to the interpretation of genome-wide peaks of indeterminate significance. In general, any replication represents a strong impetus for further analysis, even when meta-analyses of all genome-wide data fails to achieve genome-wide significance. Certainly, consensus is emerging from the published genome-wide scans for IDDM and schizophrenia (Peltonen, 1995; Todd, 1995).

5.3.7 Fine mapping within susceptibility regions

Genome-wide linkage scans define regions of excess allele sharing which remain too large (typically 10–20 cM) for conventional positional cloning procedures. In single-gene disorders, the principles of narrowing the critical recombinant region through typing of additional pedigrees are well-rehearsed (Ott, 1991). However, in complex diseases, no single recombinant can be relied on to define absolutely the position of the disease locus, and prohibitively large family resources are required to narrow the region appreciably (Kruglyak and Lander, 1995a). Even then, the critical region is defined in statistical rather than absolute terms and the formal (harrowing) possibility exists that the disease gene lies outside this narrowed region.

If additional recombinants within families are of limited value, can one use recombinants observed between pedigrees which reflect the genetic history of the underlying mutation within the population? In its broadest sense, linkage disequilibrium mapping seeks evidence of association between particular alleles and the disease of interest. Although such associations may arise for several reasons, the essential premise here is that the pathogenic mutation entered the population studied on one (or at worst, a limited number of) ancestral haplotype(s). This may have come about because of a genetic bottleneck or founder effect, as in Finns, (Hästbacka *et al.*, 1993; Norio *et al.*, 1973), through novel mutations which spread through populations by genetic drift and/or selective advantage, or through introduction of the mutation from a distinct ethnic group

(Stephens *et al.*, 1994). Whatever the cause, sequential recombinations over the generations will have removed more and more of the ancestral haplotype, leaving a progressively smaller area centred on the disease locus. Hence, finding such a region of sharing on chromosomes likely to be sharing disease mutations not only supports initial linkage findings, but may also help to fine-map the disease locus. Linkage disequilibrium mapping is of proven value in the fine-mapping of single-gene traits in founder populations (Hästbacka *et al.*, 1993), where it is no surprise to find that the overwhelming majority of cases have arisen through propagation of a single original mutation. It is less clear how helpful linkage disequilibrium mapping will be for complex traits (where multiple alleles at multiple loci are segregating) and for outbred populations (where the population dynamics should be less favourable). Under such circumstances, it may be the case that linkage disequilibrium is difficult to detect except at the disease mutation itself. However, results from the study of IDDM in British families suggest grounds for cautious optimism (Copeman *et al.*, 1995). Under some circumstances therefore, linkage disequilibrium mapping may aid identification of susceptibility regions of a size commensurate with typical positional cloning approaches (physical mapping, gene identification, etc.).

However, the need for these 'function-blind' reverse genetics approaches is likely to diminish as the biological information in databases grows. Increasingly, candidates will emerge from the list of loci mapping to the linked region which are implicated for reasons other than position alone (e.g. known biological function or through studies of animal models). Biological plausibility is increasingly used to assess positional candidates found to map to susceptibility regions uncovered in genome-wide linkage scans. Recent mapping efforts have placed about 16 000 genes on to the human chromosomal maps at high resolution, and this resource is likely to expand rapidly (Boguski and Schuler, 1995; Schuler *et al.*, 1996). Having identified a region of interest therefore, it is possible to list all the genes known to map to that location, and to prioritize those genes for further analysis using some assessment of biological relevance. The amount of functional and / or expression data available for many of these genes or gene fragments [expressed sequence tags (ESTs)] remains limited at present. Thus uncertainty about the pathophysiological basis of NIDDM, combined with limited functional information on the full pleiotropic actions of many gene products, makes it difficult to rule out loci definitively. When faced with a long list of genes known to map to a region thought to harbour a NIDDM-susceptibility locus, attempts at triage usually identify a handful of particularly 'hot' candidates and a few loci that can be rejected (e.g. collagen), leaving the majority of genes as 'possibles' for which some plausible connection to the diabetic phenotype can be constructed. Understandably, no-one wants to find out that a lack of biological imagination led them to pass over the susceptibility locus! This reticence is reinforced every time a mapping project throws up some unexpected locus (the *RET* protooncogene in Hirchsprung's, *HNF1-α* in *MODY3*) (van Heyningen, 1994; Yamagata *et al.*, 1996a). Current knowledge (as reflected in the databases) belies the pleiotropic function of many genes, and the surprises that can lurk behind the name (and implied function) assigned to a gene.

These deficits will be remedied in time. Ultimately, the Human Genome Project should deliver the sequence of all genes and their regulatory elements, and a thorough understanding of their expression patterns and functional significance. In the meantime, new technologies offer novel ways of identifying biologically relevant transcripts for study. The laborious methodology of subtractive hybridization (Reynet and Kahn, 1993) has given way to a plethora of techniques (e.g. differential display, complementary deoxyribonucleic acid- (cDNA-)representational difference analysis, cDNA-RDA) (Hubank and Schatz, 1994; Liang and Pardee, 1992) which allows the expression 'images' of tissues to be studied and compared. Reynet and Kahn (1993) were the first to use subtractive hybridization to identify a transcript (named *rad*) which was overexpressed in skeletal muscle from NIDDM subjects (but not subjects with IDDM) and therefore a strong candidate for a proximal defect (subsequent studies have not supported this promise). Fat and muscle provide accessible tissues allowing researchers to identify and characterize messages altered under relevant pathological (e.g. diabetes) or physiological (e.g. hyperinsulinaemia) situations. Such messages are likely to have some biological importance and become potential candidates. Similar application to study of the β-cell is handicapped by its inaccessibility and the consequent paucity of molecular biological resources (cell lines, cDNA libraries) (Takeda *et al.*, 1993).

Finally, the ability to sequence increasingly large tracts thought to contain susceptibility gene offers another route to gene identification. Already the *BRCA2* (breast cancer susceptibility) gene was identified from publicly deposited sequences from the Sanger Centre (Wooster *et al.*, 1995). Collaborative efforts seem likely to sequence large regions thought to harbour susceptibility genes for schizophrenia and sexual preference (Hamer *et al.*, 1993; Peltonen, 1995). A decision having been made to sequence the entire genome, it seems sensible to start with an interesting region!

5.3.8 Current status

Thus far (early 1997) the results of two small genome-wide scans have been published, and several other groups have reported interim results at meetings. Hanis *et al.* (1996) reported a locus (named *NIDDM1*) on chromosome 2q ($p < 0.0001$) amongst Mexican-American sib-pairs; to date, this has not been replicated. As described earlier, Mahtani *et al.* (1996), working on a set of pedigrees from Botnia (Finland), found evidence for excess allele sharing in the region of chromosome 12q known to harbour the *MODY3* gene. This linkage was seen predominantly in those families with the lowest fasting insulin levels, suggesting a common physiological basis with MODY (which is associated with defective β-cell secretion). This finding has not been replicated in other data sets (Lesage *et al.*, 1995) raising the possibility that this locus (designated *NIDDM2*) plays an exaggerated role amongst Finns: identification of the *MODY3* gene as *HNF1-α* will allow direct examination of the hypothesis that *MODY3* and *NIDDM2* are identical (with *NIDDM2* resulting from alleles causing lesser degrees of functional disturbance).

Other genome-wide scans are underway in a wide variety of populations; no clear picture has yet emerged from the preliminary results available at this stage (early 1997). It does seem clear, though, that there is no major gene for NIDDM with the kind of predominant effect that *HLA* plays in IDDM (Davies *et al.*, 1994).

The difficulties associated with demonstrating replication alluded to above have led to a degree of cynicism concerning the utility of this genome-wide approach (Risch and Merikangas, 1996). This is probably unwarranted. Despite the caveats above, any peak which does show replication in a second data set becomes an extremely strong candidate for involvement in NIDDM. Besides, as pointed out above, it seems likely that additional information (positional candidates, animal models, linkage disequilibrium, etc.) will be available to support the legitimacy of some findings at least. Workers in NIDDM can perhaps draw some succour from the work in schizophrenia. Several studies (but not all!) have provided qualified support for linkage to a region on chromosome 6p. Although meta-analysis of all these studies falls short of proving linkage to 6p, it must be appreciated that the studies analysed have involved a wide variety of ethnic groups and diagnostic criteria (Peltonen, 1995). The fact that that excess allele sharing on 6p survives such treatment is probably testament to a true effect (Lander and Kruglyak, 1995).

5.3.9 Many approaches – one true path?

A variety of different approaches to the identification of candidates have been described; outsiders will be tempted to ask which of these is likely to be successful. At this point, it is frankly impossible to offer a definite prediction, since the success of each strategy depends on the accuracy of the assumptions upon which each is predicated. Perversely, these assumptions will only readily be tested when the genes have been found and the genetic architecture of the disease is known.

Faced with such uncertainty, it is not surprising to find little consensus in the approaches being employed by groups across the world to address the genetic basis of NIDDM. This multiplicity of methodologies reflects differences in the assumptions made and the resources available, and has both advantages and disadvantages. On the positive side, all the research eggs are not in one methodological basket; the downside is that replication may be difficult when groups have employed different populations and diagnostic schemes.

5.4 Exploring candidate genes

5.4.1 Identifying candidate variation in candidate genes

At the core of the genetic dissection of any disease lies the identification of correlations between genomic variation and phenotypic variation relevant to the trait of interest. An essential first step, therefore, having chosen a candidate locus, lies in screening that gene to characterize candidate polymorphisms. (This is not to neglect the fact that association and linkage studies of biological candidates can be performed using anonymous microsatellites, for example, without screening the gene itself: let us assume that those experiments have been done.)

How should a gene be screened? This is not the place to discuss the multiplicity of laboratory methods advanced for the screening of mutations in the past few years – their very multiplicity is itself an indication that none enjoys the supreme confidence of the research establishment (Forrest *et al.*, 1995; Grompe, 1993). Most groups have found single-stranded conformational analysis (Orita *et al.*, 1989) quick and easy (though lacking in sensitivity), whilst others have advocated density gradient gel electrophoresis (as sensitive, though expensive to perform) (Doria *et al.*, 1994). Increasingly, direct sequencing is an option for reliable detection of sequence variation (Reeve and Fuller, 1995).

Whichever method is chosen, the screen needs to include (at least) the coding regions, splice sites and respectable lengths of the 3′-untranslated and promoter regions, for these are likely to have the greatest payoff. If regulatory changes are suspected, Northern blots or quantitative reverse transcriptase–polymerase chain reaction (RT–PCR) may provide a clue to changes in expression levels (though it may be hard to distinguish primary from secondary events). Changes in the size or sequence of cDNA fragments may identify cryptic splice sites – illegitimate transcription may be valuable when access to the tissue of interest is restricted (e.g. glucokinase in MODY) (Sun *et al.*, 1993). Finally, if DNA permits, it may be worthwhile to create a panel of Southern blots from affected individuals. These can be screened with candidate gene probes, allowing one to identify deletions or insertions; these may be rare causes of the disease in general but, if found, provide a quick pointer to the likely importance of the gene.

5.5 Assessing genotype–phenotype correlations

5.5.1 Genotype–phenotype correlations in complex trait analysis

The race to identify genes in complex traits is likely to converge on to a final common pathway – the ability: (i) to screen candidate genes (however so identified) for genetic variation, and (ii) to evaluate any variation discovered for potential relevance to the development of the trait of interest. In many ways, the second stage of this process (i.e. determining whether any variant uncovered has pathophysiological relevance) presents the greater problems. The correlation of genotype and phenotype may be a trivial matter for variants with large phenotypic effects (e.g. low-density lipoprotein receptor mutations). However, much of the phenotypic variance in complex disease may be explained by common polymorphisms with modest effects at the individual level, or effects which are evident only under particular genetic backgrounds or environmental circumstances. Evaluation of such variants is tantalizingly difficult (Gough *et al.*, 1995; Hager *et al.*, 1995; Mauriège and Bouchard, 1996) since no methodology is entirely reliable: for example, latent population stratification may lead to false positive association of a neutral variant with disease; co-segregation of a variant with disease may be due to tight linkage to a nearby mutation; the fact that a variant produces a demonstrable effect in an *in vitro* system does not prove that the gene product has any relevance to the development of the disease of interest.

How should one proceed to assess candidate polymorphisms? In many cases, one may wish to move rapidly towards functional studies conducted using appropriate *in vitro* and *in vivo* systems. However, this remains an

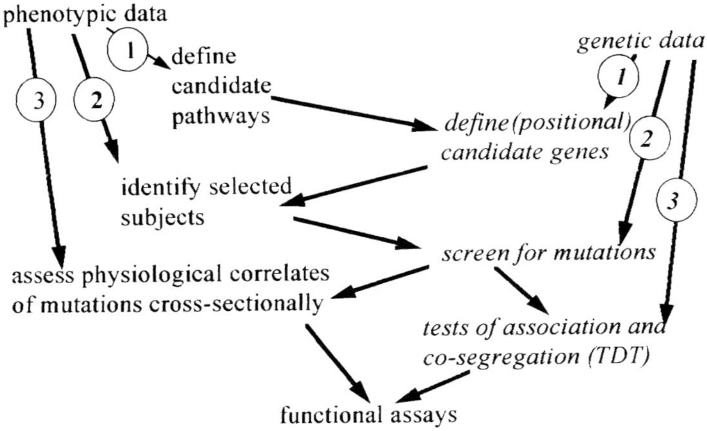

Figure 5.2. Schema for assessment of genotype–phenotype correlations by 'bootstrapping'.

impractical route for screening all candidate polymorphisms, requiring initial appraisal to use more indirect strategems. The schema summarized in *Figure 5.2* employs genetic and phenotypic analyses in concert to tease out the weak genotype–phenotype correlations that are a feature of complex diseases such as NIDDM. The success of this 'bootstrapping' strategy depends on acquiring robust genetic and phenotypic data from the same group of subjects. The physiological information is useful in three ways:

(i) It permits the identification of candidate metabolic pathways and provides a basis for evaluating positional candidates.
(ii) It allows selection a subgroup of women for candidate gene screening studies. By selecting those individuals most likely *a priori* to have a defect in the gene, of interest, the sensitivity of the mutation detection analysis is substantially improved. For example, if searching the insulin receptor gene it would be appropriate to screen individuals with the most extreme deficits in insulin sensitivity.
(iii) It allows one to examine the physiological correlates of any variants found, through study of the wider cross-sectional cohorts. This improves the specificity of the analysis.

5.5.2 Genotype–phenotype correlations in NIDDM

Armed with this outline and a collection of suitably characterized subjects, what types of analysis are likely to prove useful in assessing candidate polymorphisms? Examples follow.

Case–control analyses. The basic paradigm is simple: a cohort of cases (people with diabetes) and controls (typically subjects with no personal or family history of diabetes) are typed for a polymorphism closely linked to the gene of interest (which might be a candidate variant or simply an anonymous marker). A significant difference between the groups (an association) is usually taken to indicate that the variant has a direct pathogenic effect (or that it is in linkage disequilibrium with a nearby polymorphism that does). However, not all associations indicate linkage (Ott, 1991). The most important causes of false positive results are: (i) multiple testing (of alleles and genes) without appropriate significance corrections; and (ii) latent ethnic stratification [if cases and controls are not matched for ethnic background, then an association may reflect these ethnic differences; Lander and Schork (1994) present an entertaining example of this in their paper]. A further concern is 'survivor effects'. To take an example, an allele which in fact determines an individual's susceptibility to nicotine addiction might be observed less frequently amongst diabetic individuals than control subjects; this does not mean that the allele influences the risk of diabetes, but instead reflects the particularly poor survival prospects amongst those who smoke as well as suffering from diabetes. It is not only positive results from case-control studies that are open to suspicion. Such analyses rely on linkage disequilibrium to detect associations, so that major susceptibility effects will go completely unnoticed if the marker used is in equilibrium with the pathogenic mutation.

Family-based association methods. Given these concerns, the use of family-based control methods, such as the TDT has been strongly advocated as a means of dealing with latent ethnic stratification (Lander and Schork, 1994; Spielman *et al.*, 1993). In the TDT, the transmission of alleles from heterozygous parents is analysed and comparison made between those alleles which are transmitted to affected offspring ('cases') and those which are not. (These untransmitted alleles provide an ethnically matched 'control'.) The relative merits of the TDT over other similar methodologies [e.g. affected family-based controls (AFBAC) and haplotype relative risk (HRR)] have recently been discussed (Schaid and Sommer, 1994; Spielman and Ewens, 1996). Such methods have proved useful in the analysis of IDDM (Bennett *et al.*, 1995; Todd, 1995), but since they require knowledge of parental genotypes, they have been seen as having limited value in studies of NIDDM (where parents of affected subjects are generally dead).

Transracial analysis. Although a powerful technique which allows a more robust test of linkage disequilibrium than case–control analyses, the

TDT will not distinguish between those polymorphisms with a direct pathogenic effect and those which are merely in linkage disequilibrium with a functional variant. Several polymorphisms within the close locality of a gene may be in tight linkage disequilibrium frustrating attempts to isolate which one is functional and which are merely 'hitchhiking' (Bennett *et al.*, 1995). One solution is to compare results in a number of distinct ethnic groups. Variants positively associated with disease in diverse ethnic groups are likely to be directly pathogenic rather than reflecting ethnic stratification or linkage disequilibrium; in the latter circumstances, one would not expect any allele to show a consistent positive association (Jenkins *et al.*, 1990).

Intermediate trait analyses. The analyses above generally use a dichotomous diagnostic framework: affection status is 'diabetic' or otherwise. However, as discussed above, quantitative trait information may be valuable if the correlation between a candidate polymorphism and relevant intermediate trait measures is tighter than that between the polymorphism and the end disease phenotype. Individuals with different genotypes at the candidate polymorphism can then be compared for the distribution of these traits (Urhammer *et al.*, 1996). One practical problem arising out of this approach is the fact that available data may permit a large number of intermediate traits to be tested (e.g. different anthropometric indices, glucose and/or insulin levels and ratios at various time points; Clément *et al.*, 1995; Walston *et al.*, 1995; Widén *et al.*, 1995). Unless appropriate corrections for multiple testing are applied (and, with many of the traits showing strong intercorrelations, the degree of multiple testing may be hard to judge), it is probable that a great many spurious results will be thrown up.

The demographic and metabolic features of NIDDM previously discussed seem to frustrate many of these approaches. For example, to employ the bootstrapping approach (*Figure 5.2*), it would be desirable to have access to two things: (i) parental DNA to allow TDT analysis; and (ii) reliable intermediate trait information. However, once individuals have developed NIDDM, neither is readily available. One solution is to study individuals with a high probability of future NIDDM who have declared themselves at an early age and who are available for metabolic evaluation whilst still normoglycaemic. Groups meeting these criteria include women with gestational diabetes (Ali and Alexis, 1990) and polycystic ovarian syndrome (Franks, 1995). The congruity between the metabolic features of women with these traits and those seen in other groups at high risk of NIDDM (such as offspring of conjugal diabetic parents, and individuals with impaired glucose tolerance) suggests substantial overlap in the loci underlying susceptibility to all these conditions.

5.6 Understanding the biology of complex traits

5.6.1 After the gene-hunting

Application of the approaches above will, with luck, have identified candidate polymorphisms, and some of these will have been shown to be correlated with relevant phenotypes. The initial phase of gene discovery is over; the interesting biology can now begin.

Firstly, the relationship between the genomic variation and expressed phenotype can be examined – structure and function can be assessed at the molecular, cellular, organ, pathway and organismal levels. The ability to manipulate the genomes of model organisms (through transgenic work) permits evaluation of the consequences of loss-of-function (or gain-of-function) mutations in the whole animal. For example, knockout mice unable to express the supposedly crucial insulin receptor substrate (IRS)-1 molecule were found to show relatively modest phenotypic effects (Araki *et al*., 1994; Tamemoto *et al*., 1994): a parallel pathway mediating insulin action via IRS-2 was subsequently uncovered (Sun *et al*., 1995). In humans, such manipulations are not possible, but metabolic studies in individuals with different genotypes at the candidate polymorphism may be valuable. Studies of MODY reveal that sophisticated studies of metabolism are facilitated once subjects can be characterized in genetic terms (and genetic heterogeneity thereby minimized). Of course, this may be a more practical proposition where disease results from the action of single highly penetrant genes than in everyday NIDDM.

Secondly, as multiple susceptibility genes are found, gene–gene interactions can be examined, again at a variety of organizational levels (Clément *et al*.,1996; Takimoto *et al*.,1996). These studies are likely to be extremely interesting, not least because the dynamics of these interactions (and those with environment) will provide an indication of how successfully an individual's genotype can be used to predict their present and future disease phenotypes.

Thirdly, gene–environmental interactions will be more amenable to study once one can control (at least partly) for genotype. This may allow identification of novel environmental risk factors, and, in turn, define new preventative measures to forestall the development of diabetes.

5.6.2 Future diagnostic and therapeutic opportunities

Finally, better understanding of the pathophysiological basis of NIDDM should bring rewards in terms of improved patient care through targeted therapeutical intervention. NIDDM is a common disease, and the intermediate traits contributing to NIDDM are more frequent still. If new therapies are to achieve substantial reductions in the pathological consequences of NIDDM, the costs (both in terms of finances and side-effects) need to be low and the benefits large. Advances in genetic analysis and manipulation provide mechanisms for this targeted approach.

Targeting the presymptomatic stages of disease. Once the basic pathophysiological defects leading to NIDDM are known, pharmaceutical development will be directed towards therapies which tackle the earliest presymptomatic stages in disease development. The objective would be to identify treatments which thereby halt diabetes and associated manifestations 'in their tracks'. Even if the genetics of NIDDM turn out to be fairly complex, the situation at the metabolic level will be simpler if the susceptibility genes cluster within particular pathways (the recent work on MODY gives some support for this; Yamagata *et al.*, 1996a,b). If so, it should be possible to achieve targeted therapy with a limited number of agents, correcting the fundamental pathways even though individualized tailor-made approaches may not be feasible.

Targeting those most at risk of disease. If presymptomatic treatments are to be financially practical (particularly for less wealthy societies), they need to be prioritized to those individuals most likely to benefit. An assessment of risk would be obtained through an evaluation of an individual's genotype and exposure to relevant environmental factors.

Targeting therapies to tissue and cell types. To reduce collateral damage to other organ systems, it may be advantageous to target therapeutic agents to particular organs and cell types. Considerable work in this regard is being directed towards the targeted delivery of nucleic acids (i.e. gene therapy). In principle, the very same delivery systems (viral, liposomes) could be adapted to deliver other small molecules, such as conventional pharmacological agents.

References

Ali Z and Alexis SD. (1990) Occurrence of diabetes mellitus after gestational diabetes mellitus in Trinidad. *Diabetes Care* **13**, 527–529.

Almind K, Bjørbaek C, Vestergaard H, Hansen T, Echwald S and Pedersen O. (1993) Aminoacid polymorphisms of insulin receptor substrate-1 in non-insulin dependent diabetes mellitus. *Lancet* **342**, 828–832.

Araki E, Lipes MA, Pattl M-E, Brüning JC, Haag B, Johnson RS and Kahn CR. (1994) Alternative pathway of insulin signalling in mice with targeted disruption of the IRS-1 gene. *Nature* **372**, 186–190.

Barker DJP, Gluckman PD, Godfrey KM, Harding JE, Owens JA and Robinson JS. (1993a) Fetal nutrition and cardiovascular disease in adult life. *Lancet* **341**, 938–941.

Barker DJP, Hales CN, Fall CHD, Osmond C, Phipps K and Clark PMS. (1993b) Type 2 (non-insulin-dependent) diabetes mellitus, hypertension and hyperlipidaemia (syndrome X): relation to reduced fetal growth. *Diabetologia* **36**, 62–67.

Barnett A, Spiliopoulos A, Pyke D, Stubbs WA, Burrin J and Alberti KGMM. (1981a) Metabolic studies in unaffected co-twins of non-insulin-dependent diabetics. *Br. Med. J.* **282**, 1656–1658.

Barnett AH, Eff C, Leslie RDG and Pyke DA. (1981b) Diabetes in identical twins – a study of 200 pairs. *Diabetologia* **20**, 87–93.

Barnett M, Collier GR, Collier FMcL, Zimmet P and O'Dea K. (1994) A cross-sectional and short-term longitudinal characterisation of NIDDM in *Psammomys obesus*. *Diabetologia* **37**, 671–676.

Bell GI, Xiang K-S, Newman MV, Wu S-H, Wright LG, Fajans SS, Spielman RS and Cox NJ. (1991) Gene for non-insulin-dependent diabetes mellitus (maturity onset diabetes of the young subtype) is linked to DNA polymorphism on human chromosome 20q. *Proc. Natl Acad. Sci. USA* **88**, 1484–1488.

Bell JI and Lathrop GM. (1996) Multiple loci for multiple sclerosis. *Nature Genetics* **13**, 377–378.

Bennett ST, Lucassen AM, Gough SCL *et al.* (1995) Susceptibility to human type 1 diabetes at IDDM2 is determined by tandem repeat variation at the insulin gene minisatellite locus. *Nature Genetics* **9**, 284–292.

Boguski MS and Schuler GD. (1995) Establishing a human transcript map. *Nature Genetics* **10**, 369–371.

Bouchard C, Tremblay A, Després J-P, Nadeau A, Lupien PJ, Thériault G, Dussault J, Moorjani S, Pinault P and Fournier G. (1990) The response to long-term overfeeding in identical twins. *New England J. Med.* **322**, 1477–1482.

Bowden D, Howard T, Sale M, Qadri A, Spray B, Rich S and Freedman B. (1996) Linkage of genetic markers in the MODY1 region of chromosome 20 to NIDDM in families enriched for nephropathy (abstract). *Diabetes* (Supplement 2) **45,** 79A.

Brosseau JD, Eelkema RC, Crawford AC and Abe TA. (1979) Diabetes among the three affiliated tribes, correlation with degree of Indian inheritance. *Am. J. Pub. Health* **69**, 1277–1278.

Byrne M, Fajans S, Stoltz A, Ortiz J, Sobel R and Polonsky K. (1994) Impaired glucose priming of insulin secretion in MODY subjects with a genetic marker on chromosome 20 (abstract). *Diabetes* (Supplement 1) **43,** 69A.

Carvajal JJ, Pook MA, dos Santos M, Doudney K, Hillermann R, Minogue S, Williamson R, Hsuan JJ and Chamberlain S. (1996) The Friedreich's ataxia gene encodes a novel phosphatidylinositol-4-phosphate 5-kinase. *Nature Genetics* **14**, 157–162.

Chiu KC, Province MA, Dowse GK, Zimmet PZ, Wagner G, Serjeantson S and Permutt MA. (1992) A genetic marker at the glucokinase gene locus for type 2 (non-insulin-dependent) diabetes mellitus in Mauritian Creoles. *Diabetologia* **35**, 632–638.

Clément K, Vaisse C, Manning B StJ, Basdevant A, Guy-Grand B, Ruiz J, Silver KD, Shuldiner AR, Froguel P and Strosberg AD (1995) Genetic variation in the β3-adrenergic receptor and an increased capacity to gain weight in patients with morbid obesity. *New England J. Med.* **333**, 352–354.

Clément K, Ruiz J, Cassard-Doulcier AM, Bouillard F, Ricquier D, Basdevant A, Guy-Grand B and Froguel P. (1996) Additive effect of a A->G (-3826) variant of the uncoupling protein gene and the Trp64Arg mutation of the β_3-adrenergic receptor gene on weight gain in morbid obesity (abstract). *Diabetologia* (Supplement) **39,** A6.

Collins FS. (1992) Positional cloning: let's not call it reverse anymore. *Nature Genetics* **1**, 3–6.

Collins FS. (1995) Positional cloning moves from the perditional to traditional. *Nature Genetics* **9**, 347–350.

Cook JTE, Shields DC, Page RCL, Levy JC, Hattersley AT, Shaw JAG, Neil HAW, Wainscoat JS and Turner RC. (1994) Segregation analysis of type 2 diabetes in Caucasian families. *Diabetologia* **37**, 1231–1240.

Copeman JB, Cucca F, Hearne CM ***et al.*** (1995) Linkage disequilibrium mapping of a type 1 diabetes susceptibility gene (IDDM7) to chromosome 2q31–q33. *Nature Genetics* **9**, 80–85.

Cox NJ, Bell GI, Concannon P, Hanis CL and Spielman RS. (1996) Setting priorities for follow-up studies of a complete genome scan for NIDDM (abstract). *Am. J. Hum. Genet.* (Supplement) **59,** A45.

Curtis D. (1996) Genetic dissection of complex traits (letter). *Nature Genetics* **12,** 356–357.

Davies JL, Kawaguchi Y, Bennett ST ***et al.*** (1994) A genome-wide search for human type 1 diabetes susceptibility genes. *Nature* **371**, 130–136.

Dib C, Fauré S, Fizames C ***et al.*** (1996) A comprehensive genetic map of the human genome based on 5,264 microsatellites. *Nature* **380**, 152–154

Doria A, Warram JH and Krolewski AS. (1994) Genetic predisposition to diabetic nephropathy. Evidence for a role of the angiotensin I-converting enzyme gene. *Diabetes* **43**, 690–695.

Ebers GC, Kukay K, Bulman DE ***et al.*** (1996) A full genome search in multiple sclerosis. *Nature Genetics* **13**, 472–476.

Elston RC, Namboodiri KK, Nino HV and Pollitzer WS. (1974) Studies on blood and urine glucose in Seminole Indians : indications for segregation of a major gene. *Am. J. Hum. Genet.* **26**, 13–34.

Fajans S. (1990) Scope and heterogeneous nature of MODY. *Diabetes Care* **13**, 49–64.

Forrest S, Cotton R, Landegren U and Southern E. (1995) How to find all those mutations. *Nature Genetics* **10**, 375–376.

Frankel WN and Schork NJ. (1996) Who's afraid of epistasis? *Nature Genetics* **14**, 371–373.

Franks S. (1995) Polycystic ovary syndrome. *New England J. Med.* **333**, 853–861.

Froguel PH, Vaxillaire M, Sun F *et al.* (1992) Close linkage of glucokinase locus on chromosome 7p to early-onset non-insulin-dependent diabetes mellitus. *Nature* **356**, 162–165.

Gabbay K. (1980) The insulinopathies. *New England J. Med.* **302**, 165–167.

Galli J, Li LS, Glaser A, Östenson CG, Jiao H, Fakhrai-Rad H, Jacob HJ, Lander ES and Luthman H. (1996) Genetic analysis of non-insulin dependent diabetes mellitus in the GK rat. *Nature Genetics* **12**, 31–37.

Gardner LI, Stern MP, Haffner SM, Gaskill SP, Hazuda HP, Relethford JH and Eifler CW. (1984) Prevalence of diabetes in Mexican Americans: relationship to percentage of gene pool derived from native American sources. *Diabetes* **33**, 86–92.

Gauguier D, Froguel P, Parent V, Bernard C, Bihoreau MT, Portha B, James MR, Penicaud L, Lathrop M and Ktorza A. (1996) Chromosomal mapping of genetic loci associated with non-insulin dependent diabetes in the GK rat. *Nature Genetics* **12**, 38–43

Gough SCL, Saker PJ, Pritchard LE *et al.* (1995) Mutation of the glucagon receptor gene and diabetes mellitus in the UK: association or founder effect? *Hum. Mol. Genet.* **4**, 1609–1612.

Grompe M. (1993) The rapid detection of unknown mutations in nucleic acids. *Nature Genetics* **5**, 111–117.

Guerre-Millo M, Staels B and Auwerx J. (1996) New insights into obesity genes. *Diabetologia* **39**, 1528–1531.

Hager J, Hansen L, Vaisse C *et al.* (1995) A missense mutation in the glucagon receptor gene is associated with non-insulin-dependent diabetes mellitus. *Nature Genetics* **9**, 299–304.

Hamer DH, Hu S, Magnuson VL, Hu N and Pattatucci AML. (1993) A linkage between DNA markers on the X chromosome and male sexual orientation. *Science* **261**, 321–327.

Hanis CL, Boerwinkle E, Chakraborty R *et al.* (1996) A genome-wide search for human non-insulin-dependent (type 2) diabetes genes reveals a major susceptibility locus on chromosome 2. *Nature Genetics* **13**, 161–171.

Hanson RL, Elston RC, Pettitt DJ, Bennett PH and Knowler WC. (1995) Segregation analysis of non-insulin-dependent diabetes mellitus in Pima Indians: evidence for a major-gene effect. *Am. J. Hum. Genet.* **57**, 160–170.

Hashimoto L, Habita C, Beressi JP *et al.* (1994) Genetic mapping of a susceptibility locus for insulin-dependent diabetes on chromosome 11q. *Nature* **371**, 161–164.

Hästbacka J, de la Chapelle A, Kaitila I, Sistonen P, Weaver A and Lander E. (1993) Linkage disequilibrium mapping in isolated founder populations: diastrophic dysplasia in Finland. *Nature Genetics* **2**, 204–211.

Hattersley AT, Turner RC, Permutt MA, Patel P, Tanizawa Y, Chiu KC, O'Rahilly S, Watkins PJ and Wainscoat JS. (1992) Linkage of type 2 diabetes to the glucokinase gene. *Lancet* **339**, 1307–1310.

Hattersley AT and Turner RC. (1993) Mutations of the glucokinase gene and type 2 diabetes. *Qu. J. Med.* **86**, 227–232.

Herman WH, Fajans SS, Ortiz FJ, Smith MJ, Sturis J, Bell GI, Polonsky KS and Halter JB. (1994) Abnormal insulin secretion, not insulin resistance is the genetic or primary defect of MODY in the RW pedigree. *Diabetes* **43**, 40–46.

Hitman GA, Hawrami K, McCarthy MI, Viswanathan M, Snehalatha C, Ramachandran A, Tuomilehto J, Tuomilehto-Wolf E, Nissinen A and Pedersen O. (1995) Insulin receptor substrate-1 gene mutations in NIDDM: implications for the study of polygenic disease. *Diabetologia* **38**, 481–486.

Hubank M and Schatz DG. (1994) Identifying differences in mRNA expression by representational difference analysis of cDNA. *Nucleic Acids Res.* **22**, 5640–5648.

Huse DM, Oster G, Killen AR, Lacey MJ and Colditz GA. (1989) The economic costs of non-insulin-dependent diabetes mellitus. *J. Am. Med. Assoc.* **262**, 2708–2713.

Jenkins D, Mijovic C, Fletcher J, Jacobs KH, Bradwell AR and Barnett AH. (1990) Identification of susceptibility loci for Type 1 (insulin-dependent) diabetes by trans-racial gene mapping. *Diabetologia* **33**, 387–395.

Ji L, Yang Y, Rich SS, Warram JH and Krolewski AS. (1996) Linkage of NIDDM to the MODY 1 gene region on chromosome 20 (abstract). *Diabetes* (Supplement 2) **45**, 77A .

Kadowaki T, Kadowaki H, Mori Y *et al.* (1994) A subtype of diabetes mellitus associated with a mutation of mitochondrial DNA. *New England J. Med.* **330**, 962–968.

Kahn CR. (1994) Insulin action, diabetogenes and the cause of type II diabetes. *Diabetes* **43**, 1066–1084.

Kaprio J, Tuomilehto J, Koskenvuo M, Romanov K, Reunanen A, Eriksson J, Stengård J and Kesäniemi YA. (1992) Concordance for type 1 (insulin-dependent) and type 2 (non-insulin-dependent) diabetes mellitus in a population-based cohort of twins in Finland. *Diabetologia* **35**, 1060–1067.

King's Fund Policy Institute (1996) *Counting the Cost: The Real Impact of Non-Insulin-Dependent Diabetes.* British Diabetic Association, London.

Knowler WC, Pettitt DJ, Saad MF and Bennett PH. (1990) Diabetes mellitus in the Pima Indians: incidence, risk factors and pathogenesis. *Diabetes/Metab. Rev.* **6**, 1–27.

Kruglyak L and Lander ES. (1995a) High-resolution genetic mapping of complex traits. *Am. J. Hum. Genet.* **56**, 1212–1223.

Kruglyak L and Lander ES. (1995b) Complete multipoint sib-pair analysis of qualitative and quantitative traits. *Am. J. Hum. Genet.* **57**, 439–454.

Kruglyak L, Daly MJ, Reeve-Daly MP and Lander ES. (1996) Parametric and non-parametric linkage analysis: a unified multipoint approach. *Am. J. Hum. Genet.* **58**, 1347–1363.

Lander ES. (1989) Genetic mapping of polygenic factors causing diabetes in inbred rodent strains. In: *Genes and Gene Products in the Development of Diabetes Mellitus* (eds J Nerup, T Mandrup-Poulsen and B Hökfelt), pp. 381–385. Elsevier, Amsterdam.

Lander E and Kruglyak L. (1995) Genetic dissection of complex traits: guidelines for interpreting and reporting linkage results. *Nature Genetics* **11**, 241–247.

Lander ES and Schork NJ. (1994) Genetic dissection of complex traits. *Science* **265**, 2037–2048.

Leiter EH and Herberg L. (1996) *Genetic analysis of polygenic NIDDM in mice (abstract). Diabetes* (Supplement 2) **45,** 29A .

Lesage S, Hani EH, Philippi A, Vaxillaire M, Hager J, Passa P, Demenais F, Froguel P and Vionnet N. (1995) Linkage analyses of the MODY3 locus on chromosome 12q with late-onset NIDDM. *Diabetes* **44**, 1243–1247.

Liang P and Pardee AB. (1992) Differential display of eukaryotic messenger RNA by means of the polymerase chain reaction. *Science* **257**, 967–971.

Lillioja S, Mott DM, Spraul M, Ferraro R, Foley JE, Ravussin E, Knowler WC, Bennett PH and Bogardus C. (1993) Insulin resistance and insulin secretory dysfunction as precursors of non-insulin-dependent diabetes mellitus. Prospective studies of Pima Indians. *New England J. Med.* **229**, 1988–1992.

Mahtani MM, Widén E, Lehto M *et al.* (1996) Mapping of a gene for NIDDM associated with an insulin secretion defect by a genome scan in Finnish families. *Nature Genetics* **14**, 90–95.

Manson JE, Rimm EB, Stampfer MJ, Colditz GA, Willett WC, Krolewski AS, Rosner B, Hennekens CH and Speizer FE. (1991) Physical activity and incidence of non-insulin-dependent diabetes mellitus in women. *Lancet* **338**, 774–778.

Manson JE, Nathan DM, Krolewski AS, Stampfer MJ, Willett WC and Hennekens CH. (1992) A prospective study of exercise and incidence among US male physicians. *J. Am. Med. Assoc.* **268**, 63–67.

Matthews D, Fry L, Powles A, Weber J, McCarthy M, Fisher E, Davies K and Williamson R. (1996) Evidence for a locus for familial psoriasis mapping to chromosome 4q. *Nature Genetics* **14**, 231–233.

Mauriège P and Bouchard C. (1996) Trp64Arg mutation in b_3-adrenoceptor gene is of doubtful significance for obesity and insulin resistance. *Lancet* **348**, 698–699.

McCarthy M and Hitman GA. (1993) The genetic aspects of non-insulin dependent diabetes mellitus. In: *Causes of Diabetes: Genetics and Environmental Factors* (ed. RDG Leslie), pp. 157–183. John Wiley, London.

McCarthy MI, Hitchins M, Hitman GA, Cassell P, Hawrami K, Morton N, Mohan V, Ramachandran A, Snehalatha C and Viswanathan M. (1993) Positive association in the absence of linkage suggests a minor role for the glucokinase gene in the pathogenesis of non-insulin-dependent diabetes mellitus. *Diabetologia* **36**, 633–641.

McCarthy MI, Hitman GA, Shields DC, Morton NE, Snehalatha C, Mohan V, Ramachandran A and Viswanathan M. (1994) Family studies of non-insulin-dependent diabetes mellitus in South Indians. *Diabetologia* **37**, 1221–1230.

McCarthy MI, Kruglyak L and Lander ES. (1996) Sibpair collection strategies in complex trait analysis (abstract). *Am. J. Hum. Genet.* (Supplement) **59**, A227.

McCarty D and Zimmet P. (1994) *Diabetes 1994 to 2010: Global Estimates and Projections*. Bayer AG, Leverkusen and International Diabetes Institute, Melbourne.

Menzel S, Yamagata K, Trabb JB *et al.* (1995) Localization of MODY3 to a 5-cM region of human chromosome 12. *Diabetes* **44**, 1408–1413.

Newman B, Selby J, King M-C, Slemenda C, Fabsitz R and Friedman GD. (1987) Concordance for type 2 (non-insulin-dependent) diabetes mellitus in male twins. *Diabetologia* **30**, 763–768.

Norio R, Nevanlinna HR and Perheentupa J. (1973) Hereditary diseases in Finland. *Ann. Clin. Res.* **5,** 109–141.

Nyholm B, Mengel A, Nielsen S, Møller N and Schmitz O. (1994) The insulin resistance of relatives of type 2 diabetic subjects is significantly related to a reduced VO_2 max (abstract). *Diabetologia* (Supplement 1) **37,** A28.

Orita M, Suzuki Y, Sekiya T and Hayashi K. (1989) Rapid and sensitive detection of point mutations and DNA polymorphisms using the polymerase chain reaction. *Genomics* **5,** 874–879.

Ott J. (1990) Invited editorial: cutting a Gordian knot in the linkage analysis of complex human traits. *Am. J. Hum. Genet.* **4,** 219–221.

Ott J. (1991) *Analysis of Human Genetic Linkage* (revised edition). Johns Hopkins University Press, Baltimore.

Peltonen L. (1995) All out for chromosome six. *Nature* **378,** 665–666.

Permutt MA and Ghosh S. (1996) Rat model contributes new loci for NIDDM susceptibility in man. *Nature Genetics* **12,** 4–6.

Porte D. (1991) B-cells in type II diabetes. *Diabetes* **40,** 166–180.

Reed PW, Davies JL, Copeman JB *et al.* (1994) Chromosome-specific microsatellite sets for fluorescence-based, semi-automated genome mapping. *Nature Genetics* **7,** 390–395.

Reeve MA and Fuller CW. (1995) A novel thermostable polymerase for DNA sequencing. *Nature* **376,** 796–797.

Reynet C and Kahn CR. (1993) Rad: a member of the ras family overexpressed in muscle of type II diabetic humans. *Science* **262,** 1441–1444.

Rich SS. (1990) Mapping genes in diabetes: genetic epidemiological perspectives. *Diabetes* **39,** 1315–1319.

Rimoin D and Rotter J. (1982) Genetic syndromes associated with diabetes and glucose intolerance. In: *The Genetics of Diabetes Mellitus* (eds J Köbberling and R Tattersall), pp. 149–181, Academic Press, London.

Risch N. (1990a) Linkage strategies for genetically complex traits. I. Multilocus models. *Am. J. Hum. Genet.* **46,** 222–228.

Risch N. (1990b) Linkage strategies for genetically complex traits. II. The power of affected relative pairs. *Am. J. Hum. Genet.* **46,** 229–241.

Risch N and Botstein D. (1996) A manic depressive history. *Nature Genetics* **12,** 351–353.

Risch N and Merikangas K. (1996) The future of genetic studies of complex human diseases. *Science* **273,** 1516–1517.

Saker PJ, Barrow B, McLellan J-A, Hammersley MS, Lo Y-M D, Gillmer MD, Turner RC and Hattersley AT. (1993) A missense mutation in exon 8 of the glucokinase gene in gestational diabetic subjects (abstract). *Diabetologia* (Supplement 1) **36,** A84.

Sawcer S, Jones HB, Feakes R, Gray J, Smaldon N, Chataway J, Robertson N, Clayton D, Goodfellow PN and Compston A. (1996) A genome screen in multiple sclerosis reveals susceptibility loci on chromosome 6p21 and 17q22. *Nature Genetics* **13,** 464–468.

Schaid DJ and Sommer SS. (1994) Comparison of statistics for candidate-gene association studies using cases and parents. *Am. J. Hum. Genet.* **55,** 402–409.

Schuler GD, Boguski MS, Stewart EA *et al.* (1996) A gene map of the human genome. *Science* **274,** 540–546.

Serjeantson S, Owerbach D, Zimmet P, Nerup J and Thoma K. (1983) Genetics of Diabetes in Nauru: effects of foreign admixture, HLA antigens and the insulin-gene-linked polymorphism. *Diabetologia* **25**, 13–17.

Sheffield VC, Weber JL, Buetow K *et al.* (1995) A collection of tri- and tetranucleotide repeat markers used to generate high quality, high resolution human genome-wide linkage maps. *Hum. Mol. Genet.* **4**, 1837–1844.

Shimada F, Makino H, Hashimoto H, Taira M, Seino S, Bell GI, Kanatsuka A and Yoshida S. (1993) Type 2 (non-insulin-dependent) diabetes mellitus associated with a mutation of the glucokinase gene in a Japanese family. *Diabetologia* **36**, 433–437.

Sing CF, Haviland MB, Templeton AR, Zerba KE and Reilly SL. (1992) Biological complexity and strategies for finding DNA variations responsible for inter-individual variation in risk of a common chronic disease, coronary artery disease. *Ann. Med.* **24**, 539–548.

Spielman RS and Ewens WJ. (1996) The TDT and other family-based tests for linkage disequilibrium and association. *Am. J. Hum. Genet.* **59**, 983–989.

Spielman RS, McGinnis RE and Ewens WJ. (1993) Transmission test for linkage disequilibrium: the insulin gene region and insulin-dependent diabetes mellitus (IDDM). *Am. J. Hum. Genet.* **52**, 506–516.

Stephens JC, Briscoe D and O'Brien SJ. (1994) Mapping by admixture linkage disequilibrium in human populations: limits and guidelines. *Am. J. Hum. Genet.* **55**, 809–824.

Stone LM, Kahn SE, Fujimoto WY, Deeb SS and Porte D. (1996) A variation at position – 30 of the β-cell glucokinase gene promoter is associated with reduced β-cell function in middle-aged Japanese-American men. *Diabetes* **45**, 422–428.

Sturis J, Kurland IJ, Byrne MM, Mosekilde E, Froguel P, Pilkis S, Bell GI and Polonsky KS. (1994) Compensation in pancreatic β-cell function in subjects with glucokinase mutations. *Diabetes* **43**, 718–723.

Suarez BK, Hampe CL and van Eerdewegh P. (1994) In: *Genetic Approaches to Mental Disorders* (eds ES Gershon and CR Cloninger), pp. 23–46. American Psychiatric Press, Washington DC.

Sun F, Knebelmann B, Pueyo ME *et al.* (1993) Deletion of the donor splice site of intron 4 in the glucokinase gene causes maturity-onset diabetes of the young. *J. Clin. Invest.* **92**, 1174–1180.

Sun XJ, Wang L-M, Zhang Y, Yenush L, Myers MG, Glasheen E, Lane WS, Pierce JH and White MF. (1995) Role of IRS-2 in insulin and cytokine signalling. *Nature* **377**, 173–177.

Swai ABM, McLarty DG, Kitange HM, Kilima PM, Masuki G, Mtinangi BI, Chuwa L and Alberti KGMM. (1991) Study in Tanzania of impaired glucose tolerance. Methodological myth? *Diabetes* **40**, 516–520.

Takeda J, Yano H, Eng S, Zeng Y and Bell GI. (1993) A molecular inventory of human pancreatic islets: sequence analysis of 1000 cDNA clones. *Hum. Mol. Genet.* **2**, 1793–1798

Takimoto E, Ishida J, Sugiyama F, Horiguchi H, Murakami K and Fukamizu A. (1996) Hypertension induced in pregnant mice by placental renin and maternal angiotensinogen. *Science* **274**, 995–998.

Tamemoto H, Kadowaki T, Tobe K *et al.* (1994) Insulin resistance and growth retardation in mice lacking insulin receptor substrate-1. *Nature* **372**, 182–186.

Tattersall R and Fajans S. (1975) A difference between the inheritance of classic juvenile-onset and maturity-onset type diabetes of young people. *Diabetes* **24**, 44–53.

Tattersall RB and Mansell PI. (1991) Maturity onset-type diabetes of the young (MODY): one condition or many? *Diabetic Med.* **8**, 402–410.

Taylor SI. (1992). Lilly Lecture: molecular mechanisms of insulin resistance. Lessons from patients with mutations in the insulin-receptor gene. *Diabetes* **41**, 1473–1490.

Taylor SI, Accili D and Imai Y. (1994) Insulin resistance or insulin deficiency? Which is the primary cause of NIDDM? *Diabetes* **43**, 735–740.

Todd JA. (1995) Genetic analysis of type 1 diabetes using whole genome approaches. *Proc. Natl Acad. Sci. USA* **92**, 8560–8565.

Tokuyama Y, Sturis J, DePaoli AM, Takeda J, Stoffel M, Tang J, Sun X, Polonsky KS and Bell GI. (1995) Evolution of β-cell dysfunction in the male Zucker diabetic fatty rat. *Diabetes* **44**, 1447–1457.

Tuomi T, Groop LC, Zimmet PZ, Rowley MJ, Knowles W and Mackay IR. (1993) Antibodies to glutamic acid decarboxylase reveal latent autoimmune diabetes mellitus in adults with a non-insulin-dependent onset of disease. *Diabetes* **42**, 359–362.

Turner R, O'Rahilly S, Levy J, Rudenski A and Clark A. (1989) Does type II diabetes arise from a major gene defect producing insulin resistance or β-cell dysfunction? In: *Genes and Gene Products in the Development of Diabetes Mellitus* (eds J Nerup, T Mandrup-Poulsen and B Hökfelt), pp. 171–183. Elsevier, Amsterdam.

Turner RC and Levy JC. (1996) Notes on the GENNID study. *Diabetes Care* **19**, 892–895.

Ueda H, Ikegami H, Yamato E *et al.* (1995) The NSY mouse: a new animal model of spontaneous NIDDM with moderate obesity. *Diabetologia* **38**, 503–508.

Urhammer SA, Clausen JO, Hansen T and Pedersen O. (1996) Insulin sensitivity and body weight changes in young white carriers of the codon 64 amino acid polymorphism of the β_3-adrenergic receptor gene. *Diabetes* **45**, 1115–1120.

van Heyningen V. (1994) One gene-four syndromes. *Nature* **367**, 319–320.

Vaxillaire M, Boccio V, Philippi A, Vigouroux C, Terwilliger J, Passa P, Beckmann JS, Velho G, Lathrop GM and Froguel P. (1995) A gene for maturity onset diabetes of the young (MODY) maps to chromosome 12q. *Nature Genetics* **9**, 418–423.

Walston J, Silver K, Bogardus C *et al.* (1995) Time of onset of non-insulin-dependent diabetes mellitus and genetic variation in the β3-adrenergic-receptor gene. *New England J. Med.* **333**, 343–347.

Waterworth DM, Bennett ST, Gharani N *et al.* (1996) Association of class III alleles at the insulin gene VNTR with polycystic ovary syndrome (PCOS) (abstract). *Am. J. Hum. Genet.* (Supplement) **59**, 1701.

Widén E, Lehto M, Kanninen T, Walston J, Shuldiner AR and Groop LC. (1995) Association of a polymorphism in the β_3-adrenergic receptor gene with features of the insulin resistance syndrome in Finns. *New England J. Med.* **333**, 348–351.

Wooster R, Bignell G, Lancaster J *et al.* (1995) Identification of the breast cancer susceptibility gene *BRCA2*. *Nature* **378**, 789–793.

World Health Organisation Study Group (1985) *Diabetes Mellitus* (WHO Technical Report, no 727).

Yamagata K, Oda N, Kaisaki PJ *et al.* (1996a) Mutations in the hepatocyte nuclear factor-1a gene in maturity-onset diabetes of the young (MODY3). *Nature* **384**, 455–458.

Yamagata K, Furuta H, Oda N, Kaisaki PJ, Menzel S, Cox NJ, Fajans SS, Signorini S, Stoffel M and Bell GI. (1996b) Mutations in the hepatocyte nuclear factor-4a gene in maturity-onset diabetes of the young (MODY1). *Nature* **384**, 458–460.

Zaidi FK, McCarthy MI, Wareham NJ, Hodlstock J, Kaloo-Holein H, Krook A, Swinn RA and O'Rahilly S. (1997) Homozygosity for a common polymorphism in the islet-specific promoter of the glucokinase gene is associated with a reduced acute insulin response to glucose in pregnant women. *Diabetic Medicine* (in press).

Zhang Y, Proenca R, Maffel M, Barone M, Leopold L and Friedman JM. (1994) Positional cloning of the mouse *obese* gene and its human homologue. *Nature* **372**, 425–432.

6

Population-scale genotype assays: *APOE* gene in Alzheimer's dementia and coronary risk

Ian N.M. Day, Divya Palamand and Emmanuel Spanakis

6.1 Introduction: availability of population-scale gene testing

For research purposes, concerned with systematic studies of single nucleotide polymorphisms in candidate genes, we have devised a series of technologies making possible cheap, rapid genotype analyses of population-scale sample collections. In this context, we have, *inter alia*, established such genotype assays for the apolipoprotein E (APOE) gene, solving several complexities unique to this gene (Bolla *et al.*, 1995). From what is already known about apoE and its gene in lipoprotein metabolism, coronary disease and Alzheimer's disease, this technology opens up a cheap high-throughput gene test which could be used in a predictive mode in the general population. Here we describe this new assay system applied to the *APOE* gene and review the background of clinicobiological knowledge. The practical and ethical issues raised by the availability of such tests are considered in conjunction with possible future ways to influence *APOE*-genotype-dependent disease.

6.1.1 Microplate array diagonal gel electrophoresis (MADGE)

We are interested in large-scale molecular genetic epidemiological research into genetic variation such as single nucleotide polymorphism within genes, which is expected to underpin genetic susceptibilities to

Genetics of Common Diseases: future therapeutic and diagnostic possibilities,
edited by I. Day and S. Humphries. © 1997 BIOS Scientific Publishers Ltd, Oxford

many common diseases. Microplate array diagonal gel electrophoresis (MADGE) is a tool invented in 1993 (Day and Humphries, 1994) to enhance our ability to progress these objectives in the context of traits of coronary disease. MADGE enhances our rate of calling of single nucleotide polymorphism genotypes using electrophoresis by one to two orders of magnitude at no increase in cost. Taken together with other utilities (Bolla *et al.*, 1995), the whole process from sample acquisition to PCR template preparation to post-PCR electrophoretic analysis and validation of genotypes becomes possible and cost effective on a population scale, with no major capital expenditure on sophisticated equipment. Thus, as outlined below, it would now be quite feasible to determine the *APOE* genotypes of any large population sample, whether it be those of an epidemiological research study, all patients on a general practitioner's list, a concert hall assembly or the residents of a home for the elderly.

How does MADGE work? The MADGE method allows a very simple means to prepare open-faced, submersible or stackable polyacrylamide gels, containing an array of wells directly compatible with 96-well microplates in array and pitch, but with the array turned on a diagonal 71.6°C relative to the plane of the electrode to give a sufficient 'line of sight' for each electrophoresis track, before the next well is reached. This angle permits 2 mm cubic wells and was adopted as standard. Longer lines of sight would be possible with a different angle and narrower wells, but these would not be compatible with the spatial intolerance of standard disposable tips on usual microplate compatible pipetting devices.

In the simplest configuration, acrylamide gel mix is poured on to a plastic gel-forming device, and a 2 mm float glass plate silanized with γ-methacryloxypropyltrimethoxysilane is laid silanized-face towards the mix, directly on to the former, and a weight is then placed on top. Once the gel has set, the glass plate with attached gel is prised away from the plastic former and used submerged in a small horizontal gel tank. Such gels can be stacked directly, glass plate/gel, glass plate/gel, and so on, and, since sample identification, for example, would have been recorded for the original microplate, the whole set can be transferred using 8 × 12 or 96-channel pipetting devices direct into the gel well array for electrophoresis.

In the time that 40 samples could have been analysed on a submerged agarose gel in small tank, 500 post-PCR analyses can be completed and information recorded using the MADGE system. Thus, MADGE opens for electrophoresis the range of advantages which has become available for liquid phase handling in the form of microplates over the course of the past 30 years. Liquid phase PCR procedures have also come to rely heavily on this industry standard. While our main interest is research, we recognize that the convenience and cost-saving which this system may offer in specific genetic diagnostic tests will be considerable also, facilitating the laboratory component of tests associated with large population groups, such as screening.

6.1.2 Simultaneous analysis of hundreds of APOE genotypes

The identification of protein polymorphism for apoE (Utermann *et al.*, 1979), and the recognition of its expression in several body compartments including plasma and nervous system (Diedrich *et al.*, 1991), places the need to examine this variation in the context of a wide array of different disease states. apoE protein polymorphism can be determined by isoelectric focusing and immunoblotting, but PCR-based analysis of genotypes should now be a simpler option. However, several features of this particular genotype analysis have been problematic, which may in part explain the wide range of different methods proposed in the literature. Features which have to be addressed are:

(i) that the gene region to be analysed is %(G+C) rich, which can lead to failed PCR, insufficient yield PCR and subsequent analysis or PCR containing confounding misproducts
(ii) that there are three common alleles which represent variation at two codons, 112 and 158, resulting in six possible genotypes; this doubles the complexity for many methods such as oligonucleotide binding and allele-specific amplification
(iii) that in comparison with separation of plasma for phenotype-based assays, purification of DNA for PCR template and genotype-based assay is typically a laborious procedure; this needs to be reduced to a simplicity equivalent with centrifugation of whole blood to obtain plasma, but yielding high quality template to ensure high quality PCR.

While for historical reasons, many of the studies of apoE in hyperlipidaemias and coronary disease were hinged on determination by phenotype, some more recent studies, and all recent studies by geneticists working in other fields such as Alzheimer's dementia have hinged on PCR-based genotyping.

The preferred method of post-PCR analysis is that of Hixson and Vernier (1990), which capitalizes on the unusual feature that both the nucleotide variation in codon 112 and the nucleotide variation in codon

158 result in polymorphism for presence/absence of a *Hha*I restriction site, so that a single PCR fragment can be amplified, digested with *Hha*I, and analysed on a single electrophoresis track. Six different band patterns can be recognized, representing respectively genotypes *E2/E2, E2/E3, E2/E4, E3/E3, E3/E4* and *E4/E4*. For the scale of genetic epidemiological studies in which we are interested, polyacrylamide gel electrophoresis necessary for the Hixson and Vernier method was inconvenient.

We have addressed this problem using the MADGE system (Bolla *et al.*, 1995). In order to use the small volume sampling loading with MADGE (it is not possible to overload a 'tall well' as can be achieved with vertical electrophoresis) it was also necessary to optimize PCR yield so that all digested fragments could be identified with ethidium bromide, although we have used the more sensitive dye, SYBR-green, also. We addressed the process of sample acquisition and preparation, finding that buccal wash is a more preferable sample medium for extraction of DNA than venous blood. High quality template is more readily extracted, presumably because blood brings with it the problem that the massive excess of red cells can only contribute contaminating protein, haem, which inhibits PCR, membrane, etc. Buccal wash can be collected in groups, by post; it does not require a skilled venesector, is a less biohazardous medium for clinical staff and the laboratory and is less unpleasant for the donor. It can also be readily repeated if necessary.

The protocol streamline which we have developed means that hundreds of buccal samples can readily be processed to high quality template DNA PCR, and that a single PCR reaction and *Hha*I digest can be completed on hundreds of samples in one day. This will significantly reduce the resource costs for completing future research genetic epidemiological studies in this important candidate gene in many diseases. If large-scale diagnostic application (see below) comes into play, this system sets an economical architecture for sampling and laboratory work flow. Below we set out the knowledge of the major impact of *APOE* genotype in Alzheimer's disease, as well as its impact in hyperlipidaemias and coronary disease, and discuss the possible future role of *APOE* genotype analysis in the clinical context of Alzheimer's dementia risk.

6.2 Background

6.2.1 ApoE in type III dyslipidaemia and discovery of polymorphism

Type III hyperlipidaemia is characterized by lipoprotein particles with properties intermediate between very low density lipoprotein and low density lipoprotein. On electrophoresis, these particles appear as a smear between the low density lipoprotein (β) and very low density lipoprotein (pre-β) bands: on ultracentrifugation, these particles float with the very low density lipoprotein fraction ('beta-migrating very low density lipoprotein').

More precise definition hinges on the measurement of cholesterol and triglyceride content of the very low density lipoprotein fraction. The intermediate nature of this lipoprotein fraction reflects its pathological origin as 'remnant' particles representing partially delipidated very low density lipoprotein and chylomicrons. These particles are unusually rich in apoE (arginine-rich lipoprotein), but although apoE is the usual ligand for specific receptor mediated clearance, the apoE of patients with Type III hyperlipidaemia is unable to mediate lipoprotein clearance.

The genetic basis of Type III hyperlipidaemia remained obscure until apoE phenotyping by isoelectric focusing was initiated, and it was shown that there were several patterns marking different alleles termed initially *II, III* and *IV* (Utermann *et al.*, 1979; Zannis *et al.*, 1981). The pattern representing homozygosity for allele *IV*, was present in Type III patients, who characteristically display palmar xanthomata and an increased incidence of early coronary and peripheral arterial disease. The amino acid sequence of apoE, and demonstration (Rall *et al.*, 1982; Weisgraber *et al.*, 1982) that alleles *E2, E3* and *E4* contained at positions 112 and 158, respectively, cysteine/cysteine, cysteine/arginine and arginine/arginine was consistent with the results of isoelectric focusing.

It was further shown that most Type III patients have the E2/E2 phenotype. The absence of arginine at these codons in the epsilon-2 isoform appears to cause poor receptor binding and delayed clearance from the circulation. In support of this model, chemical conversion of the cysteines to a positively charged derivative confers upon apoE2 good receptor binding. However, the E2/E2 phenotype occurs in an estimated 1% of the population, whereas only a small (fewer than 5%) percentage of *E2/E2* individuals display Type III hyperlipidaemia. The dyslipidaemia may occur in childhood, but the hyperlipidaemia develops only later – unmasked by obesity, diabetes and ageing, all of which increase very low density lipoprotein secretion, and also by hypothyroidism or other disease. In hypothyroidism the conversion of intermediate density lipoprotein (IDL) to low density lipoprotein by hepatic lipase is poor. It seems likely that any secondary compromise, genetic or environmental, to IDL clearance which is critically rate-limited by apoE2/E2 phenotype, could precipitate Type III hyperlipidaemia.

ApoE is a 299-amino-acid polypeptide synthesized primarily in the liver and, to a minor extent, in the intestine, although recent evidence shows that it is also present in macrophages and in the brain where the major source is astrocytes (Diedrich *et al.*, 1991). ApoE is associated with a number of different lipoproteins, primarily very low density lipoprotein and chylomicron as well as a subclass of high density lipoprotein (HDL). The gene for apoE is in a 48 kb cluster on chromosome 19 which includes the gene and pseudogene of *APOC*I and the *APOC*II gene.

From the role apoE plays in lipid metabolism it is possible to predict a number of functional domains that must be present on the proteins. Two structural domains have been defined by physical biochemistry studies (Nolte and Atkinson, 1992). They consist of two regions of ordered secondary structure divided by a region of random structure. The amino terminal domain, which covers about two-thirds of the molecule, contains the low density lipoprotein receptor binding domain. apoE acts as a ligand for the low density lipoprotein receptor (B/E receptor) – binding with greater affinity than apoB. Various approaches, using monoclonal antibodies or naturally occurring apoE mutations, defective in low density lipoprotein receptor binding, have aided in the definition of this region and site-directed mutagenesis has helped to confirm this. This region is rich in basic amino acids; residues 130–150 are predicted to define the low density lipoprotein receptor binding domain. Other related receptors including a relatively brain-specific receptor have also been identified. [See Kim *et al.* (1996).]

The carboxyl-terminal domain has a region of amphipathic and helical structure of polar and non-polar residues and this represents the major lipid-binding region of apoE. In addition, apoE binds to heparin, and the heparin-binding domains have been determined by the use of monoclonal antibodies that block heparin binding, as well as ligand blotting using synthetic apoE peptides. The heparin-binding sites lie between residues 142 and 147, within the receptor-binding domain, with possible additional sites between 243–272 and 211–218.

In addition to the common variants of apoE described above, a number of rare variants have also been described, and some of these are considered below. A number of variants of apoE have been detected by isoelectric focusing, and the amino acid change established. Some are dominant in their effect on plasma lipid phenotype and are associated with Type III hyperlipidaemia. These rare defective binding mutations of apoE have extreme effects on an individual's lipid level but little impact at the population level. However, these mutations give insight into the structural importance of protein secondary structure and particularly the low density lipoprotein receptor binding domain (reviewed by Weisgraber, 1994).

Dominant variants: Substitution of Arg_{142}→Cys, L_{146}→G, or Lys_{146} → Glu all result in dominant expression of Type III hyperlipidaemia. In contrast to the apoE2 (Arg_{158} → Cys), the apoE3 (Arg_{142} → Cys) variant is expressed in childhood. Furthermore, cysteamine modification does not restore binding of the 142 variant. Thus it appears that substitutions of amino acids within the receptor-binding domain disrupt the α helical structure directly and result in expression of the disease independent of the existence of other genetic or environmental factors. The low density lipoprotein receptor binding domain repeats offer several potential apoE-binding sites per receptor molecule and a speculative molecular explanation for the dominance of these apoE variants could be the competitive interaction of the dominant variant and the normal variant of apoE on the surface of the IDL particle leading to an overall reduction in the binding of IDL to the low density lipoprotein receptor. Another dominant variant, $apoE3_{Leiden}$, is the result of a seven-amino-acid insertion representing a tandem repeat of amino acid residues 121–127. This insertion increases the size of the α helix, alters the conformation of the receptor binding domain and thus reduces binding. The 22 kDa amino-terminal fragment of $apoE3_{Leiden}$ shows normal binding to the receptor, thus the carboxyl-terminal end of the protein must exert constraint on the receptor-binding region. It appears that the seven-amino-acid insertion does not directly inhibit interaction with the receptor but alters the conformation of the apoE protein and results in dominant expression of Type III hyperlipidaemia. Since there are multiple apoE molecules per lipoprotein particle the Leiden variant could either disrupt the organization of these multiple apoE molecules on the particle thus preventing interaction with the low density lipoprotein receptor or reduce the effective concentration of receptor-active apoE molecules on the surface thus reducing the affinity for the receptor.

Clinical relevance: diagnosis and prognosis. Diagnosis means the identification of the underlying cause of a patient's clinical condition. It is thus evident that *APOE2/2* genotype will frequently, but not always, be found in type III dyslipidaemia. In the presence of a characteristic phenotype (clinical features, plasma cholesterol and triglyceride levels), genotype identification can add strong support to the diagnosis. However, a more precise phenotypic diagnosis is possible using lipoprotein electrophoresis, isoelectric focusing, apoE immunoblotting, ultracentrifugation or some combination of these. In a subset of individuals such as rarer *APOE* gene variants, and where there are confounding features in the phenotype, such as different glycation patterns of apoE in diabetes mellitus (Stavljenic-Rukavina *et al.*, 1993), the genotype and phenotype methods will yield apparently discrepant results. The resolution of such discrepancies can be highly complex, more the domain of the research laboratory than of a service laboratory. Nevertheless, in most patients, the gene test for E2/E2 genotype would concord with phenotyping.

Of what practical value is the gene test? A biochemical or genetic diagnosis of type III dyslipidaemia is useful. The clinical categorization defines likely natural history (prognosis), suggests additional investigations for possible trigger conditions, such as hypothyroidism, diabetes and obesity which may represent the primary problem, gives access to knowledge of optimal therapeutic approaches and possibility of new trials, defines potential risk to siblings, and gives the doctor and patient a clearcut disease entity to address.

6.2.2 APOE3, APOE4, *population hyperlipidaemia and coronary disease*

Coronary disease is a final common pathway of disease which can involve, in various combinations, smoking, hyperlipidaemias and atherosclerosis, hypertension, diabetes, variations in clotting, oxidation status and perhaps resistance to fatal arrhythmias. Compared with some of the other common diseases considered, it is therefore expected to be highly polygenic and complex, ranging from relatively rare single-gene disorders such as familial hypercholesterolaemia, involving low density lipoprotein receptor gene defects ($^1/_{500}$ prevalence), to cumulative contributions from many genes each having perhaps 1% impact on the genetic risk. On account of extensive research on many measured intermediate traits, particularly easy since many are plasma constituents, many genetic components have been under study for over 10 years. However, there is the prospect that the full polygenic picture will emerge only when every variation in the entire human genome has been examined; and possibly that, even then, the combinatoric nature of gene interactions (epistasis) may preclude any study size from determining the effects of every combination! Thus we can anticipate that although research into the polygenic components of coronary disease had an early start, it will be a late finisher in view of its great complexity.

How are we to establish whether a particular gene contributes to the genetic variation in a trait such as blood pressure or plasma cholesterol concentrations? Genetic epidemiology has provided statistical methods for measuring the effect of genetic variation on a phenotypic trait in a population (the so-called measured genotype approach). In this way, the overall relationship between *APOE* alleles *E2, E3* and *E4* and hyperlipidaemia and coronary disease can be examined in the general population (e.g. Zerba and Sing, 1993). In the normal population these three polymorphisms turn out to have a substantial effect on the normal variation in plasma lipid concentrations; they account for 16% of genetic variation in cholesterol concentrations in the population. Many studies and meta-analyses have added to this picture, and the reader is referred to previous reviews (e.g. Davignon, 1993; Zerba and Sing, 1993). These small changes in plasma cholesterol concentration are sufficient to produce an enrichment of the *E2* allele with advancing age and a decrease of the *E4* allele in the aging population, perhaps as a result of an increased number of deaths from myocardial infarction (but see the section on Alzheimer's dementia). Very large studies show a genotype bias also with respect to coronary events (e.g. Wilson *et al.*, 1994). However, the effect of these polymorphisms on individual lipid concentrations and risk of coronary heart disease is small and hence of very limited predictive value in any individual. It remains to be determined whether the genotype effect on coronary disease acts entirely through lipid levels, but at present considerably greater overall prediction can be made using plasma cholesterol and triglyceride assays, and these remain the basis for risk factor reduction and treatment in individual patients.

6.2.3 The apoE gene in Alzheimer's dementia

Genetic epidemiology. First recognized in 1907 by Alois Alzheimer, this disease is now known to be a common disease which is the predominant cause of dementia over the age of 65 years. Progression of mild short-term memory problems, through to massive loss of memory, language and orientation, leads to incapacitation of the individual. With increased survival to old age, society is faced with the future care of vast numbers of demented patients whose quality of life is poor.

Evidence gathered from multiplex families with late-onset dementia suggested that familial clustering was unlikely to be due to chance alone and, in 1991, Pericak-Vance *et al.* used affected pedigree member analysis to show that a region of chromosome 19 was common to affected pedigree members more frequently than random chance would predict. This linkage was confirmed, but the much earlier report in 1987 by Schellenberg *et al.* of an association of a restriction fragment length polymorphism allele of the *APOC*II gene with Alzheimer's dementia did not replicate in these data.

The *APOC*II gene is adjacent to the *APOE* gene on chromosome 19: the failed replication of the *APOC*II gene association, and the fact that apoC and apoE were regarded as lipoprotein components relevant to atherosclerosis, diverted attention from these genes as candidates for the Alzheimer locus. However, apoE was shown by Strittmatter *et al.* (1993) to bind Alzheimer amyloid β peptide, and, furthermore, apoE antisera stained senile plaques and neurofibrillary tangles, prompting investigation of *APOE* genotype in Alzheimer's disease. The three common isoproteins of apoE, designated 2, 3 and 4, have allele frequencies typically 6%, 78% and 16% but with some ethnic differences. The differences are at amino acids 112 and 158, the isoforms containing respectively, cysteine/cysteine (2), cysteine/arginine (3), or arginine/arginine (4).

The association of *APOE* genotype with susceptibility, *APOE4* marking increased risk, has now been confirmed in many laboratories, in family and sporadic late-onset disease in many studies in various racial groups (see *Table 6.1*). Many aspects of the role of apoE in Alzheimer's disease have been examined, and have been reviewed by Schellenberg (1995) and Siest *et al.* (1995). The effect of *APOE4* is dose-dependent [i.e. the *E4/E4* genotype has the most severe effect, advancing age of onset by about 8 years per allele from an age of 84 years if no *E4* allele is present, to 68 years for *E4/E4* genotype (Corder *et al.*, 1993)]. *APOE2* seems to mark reduced risk relative to *E3*. However, the observed associations do not prove that these genotypes are the etiological feature on chromosome 19; they could simply be acting as linkage disequilibrium markers for other genetic diversity in a nearby gene or in the same gene (e.g. promoter variation/expression levels) which causes the true pathological effect. However, the circumstantial evidence, the expression of apoE in the nervous system, the strong enhancement of expression in response to nerve injury, the difference in expressed isoproteins (which, at least in the cardiovascular system, confers differences in interactions, for example with lipoprotein receptors), and differential effects of isoform-specific β-very low density lipoprotein on neurite extension *in vitro*, leave *APOE* genotype with very strong candidacy as representing the etiological site.

Functional studies on an animal model may prove the assumed pathogenic role of *apoE4*, which may be an important determinant for 50% of cases of late-onset Alzheimer's dementia. *APOE* genotype seems also to influence the age of onset of dementia in the rare families developing Alzheimer's dementia at a very early age, who have an autosomal dominant inheritance caused by mutations in the *APP* gene on chromosome 21. *APOE* genotype also seems to interact with head trauma in the development of Alzheimer's dementia, and there is also the suggestion of interaction with genotype of the very low density lipoprotein receptor locus (Okuizumi *et al.*, 1995). It has been estimated that *APOE4* genotype predicts a four-fold greater risk of development of late-onset Alzheimer's dementia, but many young adults with this genotype will not go on to develop dementia, either because they will die of another disease such as coronary disease or cancer first, or because their *E4* allele is insufficient *per se* to cause Alzheimer's disease. Ultimately, refined risk prediction and understanding of pathogenesis may hinge on an array of environmental factors and genotypic variations acting in combination.

Biochemical and cellular studies. In 1991, Diedrich *et al.* showed that *apoE* expression is increased in astrocytes in Alzheimer's disease and Namba *et al.* demonstrated the presence of apoE in amyloid plaques and in neurofibrillary tangles in Alzheimer's dementia. In 1993, Strittmatter *et al.* showed the high affinity of apoE and its isoform-dependent variation, for Aβ; an affinity also demonstrated by Wiesnieski and Frangione (1992). ApoE4 binds more rapidly to Aβ, a major constituent of extracellular senile plaques, and only in a narrower pH range than apoE3. It also forms fibrils which aggregate into dense units. However, the causative role of Aβ in dementia is itself not fully proven.

It could be that it is not the Aβ-rich extracellular amyloid plaques which are most important, but the intraneuronal neurofibrillary tangles consisting of paired helical filaments formed from hyperphosphorylated TAU protein. TAU protein is usually a microtubule-associated protein. ApoE also interacts with TAU, although here it is the *E3* isoform which binds; conceivably *E3* could prevent an otherwise pathological phosphorylation. *In vivo* apoE has been observed in neurofibrillary tangles, and apoE receptor genes are expressed in neurons (Kim *et al.*, 1996). Expression is enhanced markedly at sites of injury and apoE also influences adhesion, growth and proliferation. ApoE also appears to exert allele-specific antioxidant activity and influences cytotoxicity from oxiditive insults and β amyloid peptides (Miyata and Smith, 1996).

It seems likely that a widely and highly expressed protein such as apoE, which contains significant amino acid substitutions, will be found to display great pleitropy in protein interactions and cellular effects, and may well interact in the course of other disease processes also. The problem for Alzheimer's dementia is to determine which is the major

Table 6.1. Alzheimer's disease and apoE

Serial no.	Aim	Analysis	Results	References
1	To analyse the effect of APOE ε4 gene dose and to show its correlation with increased risk and earlier onset of Alzheimer's disease (AD) N = 95, M = 34, F = 61 (age 60–91 yr) C = 139, M = 62, F = 77 (age 66–94 yr)	DNA extraction, PCR, HhaI digestion, gel electrophoresis. Statistics: allele frequency by gene counting, Mantel–Haenszel correlations.	The risk for AD significantly increased from 20% (ε2/3 or ε3/3) to 90% (ε4/4) and the mean age of onset decreased from 84 years (no ε4 allele) to 68 years (two ε4 alleles) with increasing number of apoE ε4 alleles. This suggested that ApoE ε4 gene dose is an important risk factor for the development of late-onset AD. Similar results were found by Tsai *et al.*, Harlin *et al.* and Poirer *et al.*	Corder *et al.* (1993), *Science* **261**, 921–923 Harlin *et al.* (1994), *Hum. Mol. Genet.* **3** 569–574 Poirer *et al.* (1993), *Lancet* **342**, 697–699 Tsai *et al.* (1994), Am. *J. Hum. Genet.* **54**, 643–649
2	To test the hypothesis that prevalence of apoE ε4 in very old (over age 90) cognitively normal individuals would be very small as compared with very old AD patients. Most subjects were of Eastern European Jewish ancestry N = 28, F = 20, M = 8 C = 30, F = 28, M = 2 To analyse apoE ε4 allele frequency in comparison with age in 61- to 89-year-old subjects. Sixty-six with AD (F = 48, M = 18), and 23 normal individuals (F = 10, M = 13) were identified.	DNA extraction, PCR, HhaI digestion, gel electrophoresis. Statistics: allele frequency, χ^2 between AD patients and control, regression analysis.	(General population / Oldest old normals / Oldest old AD) Oldest old normals (93.8 ± 3.4 yr); Oldest old AD (94.3 ± 3.0 yr) apoE ε4 allele frequency: General population 0.14; Oldest old normals 0.036; Oldest old AD 0.13 There was a significant reduction of apoE ε4 allele frequency in cognitively normal nonagenarians as compared with the general population of individuals less than 72 years old ($p < 0.05$) and also a four-fold decrease as compared with the age-matched AD patients. In both AD and cognitively normal controls, marked decline in apoE ε4 frequency was found with advancing age ($p < 0.003$), although there was over-representation of apoE ε4 allele ($p < 0.0001$). A few cognitively normal nonagenarians were apoE ε4 positive. This suggests that apoE ε4-associated risk of AD is age dependent. Despite this risk it is possible to reach very old age with normal cognition, thus indicating that apoE ε4 by itself is not sufficient to lead to dementia.	Rebeck *et al.* (1994), *Neurology* **44**, 1513–1516

<table>
<tr>
<td>3</td>
<td>To analyse the polymorphisms and the allele frequencies in the APOE genotype in a sample population of 446 Italian individuals
<table>
<tr><th></th><th>n</th><th>Mean age of onset (in yr)</th></tr>
<tr><td>Sporadic AD</td><td>64</td><td>60.2 ± 5.5</td></tr>
<tr><td>EOFAD</td><td>104</td><td>54.6 ± 3.4</td></tr>
<tr><td>At risk*</td><td>52</td><td>32.2 ± 9.1</td></tr>
<tr><td>Escapees**</td><td>7</td><td>70.5 ± 4.3</td></tr>
<tr><td>Centenarians (majority demented)</td><td>30</td><td>101.3 ± 1.2</td></tr>
<tr><td>Controls (apparently demented)</td><td>84</td><td>59.3 ± 6.5</td></tr>
<tr><td>Neurological diseases (pathological controls)</td><td>105</td><td>43.5 ± 7.6</td></tr>
</table>
</td>
<td>DNA extraction, PCR, HhaI digestion.

Statistics: allele frequencies determined by gene counting.</td>
<td>Both the pathological and normal controls had no differences in their apoE ε2, ε3 and ε4 allele frequencies, and this suggested a higher prevalence of ε3 (0.86) in the non-AD general Italian population.
In sporadic AD patients, there was a five-fold increase in ε4 allele frequency (0.406) as compared with controls (0.084).
In EOFAD there was no significant difference between the patients and controls in the ε4 status, suggesting that ε4 does not represent a risk factor for EOFAD. The ε2 allele frequency was significantly increased in AD (0.125) and EOFAD (0.115) patients, suggesting an association between ε2 and AD.
These findings showed a highly significant association between ε4 allele and sporadic AD. The presence of increased ε2 allele frequency in AD, EOFAD and at-risk subjects suggests that in Italian AD patients ε4 as well as ε2 alleles are involved in the etiopathogenesis of AD.</td>
<td>Sorbi et al. (1994) Neurosci. Lett. 177, 100–103</td>
</tr>
<tr>
<td>4</td>
<td>To analyse the relationship between apoE ε4 and early-onset AD patients in a Dutch population-based study. The age of onset of the disease was between 34 and 65 years

N = 175, M = 59, F = 116
C = 159, M = 64, F = 95</td>
<td>DNA extraction, PCR, HhaI digestion, gel electrophoresis, SSCP for APP mutation screening.
Statistics: allele frequency, χ^2 between AD patients and control, regression analysis and Student's t-test.
The strength of the association between a genotype and EOAD was estimated by odds ratio.</td>
<td>The frequency of apoE ε4 allele among EOAD cases (35%) was 2.3 times higher compared with that of controls (15%).
Among patients, the apoE ε4 allele frequency was 1.6 times higher in those with positive family history of dementia (41%) than those with a negative history (25%). Similarly, the apoE ε4 allele frequency in controls having a positive family history of dementia (18%) was higher than those with a negative family history (14%).
Patients homozygous for apoE ε4 allele, regardless of family history, had a significant increase in risk of EOAD.
These findings suggested that in the absence of other genetic factors apoE ε4 homozygosity is sufficient to increase the risk of EOAD.</td>
<td>Duijn et al. (1994) Nature Genet. 7, 74–78</td>
</tr>
</table>

Table 6.1. Alzheimer's disease and ApoE *continued*

Serial no.	Aim	Analysis	Results	References
5	To analyse the relationship of apoE in different neurological conditions like AD, SDLT, HC and and PD. Frozen brain tissues were obtained. The brain tissues were centrifuged and supernatant was used as source of genomic DNA HC 41 M = 30, F = 11 PD 51 M = 26, F = 25 AD 67 M = 27, F = 40 SDLT 26 M = 14, F = 12 Control 58 M = 28, F = 30	DNA extraction, PCR, *Hha*I digestion, gel electrophoresis, SSCP for APP mutation screening. apoE genotyping was done using a reverse blue dot. Statistics: allele frequency, χ^3 between AD patients and control, regression analysis and Student's *t*-test. Odds ratio for AD associated with the presence of apoE allele, ANOVA.	The apoE ε4 allele frequency was increased in SDLT (0.365) and AD (0.328) as compared with controls (0.147), PD (0.098) or HC (0.171). There was a 1.8-fold excess of amyloid β-protein in AD as compared with controls, SDLT being intermediate. Substantial accumulation of paired helical filaments *TAU* and phosphorylated *TAU* was found in AD only. There was a significant absence of *TAU* in SDLT. These findings showed that *TAU* was associated with AD and that ApoE ε4 is not an etiological factor that accounts for *TAU* pathology.	Harrington *et al.* (1994) *Am. J. Pathol.* **45** (6)
6	To determine whether apoE ε4 allele was present in AD patients with co-existing PD and to compare the allele frequency with that of pure AD patients. Fifty autopsied patients with AD were matched with 50 autopsied patients with AD + PD.	DNA extraction, PCR, *Hha*I digestion, gel electrophoresis. Statistics: allele frequency by gene counting, χ^2 test to determine the difference in the allele frequency, Yates correction and ANOVA to detect genotypic differences in age of onset.	The ε4 allele was over-represented in both AD and AD + PD cases. Seventy-two per cent of both groups had at least one apoE ε4 allele, as compared with 25% in the general population ($p < 0.005$) and the institutional autopsy population ($p < 0.001$). This suggested that some PD-related changes may be influenced by the apoE genotype in AD + PD cases.	M. Gearing *et al.* (1995) *Neurology* **45**, 1985–1990

<table>
<tr>
<td>7</td>
<td>To examine the frequency of apoE alleles in AD patients in an elderly Hong Kong Chinese population

N = 65, mean age = 76.5 yr
C = 82, mean age = 71.8 yr</td>
<td>DNA extraction, PCR, HhaI digestion, gel electrophoresis.
Statistics: allele frequency, χ^2 between AD patients and control, regression analysis and Student's t-test.</td>
<td>apoE allele frequencies:
<table>
<tr><th></th><th>N</th><th>ε2</th><th>ε3</th><th>ε4</th><th>Mean age (SD)</th></tr>
<tr><td>AD</td><td>65</td><td>0.069</td><td>0.762</td><td>0.169</td><td>76.5 (8.3)</td></tr>
<tr><td>Controls</td><td>82</td><td>0.098</td><td>0.835</td><td>0.067</td><td>71.8 (6.6)</td></tr>
</table>
The apoE ε4 allele frequency was more than twice as great in the AD group (0.169) than in the controls (0.067, $p < 0.01$). Among AD patients, those having one or two copies of apoE ε4 alleles had a lower mean age of onset than those having no copies, although it was not stastistically significant. The presence of ε4 allele lowered and that of ε2 raised the age of onset of AD, though it was not statistically significant.</td>
<td>Mak et al. (1996) Neurology 46, 146–149</td>
</tr>
<tr>
<td>8</td>
<td>Relationship between gender, ethnicity, education, a positive family history of dementia, the presence of apoE ε4 allele and age of onset in patients with AD was evaluated.
The hypothesis was that patients with the presence of apoE ε4 allele and a positive family history (FH^+) of AD would have an earlier age of onset of AD than patients having absence of apoE ε4 ($ε4^-$) allele and a negative family history (FH^-).</td>
<td>DNA extraction, PCR, HhaI digestion, gel electrophoresis.
Statistics: allele frequency, χ^2 between AD patients and control, regression analysis and ANOVA.</td>
<td>The mean age of onset was 4 years earlier in $ε4^+$ than in $ε4^-$ patients. There was no difference among the four ethnic groups in gender ratio and positive family history. Educational status was highest in Ashkenazi Jews and white non-Hispanic non-Jews.
<table>
<tr><th></th><th>Age of onset</th></tr>
<tr><td>Ashkenazi Jews</td><td>77 ± 7 yr</td></tr>
<tr><td>African Americans</td><td>72 ± 8 yr</td></tr>
<tr><td>Hispanic</td><td>72 ± 9 yr</td></tr>
<tr><td>White, non-Hispanic, non-Jews</td><td>73 ± 9 yr</td></tr>
</table>
FH^+ status was more prevalent in those patients who were $ε4^+$ (58%) than in those who were $ε4^-$ (40%), $p = 0.02$. Multiple regression analysis taking gender, education, FH^+ status and $ε4^+$ status together showed that only FH^+ and $ε4^+$ status were significant, both being associated with earlier age of onset of AD. Thus the significant predictors of age of onset of AD are FH^+ and $ε4^+$ status.</td>
<td>Duara et al. (1996), Neurology 46, 1575–1579</td>
</tr>
</table>

Table 6.1. Alzheimer's disease and apoE *continued*

Serial No.	Aim	Analysis	Results	References
9	To examine the effect of apoE ε4 on risk in a single large Amish consanguineous pedigree having late-onset AD and also to determine the frequency of the apoE alleles in the general Amish population. Samples were taken from a large inbred family having late-onset AD members N = 6 (>70 yr) (related sibships) C = 53 (20–87 yr) (unrelated non-demented individuals)	DNA extraction, PCR, *Hha*I digestion, gel electrophoresis. Statistics: allele frequency estimated by gene counting. Two-point LOD scores were calculated for apoE to check for any linkage between apoE and AD.	All the AD-affected individuals in the pedigree had apoE 3/3 or 2/2; no apoE X/4 or 4/4 was observed. The frequency of the apoE ε4 allele in the control Amish population was 0.037 ± 0.02 which was significantly less than that in the non-Amish white controls (0.16 ± 0.027, $p < 0.0002$). The LOD scores showed no evidence of linkage between AD and apoE locus. The decreased frequency of apoE ε4 allele, inbreeding, a strong clustering of cases in this family and a low frequency of dementia indicated that other genes are involved with late-onset AD.	Pericak-Vance *et al.* (1996), *Ann. Neurol.* **39**, 700–704

*offsprings of an affected member of EOFAD families and younger than the age of onset in that family; **offsprings of affected members who do not have the disease and are two standard deviations older than the mean age of onset in that family; ANOVA, analysis of variance; APP, amyloid precursor protein; C, controls; EOAD, early onset of Alzheimer's disease; EOFAD, early onset familial Alzheimer's disease; F, females; FH^-, negative family history; FH^+, positive family history; HC, Huntington's chorea; M, males; N, total number of patients; PD, Parkinson's disease; SDLT, senile dementia of the Lewy body type; SSCP, single-strand conformation polymorphism.

interaction and to find ways to alter this interaction for prevention, treatment or cure.

The history of our current understanding of the molecular genetics of late-onset Alzheimer's dementia provides reassuring evidence that the genetics of a complex disease can be unravelled, leading to the implication of a gene not obvious at the outset of the research. It also illustrates the complex interplay between family, association, and biochemical studies in such research.

6.3 Some future possibilities

6.3.1 Confirming the causal role of the apoE protein

Whether the protein gene product of the *APOE4* allele is etiological, or whether it is in linkage disequilibrium with a separate etiological site, is not yet formally established. There are significant circumstantial reasons to believe that the cysteine/arginine differences between the protein isoforms are etiological. Other *APOE* mutations causing Alzheimer's disease would strengthen the case, but these have been sought and not found (Seeman *et al.*, 1995). However, such mutations would be expected to be rare and may yet exist. Haplotype analyses have suggested that the *E4* genotype marks a haplotype with increased Alzheimer's disease risk, but between haplotype comparisons suggest that other components of the haplotype, for example, at the adjacent *APOC* gene, may contribute additional independent predictive power (Templeton, 1996). If the cysteine/arginine variations in the *APOE* gene codons are all that matters, these other components of the surrounding haplotype should have no independent predictive power.

It is of course possible that combinations of protein composition and variations in other parts of the gene or surrounding region, for example, influencing level or timing of expression, could contribute to the overall effect. More remotely, it could be a nearby gene which mattered the most, although there is no evidence to support this at present. Proving cause and effect demands a completely controllable system in which only the expressed allele differs, and then ideally in which the *E4* allele could also be switched on and off experimentally to examine influence on lesion progression. Alzheimer's disease is a complex late-onset phenotype which will be difficult to model in experimental animals, although the opportunity for pure genetic and environmental backgrounds, gene knock-outs, knock-ins and conditional gene control is very good in mouse.

6.3.2 Development of possible interventional strategies

Development of small molecules to interfere with amyloid or neurofibrillary tangle formation are obvious possibilities, and could act either on the protein interactions, on the secretases involved in the proteolysis leading to β-amyloid peptide, or on the events leading to hyperphosphorylated TAU protein. If we understood more clearly the role of apoE isoforms in enhancing the progression of Alzheimer's dementia (*E4*) or delaying it (*E2*), additional targets would be available: these might be either the targets that apoE interacts with, or apoE itself as a target. In the latter case, either protein isoform differential interactions, a protein isoform in general, its cognate mRNA or its cognate gene could be targeted. The latter would demand an approach such as modified antisense oligonucleotides with the capability to be delivered to the nervous system and then into the cell or the nucleus. However, since apoE is necessary and expressed elsewhere than the central nervous system, specificity must somehow be achieved both for organ and isoform.

6.3.3 Public demand for predictive gene tests not altering clinical management?

The importance of allele *E4* in the occurrence of late-onset Alzheimer's dementia presents new challenges. In many populations, 10–15% of individuals have an *E4* allele, and therefore an *APOE* genotype analysis makes a prediction for them for late-onset Alzheimer's risk (odds ratio approximately 4 for one *E4* allele, and estimated near 20 for *E4/E4* genotype (Tsai *et al.*, 1994), considerably greater, for example, than the predictive value of HDL-cholesterol level in predicting coronary risk. Nevertheless, many individuals with either one or two *E4* alleles will live to a great age (e.g. over 80 years) without any clinical features of Alzheimer's dementia. HDL-C is used at the primary care level in stratifying for treatment of coronary heart disease (CHD) risk.

In contrast with coronary risk, there is no opportunity at present of averting the onset of Alzheimer's dementia, and traditional clinical wisdom is that a test must offer some opportunity to influence clinical management to be justifiable. APOE genotype, in conjunction with emerging biochemical and other tests such as cerebrospinal fluid (CSF) TAU protein levels, CSF β-amyloid levels, serum levels of iron-binding protein p97 (see Kennard *et al.*, 1996), and pupillary response to a cholinergic antagonist (Scinto *et al.*, 1994), could help in the future in the diagnosis or differential diagnosis of Alzheimer's dementia in older individuals with possible incipient Alzheimer's dementia. However, some individuals consider that 'forewarned is forearmed', that is a prognostic not a diagnostic use of the assay, and will cope with their risk and modify their lives accordingly (House of Commons Science and Technology Committee, 1995) a view endorsed by the Alzheimer's Disease Society.

Additionally it is also possible that a therapeutic approach will be found. While *APOE* testing (House of Commons Science and Technology Committee, 1995) is of limited predictive value – that is, of those who do possess an *E4* allele, we cannot predict for certain who will develop Alzheimer's dementia nor when, public demand could prevail if public knowledge were greater. However, it should be remembered that after counselling, only a small number of patients at risk for the devastating consequences of Huntington's chorea elect to have genetic testing, but nevertheless for some it brings immense relief and for others effective life planning and reproductive choices. Needless to say, *E4* genotype should not be reported to a patient tested for *E2/E2* genotype in the context of dyslipidaemia, but here we have a prognostic test which might be applicable in conjunction with prior counselling. In a patient with Alzheimer's dementia, treatable or untreatable cause, it could add to the picture of diagnostic evaluation. In the seemingly remote future of a preventive therapy, it could also become a screening test in later life, for example part of a 'retirement screening programme'; *E4* positive individuals would then receive the preventive therapy. These ideas are pure conjecture, but it is hoped that even the most conservative reader will appreciate that *E4* genotyping could become a massive demand on future laboratory services, compared with the highly selective instances in which *E2* genotype is currently considered.

Acknowledgements

Ian Day is a Lister Institute Research Fellow. Emmanuel Spanakis is supported by Medical Research Council grant G9605150MB.

References

Bolla M, Haddad L, Winder AF, Humphries SE and Day INM. (1995) High-throughput method for determination of apolipoprotein E genotypes with use of restriction digestion analysis by microplate array diagonal gel electrophoresis (MADGE). *Clin. Chem.* **41**, 1599–1604.

Corder EH, Saunders AM, Strittmatter WJ *et al.* (1993) Gene dose of apolipoprotein-E type-4 allele and the risk of Alzheimer's disease in late onset families. *Science* **261**, 921–923.

Davignon J. (1993) Apolipoprotein E polymorphism and atherosclerosis. In: *New Horizons in Coronary Heart Disease* (eds CVR Born and CJ Schwartz), pp. 5.1–5.21. Xerox Education Publications, Middleton, CT. .

Day INM and Humphries SE. (1994) Electrophoresis for genotyping: microtitre array diagonal gel electrophoresis (MADGE) on horizontal polyacrylamide (H-PAGE) gels, Hydrolink or agarose. *Anal. Biochem.* **222**, 389–395.

Diedrich JF, Minnigan H, Carp RI, Whitaker JN, Race R, Frey W and Haase AT. (1991) Neuropathological changes in scrapie and Alzheimer's disease are associated with increased expression of apolipoprotein E and cathepsin in astrocytes. *J. Virol.* **65**, 4759–4768.

Hixson JE andVernier DT. (1990) Restriction isotyping of human apolipoprotein E by gene amplification and cleavage with HhaI. *J. Lipid Res.* **31**, 545-548.

House of Commons Science and Technology Committee (1995) *Human Genetics: The Science and its Consequences,* Vol. 2, p. 123. HMSO, London.

Kennard ML, Feldman H, Yamada T and Jefferies WA. (1996) Serum levels of the iron binding protein p97 are elevated in Alzheimer's disease. *Nature Med.* **2**, 1230–1235.

Kim D-H, Iijima H, Goto K *et al.* (1996) Human apolipoprotein E receptor 2. A novel lipoprotein receptor of the low density lipoprotein receptor family predominantly expressed in brain. *J. Biol. Chem.* **271**, 8373–8380.

Miyata M and Smith JD. (1996) Apolipoprotein E allele-specific antioxidant activity and effects on cytotoxicity by oxidative insults and beta-amyloid peptides. *Nature Genetics* **14,** 55–61.

Namba Y, Tomonaga M, Kawasaki H, Otomo E and Ikeda K. (1991) Apolipoprotein E immunoreactivity in cerebral amyloid deposits and neurofibrillary tangles in Alzheimer's disease and kuru plaque amyloid in Creutzfeldt-Jacob disease. *Brain Res.* **541**, 163–166.

Nolte RT and Atkinson D. (1992) Conformational analysis of apolipoprotein A-I and E-3 based on primary sequence and circular dichroism. *Biophys. J.* **63**, 1221–1239.

Okuizumi K, Onodera O, Namba Y *et al.* (1995) Genetic association of the very low density lipoprotein (VLDL) receptor gene with sporadic Alzheimer's disease. *Nature Genetics* **11**, 207–209.

Pericak-Vance MA, Bebout JL, Gaskell PC *et al.* (1991) Linkage studies in familial Alzheimer's disease: evidence for chromosome 19 linkage. *Am. J. Hum. Genet.* **48**, 1034–1050.

Rall SCJ, Weisgraber KH, Innerarity TL and Mahley RW. (1982) Structural basis for receptor binding heterogeneity of apolipoprotein E from type III hyperlipoproteinemic subjects. *Proc. Natl Acad. Sci. USA* **79**, 4696–4700.

Schellenberg GD. (1995) Genetic dissection of Alzheimer disease, a heterogeneous disorder. *Proc. Natl Acad. Sci. USA* **92**, 8552–8559.

Schellenberg GD, Deeb S, Boehnke LM *et al.* (1987) Association of apolipoprotein CII allele with familial dementia of the Alzheimer type. *J. Neurogenetics* **4**, 97–108.

Scinto LFM, Daffner KR and Dressler D. (1994) A potential noninvasive neurobiological test for Alzheimer's disease. *Science* **266**, 1051–1053.

Seeman P, Laccone F, Reiss J and Stoppe G. (1995) Absence of mutations in the apolipoprotein E (APOE) gene of patients with Alzheimer disease. *Hum. Mutat.* **5**, 103–104.

Siest G, Pillot T, Regis-Bailly A, Leininger-Muller B, Steinmetz J, Galteau M-M and Visvikis S. (1995) Apolipoprotein E: an important gene and protein to follow in laboratory medicine. *Clin. Chem.* **41**, 1068–1086.

Stavljenic-Rukavina A, Sertic J, Salzer B, Dumic M, Radica A, Fumik K and Krajina A. (1993) Apolipoprotein E phenotypes and genotypes as determined by polymerase chain reaction using allele-specific oligonucleotide probes and the amplification refractory mutation system in children with insulin-dependent diabetes mellitus. *Clin. Chem. Acta* **216**, 191–198.

Strittmatter WJ, Weisgraber KH, Huang DY *et al.* (1993) Binding of human apolipoprotein E to synthetic amyloid beta-peptide: isoform-specific effects and implications for late-onset Alzheimer's disease. *Proc. Natl Acad. Sci. USA* **90**, 8098–8102.

Templeton AR. (1996). Cladistic approaches to identifying determinants of variability in multifactorial phenotypes and the evolutionary significance of variation in the human genome. In: *Variation in the Human Genome. Ciba Foundation Symposium 197*, pp. 259–283. John Wiley & Sons, Chichester.

Tsai M-S, Tangalos EG, Petersen RC *et al.* (1994) Apolipoprotein E: risk factor for Alzheimer disease. *Am. J. Hum. Genet.* **54**, 643–649.

Utermann G, Pruin N and Steinmetz A. (1979) Polymorphism of apolipoprotein E. III. Effect of a single polymorphic gene locus on plasma lipid levels in man. *Clin. Genet.* **15**, 63–72.

Weisgraber KH. (1994) Apolipoprotein E: structure–function relationships. In: *Advances in Protein Chemistry. Vol. 45* (ed. VN Schumaker), pp. 249–302. Academic Press, New York.

Weisgraber KH, Innerarity TL and Mahley RW. (1982) Abnormal lipoprotein receptor-binding activity of the human E apoprotein due to cysteine-arginine interchange at a single site. *J. Biol. Chem.* **257**, 2518–2521.

Wisniewski T and Frangione B. (1992) Apolipoprotein E: a pathological chaperone protein in patients with cerebral and systemic amyloid. *Neurosci. Lett.* **135**, 235–238.

Wilson PWF, Myers RH, Larson MG, Ordovas JM, Wolf PA and Schaefer EJ. (1994) Apolipoprotein E alleles, dyslipidemia, and coronary disease: the Framingham offspring study. *J. Am. Med. Assoc.* **272**, 1666–1671.

Zannis VI, Just PW and Breslow JL. (1981) Human apolipoprotein E isoprotein subclasses are genetically determined. *Am. J. Hum. Genet.* **33**, 11–24.

Zerba KE and Sing CF. (1993) The role of genome type–environment interaction and time in understanding the impact of genetic polymorphisms on lipid metabolism. *Curr. Op. Lipidol.* **4**, 152–162.

7

Genetic tests for coronary artery disease risk: the fibrinogen and stromelysin genes as examples

Steve E. Humphries, Hugh Montgomery, Shu Ye and Adriano Henney

7.1 Introduction

The critical role of genes is to code for structural proteins and enzymes which enable the cell, organ or organism to maintain homeostasis in the face of the environmental challenges experienced. The clinical features of a disorder with a late age of onset, such as coronary artery disease (CAD), can therefore be thought of as being caused in large measure by the failure of the individual to maintain homeostasis. However, CAD is a multifactorial disorder, with both genetic and environmental factors being involved to varying extents in causing the disease in different individuals in the general population. Over recent years, epidemiological studies have identified a number of factors that are associated with increased risk of CAD, including high blood pressure, smoking, high dietary fat intake, obesity and the development of diabetes. From these studies, a number of plasma risk factors have been identified such as elevated levels of cholesterol and high levels of the clotting factor, fibrinogen.

In general, there are two ways where information at the level of the genotype may give added information over and above that of measurement of an individual's current level of a CAD risk factor. The first is where genotype predicts a level of measurable risk factor at some future time. For example, it is well known that plasma lipid levels change

Genetics of Common Diseases: future therapeutic and diagnostic possibilities,
edited by I. Day and S. Humphries.

with the development of obesity and diabetes, as well as with increasing age. The levels of some clotting factors change in a similar way, and the menopause has a major impact on plasma levels of such risk factors. Thus, if individuals with a certain genotype experience a greater than average increase in risk factors with such life changes, such individuals would benefit from this knowledge and could be given advice to avoid this dangerous lifestyle change, for example by more strenuous efforts to prevent becoming obese. The second situation where a genotype may give extra predictive information over classical risk factor traits is where the gene codes for a protein or enzyme which is not easily measurable. Thus, if the protein is expressed only in certain cells which are not easily biopsied, for example heart muscle, the intestine, the liver cells or the vessel wall, then identification of a high activity variant protein at the DNA level may be extremely informative. In this review, examples of both of these types of genotype information will be presented.

7.2 Molecular biology of variability

For any selected individual, the measured level of a risk factor in the blood, such as fibrinogen, is determined by an individual's genetic predisposition that 'sets' their level at a particular point plus their genetically determined ability to maintain homeostasis in response to the environment being experienced. Under this model, the current epidemic of CAD being seen in Western societies is not due to an increase in the frequency of mutations in important genes, but rather to an inability in some, but not all, individuals to maintain optimum blood levels of these risk-factor components in the environment experienced as a result of 'affluent' lifestyle.

Failure to maintain homeostasis has been previously proposed (Berg, 1987), in terms of a 'variability gene' concept, with a distinction between genes (or genetic variation) affecting the 'level' of a risk factor and other genes (or genetic variation) determining 'variability', with an individual's risk of CAD being a combination of the two. Thus, an individual with genetically determined high risk-factor levels plus a 'restrictive' genotype would be at greater risk of CAD than one with a 'permissive' genotype, who could potentially lower risk by modifying lifestyle. Conversely, an individual with genetically determined low risk factor levels and a 'restrictive' genotype may not experience an elevation of risk by entering a high risk environment, whereas one who carries a 'permissive' genotype would suffer a large and clinically important increase in risk and succumb to premature CAD. An example of this for a genotype determining variability in plasma fibrinogen levels over time is shown in *Figure 7.1*. If DNA-based tests could be developed that would distinguish between individuals with the high fibrinogen variability pattern compared with the low fibrinogen variability pattern, this, plus an individual's current levels, might have a better predictive value for an individual's future risk of having a thrombotic ischaemic attack than current levels alone.

The concept of a 'variability gene' can be formulated in molecular terms in a number of different ways. One possibility is that variation in the sequences involved in control of expression of the gene (e.g. transcription, RNA processing, translation or mRNA stability) may respond to particular environmental factors. It is well established that control of the rate of transcription is affected by the interaction of a large number of different *trans*-acting DNA-binding proteins that bind close to, and mainly but not exclusively, upstream from the start of transcription of the gene. For many genes , this 'proximal promoter region' of 200–300 base pairs (bp) contains a number of different elements of 6–12 bp in length that are common to many genes, for example the TATA, CAAT and SP-1 boxes and the IL-6 element. Proteins that bind to these elements have been identified, and many of the genes encoding them have been cloned. Some of these elements have a positive effect on transcription, whilst others have a negative effect (repressor), so that the overall rate of transcription is under multifactorial control (Frankel and

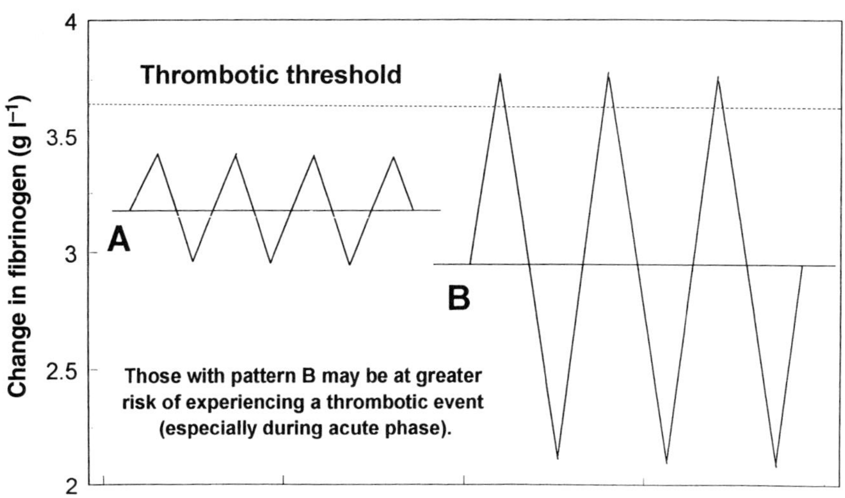

Figure 7.1. Change in plasma fibrinogen in those with 'restrictive' or 'plastic' genotype.

Kim, 1991; Latchman, 1996). It is clear that the affinity of a particular *trans*-acting protein is highly dependent on the sequence of the DNA, and single base changes may reduce or abolish binding, leading to severely reduced transcription, as has been found in some examples of thalassaemia (Treisman *et al.*, 1983). Conversely, other sequence changes within the promoter may increase the strength of the promoter, for example causing hereditary persistence of fetal haemoglobin (Ronchi *et al.*, 1989). Similar conclusions have been reached on other promoters by *in vitro* expression studies using site-directed mutagenesis (e.g. reviewed in Maniatis *et al.*, 1987).

It is of relevance that many of these DNA-binding proteins function as homodimers, heterodimers or multimers, and some of these proteins are modified post-translationally, such as by phosphorylation (Gomes and Cohen, 1991), influencing affinity for DNA or for each other and, thus, their activity. Environmental changes might then affect these processes in several ways; for example, changes in dietary lipid composition, or other components such as drugs, may have an effect on cellular membrane fluidity or structure which could alter the activity of membrane-associated enzymes such as protein kinases, and thus, in turn, alter the modification state and DNA affinity of the nuclear proteins. This modification may result in a large effect on transcription from one (polymorphic) variant promoter sequence, but not another. In addition to sequence variation in the promoter region of the candidate gene itself, there is also the possibility that changes in the levels or function of the *trans*-acting DNA-binding proteins themselves may affect 'variability'. The effect of such changes would be likely to be wide ranging (pleiotropy) since many of these sequence elements are common to many genes and, thus, any one of these proteins is likely to be involved in control of expression of several genes.

7.3 Sequence polymorphism at the fibrinogen locus, and variability in plasma fibrinogen levels

Several prospective studies have shown a direct association between plasma fibrinogen concentration and the subsequent incidence of CAD (e.g. Meade *et al.*, 1986; reviewed in Cook and Ubben, 1990). For men in the Northwick Park Heart Study (NPHS), an elevation of one standard deviation in fibrinogen (about $0.6\ \mathrm{g\ l^{-1}}$) was associated with an 84% increase in the risk of CAD within the next 5 years. This association is probably mediated through a number of different pathophysiological processes. Individuals with elevated fibrinogen levels may have an increased propensity for coagulation, and thrombosis formation in an artery that is already narrowed by atherosclerosis is a frequent cause of acute symptoms such as myocardial infarction or stroke. High levels of fibrinogen increase blood viscosity, which itself may partly be involved. Fibrinogen interacts with specific receptors on activated platelets and increases platelet aggregability *in vitro*. Elevated levels of fibrinogen may also be having a direct effect on the development of the atherosclerotic lesion, and fibrinogen and its degradation products can be detected histochemically and immunologically in the intima of diseased artery walls and within atherosclerotic plaques (Smith and Staples, 1981). Animal studies have also shown that intravascular fibrin deposition is related to plasma fibrinogen levels. Finally, fibrinogen has been shown to have a mitogenic effect on haemopoetic cells, through apparently specific cell-surface receptors, although these studies have not been extended to endothelial cells.

Fibrinogen is synthesized in the liver and, since it is an acute phase protein, its plasma level is raised following infection or injury. Because of this sensitivity to environmental factors, the within-individual variation of fibrinogen levels is high, accounting for up to 26% of the sample variance in standardized assays (Thompson *et al.*, 1987). Each plasma fibrinogen molecule comprises two each of the Aα-, Bβ- and γ-fibrinogen polypeptide chains, and the complex is held together by a number of inter- and intra-chain disulphide links. The cDNA sequence and gene structure for all three fibrinogen genes have now been determined. The genes are in a cluster of less than 50 kb on the long arm of chromosome 4, and each chain is synthesized as a separate mRNA, with the levels of all three mRNAs being co-ordinatedly controlled at least in response to acute stimulation such as defibrination. The rate-limiting step in the production of the mature fibrinogen molecule in the human hepatoma cell line, HepG2, is the synthesis of the Bβ-polypeptide chain (Roy *et al.*, 1990), which in turn is influenced by the amount of its mRNA available. It is therefore likely that an alteration in the level of synthesis of the Bβ-chain may have an effect on the amount of fibrinogen secreted by the liver.

A cartoon of the β-gene promoter is shown in *Figure 7.2*, and the region from –150 base pairs to the start of transcription contains a number of elements of interest (Huber *et al.*, 1990). Only in this region is there significant sequence homology between the rat and human genes, suggesting that this sequence has important conserved functions. This region has homology with other 'acute phase' genes such as α-1-antitrypsin (Courtois *et al.*, 1987). In addition, this region has been reported to contain all the information required to act as a promoter in

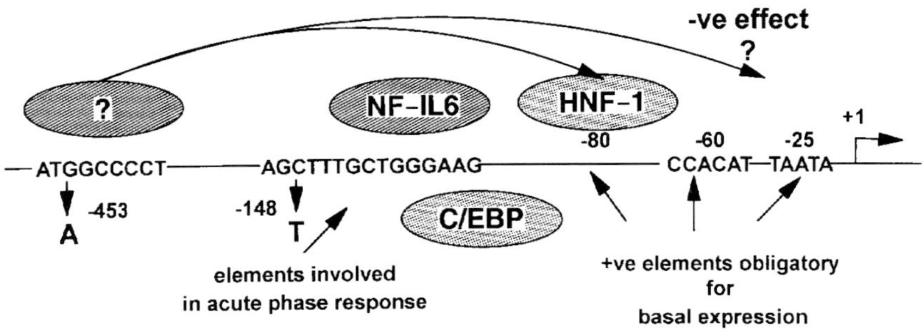

Figure 7.2. Promoter region of the β-fibrinogen gene.

HepG2 cells and has been shown to bind proteins from a HepG2 cell nuclear extract. The sequence from –89 to –76 contains a conserved liver-specific transcription element which binds hepatic nuclear factor 1 (HNF1), and deletion mapping shows that just upstream lies an interleukin 6 (IL-6) responsive element, which has been identified in other genes as the motif CTGGGA (Dalmon *et al.*, 1993). It is therefore possible that sequence changes in this region of the gene may have a direct effect on the rate of transcription and thus on plasma fibrinogen levels.

In studies of the β-fibrinogen promoter, a common G/A sequence variation at position –455 was detected by use of the enzyme, *Hae*III, with a loss of the predicted *Hae*III site in roughly 20% of alleles examined (Thomas *et al.*, 1991). In several samples of healthy individuals it has been reported that the A^{-455} allele (lack of *Hae*III cutting site) is consistently associated with higher fibrinogen levels (*Table 7.1*), with those with one or more copies of the A^{-455} allele having higher fibrinogen levels than those with the genotype *G/G* (0.28 g l^{-1} higher in healthy men in the UK; Thomas *et al.*, 1991). The magnitude of this genotype effect indicates that it is likely to be of biological significance in causing an elevated risk of thrombosis and, by extrapolation from the NPHS data of the relationship between fibrinogen and CAD risk (0.6 g l^{-1} associated with 84% greater risk), men with the *A* allele would be at 40% higher risk of a thrombotic event. Of course, this estimate is based on healthy middle-aged men from north London, and may not be the same in other groups. However, in support of the relationship between fibrinogen genotype and risk of disease, polymorphisms at the fibrinogen locus have been reported to be associated with risk of peripheral vascular disease (Fowkes *et al.*, 1992), although not with risk of myocardial infarction (MI) in a case– control study (Scarabin *et al.*, 1993).

7.4 Fibrinogen and the acute phase response

The 'acute phase' response that accompanies severe inflammation is characterized by changes in the plasma level of a number of proteins, with some positive acute phase reactants increasing two-to four-fold (e.g. fibrinogen and α-1-antitrypsin), others several hundred-fold (e.g. C-reactive protein), and some decreasing (negative acute phase reactants, including albumin and apoAl). Monocytes play a central role in the acute phase. They migrate to sites of tissue damage and respond to various external stimuli by secreting a number of cytokines and growth factors, the most important of which appear to be IL-1, IL-6, tumour necrosis factor (TNF) and transforming growth factor-β (TGF-β). In CAD it is

Table 7.1. Mean fibrinogen levels in men with different 'G/A' genotypes

Sample (ref.)	Genotype	No.	Fibrinogen (g l^{-1})
Healthy men, UK (Thomas, 1991)	GG GA + AA	188 101	2.71 2.99[b]
Peripheral vascular disease (PVD) + Healthy, UK (Conner *et al.*, 1992)[c]	GG GA + AA	165 82	2.92 2.96
M1 + Healthy, Sweden (Green *et al.*, 1993)	GG GG + AA	44 32	2.9 3.30[b]
Young males, European Arteriosclerosis Research Study (EARS) (Humphries *et al.*, 1995)	GG GG + AA	326 188	2.24 2.34[b]
Etude Cas Témoin sur l'Infarctus du Myocarde (ECTIM) control group (Scarabin *et al.*, 1993)	GG GG + AA	410 238	2.97 3.06[b]
ECTIM, MI group (Scarabin *et al.*, 1993)	GG GA + AA	352 181	3.38 3.54[b]
Healthy men, Germany (Heinrich *et al.*, 1995)	GG GA + AA	160 87	2.63 2.84[b]
Healthy men, USA (Iso *et al.*, 1995)	GG GA + AA	35 30	2.80 2.98
Healthy men, UK (Thomas *et al.*, 1996)	GG GA + AA	232 152	3.17 3.36[a]

[a]$p < 0.05$, [b]$p < 0.025$
[c]Data not adjusted for the effect of age/smoking, etc.

believed that macrophages, recruited particularly by smoking-induced damage to the lung tissues, as well as those present in atherosclerotic tissues as foam cells, are responsible for the secretion of cytokines into the blood, and these stimulate the liver to make a 'low grade' acute response. Hepatocytes have a specific IL-6 receptor on their surface which comprises two proteins, one of which binds IL-6 and, through interaction with the second (a transmembrane tyrosine kinase; Kishimoto *et al.*, 1992), stimulates the phosphorylation of specific cytoplasmic proteins. This initiates a cellular cascade of events which results, among other things, in the rapid modification of a nuclear transcription factor NF-IL-6, which significantly enhances the DNA-binding ability of the protein (Akira *et al.*, 1992). NF-IL-6 is a 'leucine zipper' - containing protein which has homology to the CCAAT enhancer-binding protein (C/EBP) transcription factor. The transcription of a number of liver-specific genes is controlled by C/EBP binding, due to the presence of a sequence element (consensus TGTGGAAA) in the promoter region of both positive and negative acute phase genes; such an element is found in both the albumin and apoAI promoter. It appears that NF-IL-6 competes for C/EBP binding in these genes, and this has the effect of suppressing the transcription of negative acute phase proteins. By contrast, positive acute phase proteins have related sequence elements which are recognized only by NF-IL-6, and binding results in strong transcription; such elements have been identified in the fibrinogen gene promoter amongst others.

Thus although the G/A^{-455} polymorphic sequence is outside the region of the reported promoter sequence, it is possible that it is having a direct effect on transcription of the gene, and preliminary studies have demonstrated binding of a hepatic nuclear protein to the *G* but not the *A* sequence (Green F. and Humphries SE., unpublished). However, recently it has been found that the A^{-455} sequence change is in complete allelic association in all Caucasian populations studied to date with a C^{-148}/T change located close to the consensus of the IL-6 element (*Figure 7.2*). This raises the possibility that the *G/A* change is acting as a neutral marker for the *C/T* change, which is the functional change working through effects on transcription of the β-fibrinogen gene that are mediated by IL-6. One possibility is that variation in the IL-6-responsive element in the β-promoter may increase the affinity of NF-IL-6, leading to enhanced transcription. This hypothesis is supported by recent data from our laboratory (Lane *et al.*, 1993) and others (Baumann and Henschen, 1993) using band shift assays that the T^{-148} sequence binds a nuclear protein which the C^{-148} sequence does not. In order to test this hypothesis, experiments are in progress to insert this fragment of the gene into the appropriate vector to test promoter strength.

7.5 Acute phase stimulation of fibrinogen by exercise

As shown in *Figure 7.1*, it is possible that, in the general population, some individuals may have a 'plastic' fibrinogen genotype that responds easily to environmental factors, and such individuals would thus experience a greater than average increase in plasma fibrinogen levels in response to a moderate environmental stimulus. These individuals would then be at greater than average risk of a thrombotic event at the peak of fibrinogen levels, and thus be at greater risk of CAD. To investigate this possibility, we have examined the effects of acute intensive exercise on plasma fibrinogen levels and the relationship of these responses to genotype (Montgomery *et al.*, 1996). One hundred and fifty-six male British Army recruits were studied at the start of their 10-week basic training (which emphasizes physical fitness) and between 0.5 and 5 days after a major 2-day strenuous military exercise (ME). Compared with baseline values, fibrinogen concentrations were significantly lower (12%, $p = 0.04$) at day 5 after ME, consistent with beneficial effect of training (Stratton *et al.*, 1991). However, within 12 h of the completion of the 2-day ME, fibrinogen levels were 14.5% higher than those at 5 days, suggesting that the acute response had already begun. Levels were significantly higher on days 1–3 after ME (suggesting an 'acute phase' response to strenuous exercise) and were maximal on days 1 and 2 (27% and 37%, $p < 0.001$ respectively).

The time course of this response was similar to that seen after the physiological stress of MI (Haines *et al.*, 1983), suggesting strongly that the men in this situation were experiencing an acute response cytokine-mediated change in plasma fibrinogen levels. However, as shown in *Figure 7.3*, the degree of rise was strongly related to the presence of the *A* allele, being 27, 37, and 89% for those of *GG*, *GA* and *AA* genotype respectively ($p = 0.01$). Although the group was small, the differences were statistically significant, and the size of the effect is sufficient to be of biological significance. In particular, 2 days after the exercise the men homozygous for the *A* allele (representing 4% of the population based on the observed allele frequencies) had fibrinogen levels which had risen by more than 105% compared with their 'untrained' levels, and those were amongst the highest levels in the whole sample. These findings support an association of the *A* allele with greater fibrinogen level 'responsiveness' after a cytokine-inducing event.

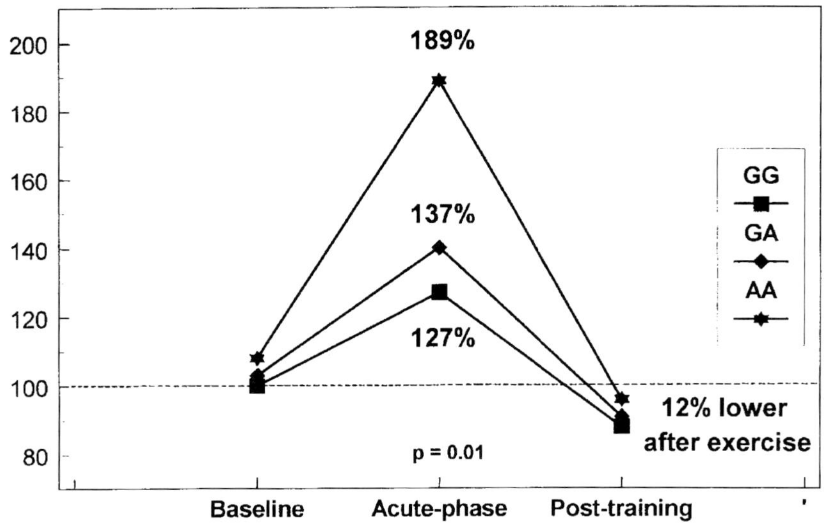

Figure 7.3. Mean percentage change in plasma fibrinogen on days 1+2 and day 5 after severe exercise in men with different *G/A* genotype.

The young fit men in this study showed no adverse effects owing to the acute rise in fibrinogen levels, and it is likely that compensating increases in fibrinolytic potential also occurred. However, the observations raise the possibility that, in a man with advanced atherosclerosis with a risk of plaque rupture, an acute phase reaction may result in a greater risk of a thrombotic event if he has the genotype *A/A* than the genotype *G/G*. This aspect of genotype-associated risk remains to be explored.

7.6 Genes determining development of atherosclerosis and plaque rupture

In recent years, a much greater understanding has been obtained of the key role of cell-mediated processes in the development of the atherosclerotic plaque and in plaque rupture, and it is highly likely that genetic variation is also affecting these processes. The importance of the macrophage-derived foam cells that accumulate in the plaque is now well established, and fluctuations in the type of macrophages or in their activation status as a result of changes in the use of drugs, hormonal status or injury is likely to be genetically determined. The importance of immune mechanisms and of growth factors such as a platelet-derived growth factor (PDGF) has also been established, and their action studied on the regulation of expression of genes, and proliferative processes, in macrophages, smooth muscle cells (SMC) and endothelial cells. It is also likely that variability in the levels of growth factors present in the plaque as a result of such environmental changes will be under genetic control.

The acute event which precipitates an MI is usually the rupture of an advanced foam-cell and lipid-laden atherosclerotic plaque, thus exposing a thrombogenic surface, with the resulting thrombosis causing occlusion of the artery (Davies *et al.*, 1989). Thus the variability of expression of proteinases that could weaken and destroy the plaque is a process that is likely to be critical. Any or all of the cells found in the normal or atherosclerotic vessel wall may synthesise and secrete such proteinases; for example, as a result of stimulation by growth factors found in SMC (James *et al.*, 1993) or of accumulation of lipids in foam cells (Rouis *et al.*, 1990). Variability in the tight control of these processes or the structural proteins themselves could also contribute to the development of vascular disease. Since the level of expression of these enzymes in the plaque is essentially impossible to measure, and plasma levels (even if assays were available) may not show a strong correlation with levels in the plaque, this is a second area of cardiovascular disease where a specific genotype, if it predicted high enzyme levels, may add information to measurement of plasma risk factors for estimating an individual's CAD risk.

7.7 Stromelysin genotype and progression of atherosclerosis

Stromelysin-1 is a key member of the matrix metalloproteinase (MMP) family, with a broad substrate specificity. It can degrade types II, IV and IX collagen, proteoglycans, laminin, fibronectin, gelatins and elastin (Murphy and Reynold, 1993). In addition, stromelysin-1 can also activate other MMPs such as collagenase, matrilysin and gelatinase B, rendering stromelysin-1 crucial in connective tissue remodelling (Imai *et al.*, 1995). Expression of stromelysin-1 is primarily regulated at the level of transcription, where the promoter of the gene responds to various stimuli, including growth factors, cytokines, tumour promoters and oncogene products (Frisch and Ruley, 1987; Hanemaaijer *et al.*, 1993; Kerr *et al.*, 1988). The regulatory effects of such stimuli are mediated through a number of *cis*-elements located in the stromelysin-1 promoter, the location of some of which are shown in *Figure 7.4*. For instance, the activator protein-1 binding site at positions –63 to –70 is necessary for the basal expression of the gene and is also involved in IL-1 induction (Quinones *et al.*, 1994). A promoter element located between –1218 and –1202 is responsible for the induction of stromelysin-1 expression by

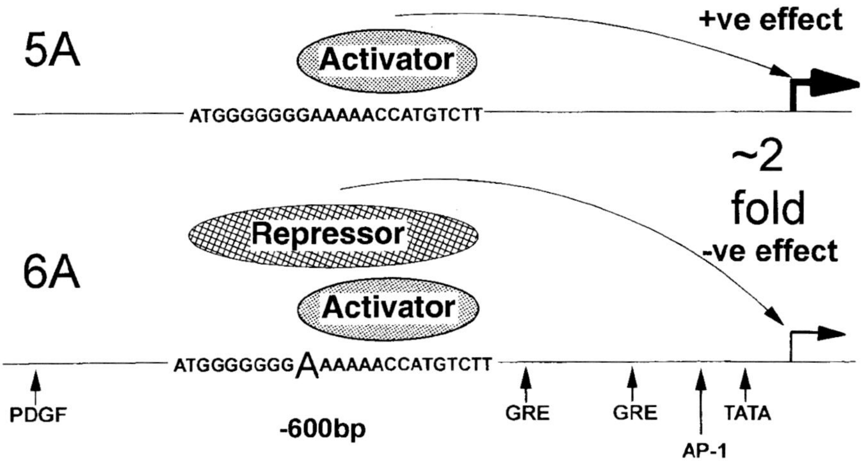

Figure 7.4. Promoter region of the stromelysin gene.

PDGF B/B (Diaz-Meco *et al.*, 1991; Sanz *et al.*, 1994), whereas three sequences that share strong homology with the glucocorticoid-responsive consensus element have been implicated in dexamethasone suppression (Quinones *et al.*, 1989).

Over the last few years, MMPs have also been implicated in connective tissue remodelling during atherogenesis (Newby *et al.*, 1994). By *in situ* mRNA hybridization, we originally demonstrated the presence of stromelysin-1 in coronary atherosclerotic plaques (Henney *et al.*, 1991). Extensive expression of the *stromelysin-1* gene was localized particularly to the regions considered prone to rupture, such as the cap and its adjacent tissues. These observations have been supported subsequently by studies using *in situ* zymography, that demonstrated gelatinolytic and caseinolytic activity in atherosclerotic but not in the uninvolved arterial tissues. Additional reports of MMP expression in atheroma have also appeared (Brown *et al.*, 1995; Nikkari *et al.*, 1995).

Using single-strand conformation polymorphism analysis, we identified a common polymorphism in the stromelysin gene promoter, which, as is shown in *Figure 7.4*, is located roughly 600 bp upstream from the start of transcription, where one allele has a run of six adenosines (*6A*) and another has five adenosines (*5A*). To investigate the relationship between this polymorphism and progression of the disease, we genotyped a small group of men with CAD, who participated in the St Thomas' Atherosclerosis Regression Study (STARS) and who were randomized to receive usual care (UC), dietary intervention (D) or diet plus cholestyramine (DC), with angiography at baseline and at 39 months. In these patients, the frequency of the *5A* allele was 0.49 (95% CI from 0.41 to 0.57) and was not significantly different from that in a sample of 155 healthy UK men. The clinical and demographic characteristics of the patients, as well as the study design, methodologies of coronary angiography and other clinical measurements in STARS have been described in detail previously (Watts *et al.*, 1992). Briefly, 90 men (aged under 66 years, with plasma cholesterol level above 6.0 mmol l^{-1} but not exceeding 10 mmol l^{-1} who had not been treated with lipid-lowering drugs) referred for coronary angiography to investigate angina pectoris or other findings suggestive of CAD were recruited, with 74 of them completing the study. The patients had coronary angiography and plasma lipids and lipoproteins measured prior to being randomly assigned to one of the three treatment groups, and again approximately 3 years after randomization. Paired measurements were made using a computerized method on 489 segments, with a mean (SD) of 6.4 (2.0) segments analysed per patient. The study end-points were the overall change (per patient) in mean and minimum absolute width of coronary segments (ΔMAWS and ΔMinAWS, respectively).

As previously reported, the DC group showed regression of coronary artery disease and the UC group showed progression of disease during the trial period, with the D group being intermediate. A strong association between the polymorphism and change in luminal diameter was seen in

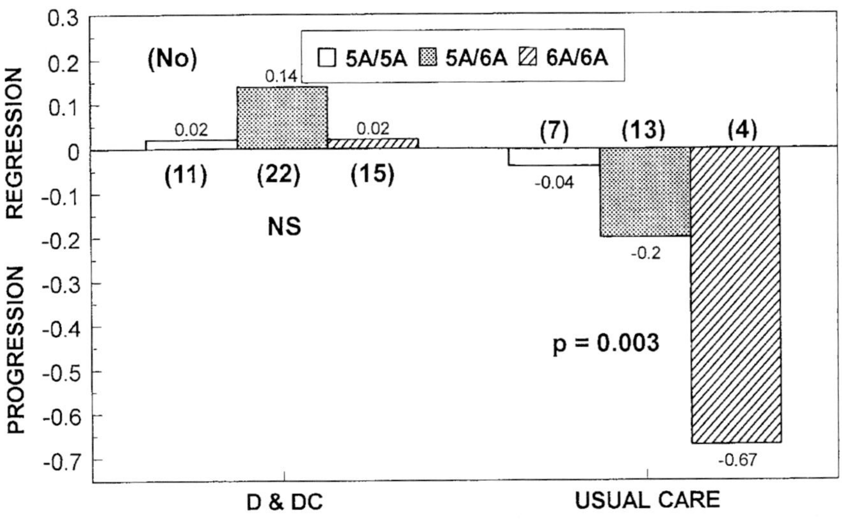

Figure 7.5. Effect of SEL genotype on progression of CAD.

the UC group in which the decrease in MinAWS in patients homozygous for the *6A* allele was 4.8-fold greater than that in those with other genotypes ($p < 0.01$). The findings were similar in measures of the change in MAWS, with a decrease 3.7-fold greater in individuals with the genotype *6A6A* than those with other genotypes ($p < 0.05$, *Figure 7.5*). No significant associations were observed in patients in the D or DC treatment groups.

This observation supports the findings by others that the MMPs are involved in the connective tissue remodelling during atherogenesis (Galis *et al.*, 1994, 1995; Henney *et al.*, 1991). These results provide the first evidence of a link between genetic variation in stromelysin and progression of coronary atherosclerosis, and support the hypothesis that connective tissue remodelling mediated by metalloproteinases contributes to the pathogenesis of atherosclerosis.

It was, however, unclear whether the *5A/6A* polymorphism plays a role in the regulation of stromelysin-1 expression, or if it is in linkage disequilibrium with variants elsewhere at the gene locus which are functional. Transient gene expression experiments were therefore carried out (Ye *et al.*, 1996) to investigate whether the *5A/6A* polymorphism has an effect on the strength of the stromelysin-1 promoter. For these experiments, two reporter gene constructs were made, in which the stromelysin-1 promoter with either *5A*s or *6A*s at the polymorphic site was placed upstream of the reporter gene *CAT*. The resultant plasmids, *5A-CAT* and *6A-CAT*, were then separately transfected into cultured human fetal foreskin fibroblasts (HFFF2), and the *CAT* expression measured. As shown in *Figure 7.4*, the results of these experiments demonstrated that the *5A-CAT* transfectants expressed higher *CAT* activity than the cells transfected with *6A-CAT*. The difference between the two types of transfectants was approximately two-fold and was statistically significant by a Student's *t*-test ($p = 0.013$). A similar trend was also observed in transiently transfected vascular SMC (data not shown).

Having found a significant difference in the stromelysin-1 promoter activity between the *6A* and *5A* alleles, an electrophoretic mobility shift assay (EMSA) was carried out to investigate whether the sequence at the *5A/6A* polymorphic site is a binding site for nuclear proteins. For these experiments, two 27/26mer oligonucleotide probes corresponding to the stromelysin-1 promoter sequence from –1189 to –1161 were synthesized, with either *6A* or *5A* in the polymorphic site. These probes were then reacted with crude nuclear extracts prepared from both HFFF2 and vascular SMC. From both cell types, two shifted bands designated A and

B were observed, regardless of whether the 5A or 6A probes were used. These bands may represent two different DNA-binding proteins or different forms of the same protein. Although bands A and B are common to both alleles, their relative intensities were observed to differ with the two probes. The signals of band B were equally intense in the assays with either the 5A or 6A probes, suggesting a similar binding affinity of the nucleoprotein factor with either allele. In contrast, the intensity of band A was significantly higher with the 6A probe than with the 5A probe, suggesting that this factor binds preferentially to the *6A* allele. In agreement with this, the binding of factor A to the 5A probe was readily blocked by unlabelled competitor oligos, while the same molarity of specific competitor oligos failed to deplete entirely the nucleoprotein in the assay with the 6A probe. A similar mobility shift pattern was also observed in experiments with nuclear extracts prepared from HUVEC and HepG2 cells, suggesting that the nuclear protein(s) are expressed in a wide variety of cell types.

Since the *6A* allele of the stromelysin polymorphism is associated with a more rapid progression of coronary stenosis due to atherosclerosis (Ye *et al.*, 1995), and the results of the molecular analysis suggest a lower promoter activity of the *6A* allele, a reduced stromelysin expression seems to favour progression of atherosclerosis. However, what are the mechanisms? A typical atheroma contains a lipid-rich core and, on its luminal aspect, a fibrous cap which is composed of vascular smooth muscle cells and extracellular matrix macromolecules including interstitial collagens, elastin and proteoglycans (Stary *et al.*, 1995). During atherogenesis, an atheroma can develop into different types of lesions. At one end of the spectrum are the atherosclerotic plaques with a thick fibrous cap containing a substantial amount of extracellular matrix. The bulk of the plaque can protrude into the lumen of the artery and, in the case of coronary atherosclerosis, encroach into the coronary flow, causing clinical manifestations like angina. At the other end of the spectrum are the plaques with a thin fibrous cap and a relatively well preserved lumen. The thin cap in this type of plaque makes it less resistant to mechanical stress and therefore more prone to plaque rupture, the commonest pathological event initiating acute coronary ischaemic episodes including unstable angina and myocardial infarction (Davies *et al.*, 1993). It has been suggested by recent pathological studies that excessive MMP activity may be partially responsible for the weakening of the fibrous cap. It is also possible that inadequate production of MMPs could result in reduced matrix degradation, favouring the deposition of connective tissue in the lesion and contributing to its growth.

Thus, taken together, the results suggest that, compared with other genotypes, individuals homozygous for the *6A* allele would have lower stromelysin-1 levels in their arterial walls because of reduced transcription of the gene. The lower stromelysin-1 activity in such individuals would favour deposition of extracellular matrix in the atherosclerotic lesions, as a result of an imbalance between the activities of matrix-degrading enzymes and their inhibitors. The excessive accumulation of connective tissue, in addition to other genetic and environmental factors, would lead to the development of the first type of atherosclerotic plaques described above (i.e. those with a thick cap) and result in a more rapid progression of angiographically defined stenosis. Further studies are required to test this hypothesis.

7.8 Conclusion

We have described two genetic predictors for an individual's future CAD risk, which may be useful over and above that of measures of classical risk factors. Both polymorphisms are in promoters, and we have evidence that they affect binding of nuclear proteins and thus influence the rate of transcription of the gene, with resulting effects on levels of the protein. We believe that this class of genetic variation may be common in the human population, and may be of particular relevance in understanding the genetic contribution to multifactorial disorders. This is because such changes are, in general, of modest impact on transcription, and thus will be of pathological consequence only in the presence of genetic or other environmental risk factors. For both of these promoter changes, we also believe that a better understanding of the detailed molecular mechanisms of their effect, including characterization of the nuclear factors whose binding and thus activity they influence, should allow the development of novel therapeutic strategies, where levels of transcription may be modulated by specific antagonists of such nuclear protein binding.

Once the mechanisms controlling changes in CAD risk factors in response to personal environmental changes are better understood, it may be possible to develop directed therapeutic strategies that will reduce risk in a genotype-specific manner, an approach which is not possible at present.

Acknowledgements

We are particularly grateful to Gina Shoesmith for help in preparing this manuscript. This work was supported by grants from the British Heart Foundation (PG007 and 86-77).

References

Akira S, Isshiki H, Nakajima T, Kinoshita S, Nishio Y, Natsuka S and Kishimoto T. (1992) Regulation of expression of the interleukin 6 gene: structure and function of the transcription factor NF-IL6. In: *Ciba Foundation Symposium 167. Polyfunctional Cytokines: IL6 and LIF* (eds GR Block, J March and K Widdows), pp. 47–67. John Wiley & Sons, Chichester.

Baumann RE and Henschen AH. (1993) Genetic variation in human Bβ fibrinogen gene promoter influences formation of a specific DNA–protein complex with the interleukin 6 response element (abstract). *Thromb. Haemostasis* **69**, 1515.

Berg K. (1987) Genetics of coronary heart disease and its risk factors. In: *Ciba Foundation Symposium 130. Molecular Approaches to Human Polygenic Disease* (ed. K Berg), pp. 14–133. John Wiley & Sons, Chichester.

Brown DL, Hibbs MS, Kearney M, Loushin C and Isner JM. (1995) Identification of 92-kD gelatinase in human coronary atherosclerotic lesions. *Circulation* **91**, 2125–2131.

Conner JM, Fowkes FGR, Wood J, Smith FB, Donnon PT and Lowe GDO. (1992) Genetic variation at fibrinogen loci and plasma fibrinogen levels. *J. Med. Genet.* **29**, 480–482.

Cook NS and Ubben D. (1990) Fibrinogen as a major risk factor in cardiovascular disease. *Trends Pharmocol. Sci.* **11**, 444–451.

Courtois G, Morgan JG, Campbell LA, Fourel G and Crabtree GR. (1987) Interaction of a liver-specific nuclear factor with the fibrinogen and α-1-antitrypsin promoters. *Science* **238**, 688–692.

Dalmon J, Laurent M and Courtios G. (1993) The human β fibrinogen promoter contains a hepatocyte nuclear factor 1-dependent interleukin-6 responsive element. *Mol. Cell. Biol.* **13**, 1183–1193.

Davies MJ, Bland MJ, Hangartner WR, Angelini A and Thomas AC. (1989) Factors influencing the presence or absence of acute coronary thrombi in sudden ischemic death. *Eur. Heart J.* **10**, 203–208.

Davies MJ, Richardson PD, Woolf N, Katz DR and Mann J. (1993) Risk of thrombosis in human athrosclerotic plaques: role of extracellular lipid, macrophage, and smooth muscle cell content. *Br. Heart J.* **69**, 377–381.

Diaz-Meco MT, Quinones S, Municio MM, Sanz L, Bernal D, Cabrero E, Saus J and Moscat J. (1991) Protein kinase C-independent expression of stromelysin by platelet-derived growth factor, *ras* oncogene, and phosphatidylcholine-hydrolyzing phospholipase C. *J. Biol. Chem.* **266**, 22597–22602.

Frisch SM and Ruley HE. (1987) Transcription from the stromelysin promoter is induced by interleukin-1 and repressed by dexamethasone. *J. Biol. Chem.* **262**, 16300–16304.

Fowkes FGR, Conner JM, Smith FB, Wood J, Donnan PT and Lowe GDO. (1992) Fibrinogen genotype and risk of peripheral atherosclerosis. *Lancet* **339**, 693–696.

Frankel AD and Kim PS. (1991) Modular structure of transcription factors: implications for gene regulation. *Cell* **65**, 717–719.

Galis ZS, Sukhova GK, Lark MW and Libby P. (1994) Increased expression of matrix metalloproteinases and matrix degrading activity in vulnerable regions of human atherosclerotic plaques. *J. Clin. Invest.* **94**, 2493–2503.

Galis ZS, Muszynski M, Sukhova GK, Simon-Morrissey E and Libby P. (1995) Enhanced expression of vascular matrix metalloproteinases induced *in vitro* by cytokines and in regions of human atherosclerotic lesions. *Ann. NY Acad. Sci.* **748**, 501–507.

Gomez N and Cohen P. (1991) Dissection of the protein kinase cascade by which nerve growth factor activates MAP kinases. *Lett. Nature* **353**, 170–172.

Green F, Hamsten A, Blomback M and Humphries SE. (1993) The role of β-fibrinogen genotype in determining plasma fibrinogen levels in young survivors of myocardial infarction and healthy controls from Sweden. *Thromb. Haemost.* **70,** 915–920.

Haines AP, Howarth D, North WRS, Goldenberg E, Stirling Y, Meade TW, Raftery EB and Craig MWM. (1983) Haemostatic variables and outcome of acute myocardial infarction. *Thromb. Haemost.* **50**, 800–803.

Hanemaaijer R. (1993) Regulation of matrix metalloproteinase expression in human vein and microvascular endothelial cells. Effects of tumour necrosis factor alpha, interleukin 1 and phorbol ester. *Biochem. J.* **296**, 803–809.

Henney AM, Wakeley P, Davies MJ, Foster K, Hembry R, Murphy G and Humphries SE. (1991) Localization of stromelysin gene expression in atherosclerotic plaques by *in situ* hybridization. *Proc. Natl Acad. Sci. USA* **88,** 8154–8158.

Heinrich J, Funke H, Rust S, Schulte H, Schonfeld R, Kohler E and Assman G. (1995) Impact of polymorphisms in the alpha- and beta-fibrinogen gene on plasma fibrinogen concentrations of coronary heart disease patients. *Thromb. Res.* **77**, 209–215.

Huber P, Laurent M and Dalmon J. (1990) Human beta-fibrinogen gene expression: upstream sequences involved in its tissue specific expression and its dexamethasone and interleukin 6 stimulation. *J. Biol. Chem.* **265**, 5659–5701.

Humphries S, Ye S, Talmud P, Bara L, Wilhelmsen L and Tiret L. (1995) European Atherosclerosis Research Study: Genotype at the fibrinogen locus (G_{-455} - A beta-gene) is associated with differences in plasma fibrinogen levels in young men and women from different regions in Europe: evidence for gender – genotype – environment interaction. *Arterioscler. Thromb. Vasc. Biol.* **15**, 96–104.

Imai K, Yokohama Y, Nakanishi I, Ohuchi E, Fujii Y, Nakai N and Okada Y. (1995) Matrix metalloproteinase 7 (matrilysin) from human rectal carcinoma cells. Activation of the precursor, interaction with other matrix metalloproteinases and enzymic properties. *J. Biol. Chem.* **270**, 6691–6697.

Iso H, Folsom A, Winkelmann JC, Kioke K, Harada S, Greenberg B, Sato S, Shimamoto T, Lida M and Komachi Y. (1995) Polymorphisms of the beta fibrinogen gene and plasma fibrinogen concentration in Caucasian and Japanese population samples. *Thromb. Haemost.* **73**, 106–111.

James TW, Wagner R, White LA, Zwolak RM and Brinkerhoff CE. (1993) Induction of collagenase and stromelysin gene expression by mechanical injury in a vascular smooth muscle-derived cell line. *J. Cell. Physiol.* **157**, 426–437.

Kerr LD, Holt JT and Matrisian LM. (1988) Growth factors regulate transin gene expression by c-fos-independent pathways. *Science* **242**, 1424–1427.

Kishimoto T, Hibi M, Murakami M, Narazaki M, Saito M and Taga T. (1992) The molecular biology of interleukin 6 and its receptor. In: *Ciba Foundation Symposium 167. Polyfunctional Cytokines: IL6 and LIF* (eds GR Bock, J March and K Widdows), pp. 5–23. John Wiley & Sons, Chichester.

Lane A, Humphries SE and Green FR. (1993) Effect on transcription of two common genetic polymorphisms adjacent to the promoter region of the B-fibrinogen gene. *Thromb. Haemost.* **69**, 962.

Latchman DS. (1996) Transcription factors mutations and disease. *N. Engl. J. Med.* **334**, 28–33.

Maniatis T, Goodburn S and Fischer JA. (1987) Regulation of inducible and tissue-specific gene expression. *Science* **236**, 1237–1244.

Meade TW, Mellows S, Brozovic M *et al.* (1986) Haemostatic function and ischaemic heart disease: principle results of the Northwick Park Heart Study. *Lancet* **2**, 533–537.

Montgomery HE, Clarkson P, Nwose OM *et al.* (1996) The acute rise in serum fibrinogen concentration with exercise is influenced by the G-453-A polymorphism of the beta-fibrinogen gene. *Arterioscl. Thromb. Vasc. Biol.* **16**, 385–391.

Murphy G and Reynolds JJ. (1993) Extracellular matrix degradation. In: *Connective Tissue and Its Heritable Disorders*, pp. 287–316. Wiley-Liss, New York.

Newby AC, Southgate KM and Davies M. (1994) Extracellular matrix degrading metalloproteinases in the pathogenesis of arteriosclerosis. *Basic Res. Cardiol.* **89**, 59–70.

Nikkari ST, O'Brien KD, Ferguson M, Hatsukami T, Welgus HG, Alpers CE and Clowes AW. (1995) Interstitial collagenase (MMP-1) expression in human carotid atherosclerosis. *Circulation* **92**, 1393–1398.

Quinones S, Saus J, Otani Y, Harris ED and Kurkinen M. (1989) Transcriptional regulation of human stromelysin. *J. Biol. Chem.* **264**, 8339–8344.

Quinones S, Buttice G and Kirkinen M. (1994) Promoter elements in the transcriptional activation of the human stromelysin-1 gene by the inflammatory cytokine, interleukin 1. *Biochem. J.* **302**, 471–477.

Ronchi A, Nicolis S, Santor C and Ottolenghi S. (1989) Increased Spl binding mediates erythroid-specific overexpression of a mutated (HPFH) γ-globulin promoter. *Nucleic Acids Res.* **17**, 10231–10241.

Rouis M, Nigon F, Lafuma C, Hornebeck W and Chapman MJ. (1990) Expression of elastase activity by human monocyte-macrophages is modulated by cellular cholesterol content, inflammatory mediators, and phorbol myristate acetate. *Arteriosclerosis* **10**, 246–255.

Roy SN, Mukhopadhyay G and Redman CM. (1990) Regulation of fibrinogen assembly. Transfection of HepG2 cells with Bβ cDNA specifically enhances synthesis of the three component chains of fibrinogen. *J. Biol. Chem.* **265**, 6389–6393.

Sanz L, Berra E, Municio MM, Dominquez I, Lozano J, Johansen T, Moscat J and Diaz-Meco MT. (1994) Zeta PKC plays a critical role during stromelysin promoter activation by platelet-derived growth factor through a novel palindromic element. *J. Biol. Chem.* **269**, 10044–10049.

Scarabin P-Y, Bara L, Ricard S, Poirer O, Cambou JP, Arveiler D, Luc G, Evans AE, Samama MM and Cambien F. (1993), Genetic variation at the b-fibrinogen locus in relation to plasma fibrinogen concentrations and risk of myocardial infarction. The ECTIM Study. *Arterioscl. Thromb.* **13**, 886–891.

Smith EB and Staples EM. (1981) Haemostatic factors in human aortic intima. *Lancet* **1**, 1171–1174.

Stary HC, Chandler B, Dinsmore RE *et al.* (1995) A definition of advanced types of atherosclerotic lesions and a histological classification of atherosclerosis. A report from the Committee on Vascular Lesions of the Council on Arteriosclerosis, American Heart Association. *Arterioscl. Thromb. Vasc. Biol.* **15,** 1512–1531.

Stratton JR, Chandler WL, Schwartz RS *et al.* (1991) Effects of physical conditioning on fibrinolytic variables and fibrinogen in young and old healthy adults. *Circulation* **83,** 1692–1697.

Thomas A, Kelleher C, Green F, Meade TW and Humphries SE. (1991) Variation in the promoter region of the β-fibrinogen gene is associated with plasma fibrinogen levels in smokers and non-smokers. *Thromb. Haemost.* **65**, 487–490.

Thomas AE, Kelleher CC, Green F and Humphries SE. (1996) Association of genetic variation at the β-fibrinogen gene locus and plasma fibrinogen levels: interaction between allele frequency of the G/A^{-455} polymorphism, age smoking. *Clin. Genet.* **50**, 184–190.

Thompson SG, Martin JC and Meade TW. (1987) Sources of variability in coagulation factor assays. *Thromb. Haemost.* **58**, 1073–1077.

Treisman R, Orkin SH and Maniatis T. (1983) Specific transcription and RNA splicing defects in five cloned β-thalassaemia genes. *Nature* **302**, 591–596.

Watts GF, Lewis B, Brunt JNH, Lewis ES, Coltart DJ, Smith LDR, Mann JI and Swan AV. (1992) Effects on coronary artery disease of lipid-lowering diet plus cholestyramine, in the St Thomas' atherosclerosis regression study (STARS). *Lancet* **339**, 563–569.

Ye S, Watts GF, Mandalia S, Humphries SE and Henney AM. (1995) Preliminary report: genetic variation in the human stromelysin promoter is associated with progression of coronary atherosclerosis. *Br. Heart J.* **73**, 209–215.

Ye S, Eriksson P, Hamsten A, Kirkinen M, Humphries SE and Henney AM. (1996). Progression of coronary atherosclerosis is associated with a common genetic variant of the human stromelysin-1 promoter which results in reduced gene expression. *J. Biol. Chem.* **271**, 13055–13060.

8

HLA genes and rheumatoid arthritis susceptibility

W. Ollier

8.1 Introduction

Rheumatoid arthritis (RA) is an inflammatory disease, which predominantly affects the articulated joints of the body where there is a synovial membrane. This is usually a chronic inflammatory process within the joint which leads to pain, swelling and the gradual destruction of the cartilage, bone, tendons and surrounding tissues. This often results in deformity and accompanying disability. A key feature of the disease is the development of 'pannus' a proliferative outgrowth of synovial tissue, which takes on invasive properties and largely contributes to the bone and cartilage erosion observed on X-ray. Another hallmark of this condition is the production of rheumatoid factors (RFs) which are autoantibodies directed to the Fc part of IgG molecules. They are present in over three-quarters of all RA patients and are usually of IgM isotype although both IgG and IgA RFs can occur.

RA is considered to be an autoimmune disease although, as yet, no joint-specific autoantigen has been identified. A relatively small proportion of patients develop autoantibodies to type II collagen although these probably represent a distinct etiopathogenic subset (Morgan, 1990). Many cells infiltrate into the synovial fluid and the synovium. A proportion of these are T-lymphocytes and RA is thought by many to be a T-cell driven disease (Panayi *et al.*, 1992). However, the majority of these cells appear to

Genetics of Common Diseases: future therapeutic and diagnostic possibilities,
edited by I. Day and S. Humphries.

be annergized, and another school of thought exists which attributes RA etiopathogenesis to macrophage activity and the pro-inflammatory cytokine proteins they produce (Firestein and Zvaifler, 1990).

RA is clinically heterogeneous and highly variable with respect to its expression, progression and outcome. Some cases can be mild or non-erosive; others can be severe and progressive. Patients can go into remission or the disease can 'burn itself out'. Occasionally patients can follow a relapse-and-remit or 'palindromic RA' course. Furthermore, patients can often develop a range of extra-articular or systemic disease complications, such as secondary Sjögren's syndrome, pulmonary fibrosis, rheumatoid nodules or Felty's syndrome. The diagnosis of RA can therefore sometimes be difficult to make, and although diagnostic criteria have been devised and regularly updated (Arnett *et al.*, 1988) a level of uncertainty can exist. This tends to bias confident diagnosis towards those who are at the more severe end of the clinical spectrum.

8.2 Epidemiology of RA

RA is a common condition affecting approximately 1% of the adult population (Silman and Hochberg, 1993). It has been documented in nearly all populations worldwide, although some variations in prevalence have been described (Harvey *et al.*, 1981; Silman *et al.*, 1993a). In most populations, more women than men are affected, although this moves towards equality with an increasing age at disease onset (Silman and Hochberg, 1993). RA tends to have a peak onset in the fourth and fifth decades of life, although it can occur at any age. No seasonal variation has been found and there is little, evidence to suggest clear bacterial, viral or environmental triggers (Silman and Hochberg, 1993). In females, pregnancy represents a risk factor for RA onset in women (Del Junco *et al.*, 1989; Silman *et al.*, 1992), although if patients with existing RA become pregnant, the disease usually goes into temporary remission.

8.3 Defining the genetic basis of RA

RA is considered to be a complex disease where both genetic and environmental components contribute to the disease. This has largely been concluded from twin studies where estimates of RA concordance have been made in monozygotic (MZ) and dizygotic (DZ) pairs. Although a number of such studies have been performed, only four have examined reasonable sample sizes (Aho *et al.*, 1986; Harvald *et al.*, 1965; Lawrence, 1970; Silman *et al.*, 1993b). RA concordance in MZ twins ranges from 12.3 to 34%, although the relative concordance risk in MZ versus DZ is approximately the same in each study (3.0–5.3). Whilst these figures may initially suggest a relatively small genetic component in RA, it should be emphasized that crude concordance rates are poor measures of genetic contribution (for a fuller criticism, see Ollier and MacGregor, 1995) and, using alternative methods, such as Falcolner's heritability index (Falconer, 1989), a figure of 60% is calculated as the inherited component (MacGregor, 1994).

Additional evidence for establishing a genetic component in RA comes from the quantification of familial disease recurrence risks in relatives of RA patients. Again a range of estimates has been observed in different studies. The relative risk of a first degree relative of a disease proband developing RA ranges from 1.2 (Lawrence and Wood, 1968) to 10 (Wasmuth *et al.*, 1978). These differences to some extent reflect the variation in methodological approach taken by these studies. An alternative measure for the genetic component has been suggested by Risch (1987). This is referred to as the λs and can be determined by dividing the sibling recurrence risk by the population prevalence. In RA this figure could be as high as 10; for comparison, the λs for juvenile onset diabetes is approximately 15.

The consensus view now is that the genetic component in RA is oligogenic in nature, and this may be composed of six or more genes. Major worldwide initiatives are presently going ahead to try to identify these genetic components. These are concentrating on whole genome screening approaches where highly polymorphic microsatellite markers are being tested for RA linkage at 10–15 cM intervals in large collections of affected sibling pairs. Preliminary reports have been made in both French (Cornelis *et al.*, 1996) and UK (Hardwick *et al.*, 1996) RA families, although results suggest that larger sample sizes will be required to detect relatively small genetic effects. In contrast, many previous studies of relatively small sized samples have reproducibly detected a large genetic effect linked with the HLA complex on the short arm of chromosome 6 (Ollier *et al.*, 1986).

8.4 HLA associations with RA

The first reports of an association between HLA and RA were made in the mid 1970s (Stastny, 1978; Winchester, 1977). These suggested that the mixed lymphocyte culture-defined HLA specificity, Dw4, and the serologically defined HLA class II antigen, DR4, were strongly associated with RA susceptibility. This was followed by numerous reports confirming the associations in other populations, with odds ratios ranging from 14.7 (Mehra *et al.*, 1982) to 1.8 (Sanchez *et al.*, 1990). It was gradually realized that not all *DRB1* variant alleles previously recognized as being of the same DR4 serotype were associated with RA susceptibility in Caucasoid populations. *HLA-DRB1*0402* (previously *DR4/Dw10*), *DRB1*0403* and *DRB1*0407* (previously *DR4/Dw13.1* and *DR4/Dw13.2* respectively) were not associated with risk (Brautbar *et al.*, 1986; Gao *et al.*, 1990; Ollier *et al.*, 1988; Wordsworth *et al.*, 1989). However, *DRB1*0401* (*DR4/Dw4*), *DRB1*0404* and *DRB1*0408* (previously *DR4/Dw14.1* and *DR4/Dw14.2* respectively) were strongly RA associated.

Furthermore, as more studies were done, it was realized that some additional *DRB1* alleles were also associated with RA. These included *DR1*, *DR10* and *DRB1*0405* in a range of Mediterranean and Jewish populations, the *DRB1*0405* variant of *DR4* (previously *Dw15*) in Oriental and Japanese populations and the *DRB1*1402* allele in native North Americans (Wilkens *et al.*, 1991). A review of HLA associations with RA in populations has been given elsewhere (Ollier and Thomson, 1992).

8.5 The shared epitope hypothesis

Although these multiple HLA associations with RA initially confused the issue of HLA-encoded risk, an examination of DRB1 amino acid sequence suggested that a conserved 'epitope' of basic charged residues (QKRAA/QRRAA/RRRAA) existed in the third hypervariable region of RA-associated alleles but not in non-associated ones (Gregersen *et al.*, 1987). This was termed the 'RA shared epitope' hypothesis and has now been widely accepted as being a plausible explanation for HLA-encoded RA risk.

The third hypervariable region structurally resides in a part of the DRB1 molecule which potentially affects both peptide binding and its interaction with T-cell receptor. This has suggested that the RA shared epitope (SE) may convey risk by either exhibiting a particularly high affinity for an arthritogenic peptide or by modelling the T-cell repertoire during thymic selection towards a potential for arthritogenic response. Other explanations have been put forward, the most plausible perhaps being based on the sequence homology between the SE (QKRRA) and prokaryotic proteins from Epstein–Barr virus (EBV) (gp110) (Roudier *et al.*, 1989) and *Escherichia coli* (dnaJ) (Albani *et al.*, 1992). Data has been presented to suggest that such molecular mimicry may be responsible for the selection of low affinity self-reactive T cells which can lead to the development of an inflammatory response in arthrodial joints (Albani *et al.*, 1995).

8.6 Is HLA associated with RA severity or susceptibility?

A number of studies have suggested that *HLA-DR4* and *DR4* homozygosity are more strongly associated with RA severity or disease progression than with RA *per se* (Jaraquemada *et al.*, 1986, 1989; Weyand *et al.*, 1992a,b). This particularly appears to be the case for the development of erosions (Young *et al.*, 1984) and extra-articular forms of the disease such as major vasculitis (Scott *et al.*, 1981) and Felty's syndrome (Kluda *et al.*, 1986). This is further supported by community-based studies of RA prevalence (DeJongh *et al.*, 1984) and incidence (Thomson *et al.*, 1993) where DR4 exerts at best only a weak effect. This is expected, as prevalent cases represent a more random sample of RA patients and incident cases are individuals with early disease. In contrast, the majority of other studies have examined HLA associations in hospital-based series which are biased towards greater severity and progression.

More recently, studies have demonstrated that the association between HLA and RA severity holds true for the presence of the SE. Although homozygosity for the SE was associated with a higher risk for severe RA, the relationship between severity and the different *DRB1* alleles carrying the SE sequence was not identical.

Patients with the *DRB1*0401/*0404* genotype particularly appear to be at highest risk for developing severe disease (Wordsworth *et al.*, 1992), and this is even more apparent in male patients and those with a younger disease onset (MacGregor *et al.*, 1995). As both males and young individuals represent groups who are normally protected from RA, they are presumed to have a higher threshold for developing disease and thus require a stronger genetic input. Such variations in RA risk for different SE-bearing alleles and genotypes are inconsistent with the SE representing a discrete risk factor for RA. One possibility may be that some SE-bearing alleles are associated with susceptibility and others with severity. The latter could be explained by linkage disequilibrium of some *DRB1* alleles carrying the SE with high producing genetic variants of the TNF-α gene which is located in the HLA class III region. A good case can be made for TNF-α having an inflammatory and erosive role in RA joint pathology (Brennan *et al.*, 1991; Elliott *et al.*, 1994; Vassalli, 1992). Polymorphism has been found in the promoter region of TNF and this appears to correlate with the level of gene expression (Wilson *et al.*, 1994). Furthermore, a relationship exists between some HLA haplotypes and the level of *in vitro* TNF production (Pociot *et al.*, 1993). Recently we have used polymorphic microsatellite markers in the TNF region to establish whether the DR4-bearing haplotypes found in RA patients have characteristic TNF microsatellite profiles. Two such haplotypes were identified (Hajeer *et al.*, 1996).

- *DRB1*0401-TNFd4-TNF a6-TNF b5-HLA-B44*
- *DRB1* 0401-TNFd5-TNFc2-TNFb1-HLA-B62*

In addition the *TNFa2* allele was found at increased frequency in *DRB1*0404* patients. This *TNF* allele has previously been associated with high *in vitro* TNF-α production (Pociot *et al.*, 1993).

A recent study has also presented evidence for *DRB1*0401* and *DRB1*1001* alleles directly influencing the exogenous peptide antigen processing pathway (Auger *et al.*, 1996). They found that the shared epitope sequences of *DRB1*0401* (QKRAA) and *DRB1*1001* (RRRAA) specifically bound HSP73, a constitutive 70 kDa heat shock protein. This association takes place at an early stage of the HLA-DR α/β heterodimer formation and appears to affect intracellular trafficking and how the DR molecule interacts with peptide fragments. These effects may have relevance to RA susceptibility as they could:

- competitively inhibit the lysosomal transport of any proteins containing the QKRAA or RRRAA sequence (for example, EBV gp110 and dna J);
- enhance proteolysis in lysosomes and allow peptides carrying the shared epitope to be presented to the immune system, thus influencing thymic selection;
- lead to HSP73 autoimmunization;
- impair HLA-DR cell surface expression.

Any of the above mechanisms could be implicated in RA etiopathology and provide an explanation why these alleles are associated with susceptibility or severity.

8.7 Is HLA-encoded RA susceptibility due to a single or multiple genetic factors?

The shared epitope hypothesis has now largely entered into immunological dogma although as yet there is not categorical evidence to support its validity. Indeed there is some genetic analysis which contradicts the existence of the SE (Dizier *et al.*, 1993). Whilst the SE remains an attractive explanation and may account for some HLA-encoded risk, it is not compatible with all observations. It is possible that other HLA-encoded genes play a role in RA, either independently of, or in concert with, a DRB1-shared epitope.

Linkage disequilibrium between genes in the HLA complex is known to be very strong and operates over considerable distances. In RA, many of the disease-associated *DRB1* alleles are known to be present on established and conserved HLA haplotypes. Thus linkage disequilibrium could be responsible for several RA susceptibility genes being found together and explain both some of the clinical heterogeneity seen and why certain shared epitope-bearing alleles contribute differentially to RA. One prime candidate is TNF-α, which has been discussed earlier in the context of RA severity. However, it is possible that other genes within the HLA complex are making a contribution.

RA susceptibility in males and females is different, with the disease being at least twice as prevalent in women (Symmons *et al.*, 1994). Much of this has been attributed to the influences of hormonal factors, although these are poorly understood. Epidemiological studies have implicated pregnancy (Del Junco *et al.*, 1989; Silman *et al.*, 1993a), breast feeding (Brennan and Silman, 1994), oral contraceptive use and menopause as contributory factors (Hazes and Van Zeben, 1991).

It is perhaps less well known that HLA associations with RA differ significantly between males and females. A higher frequency of the *HLA-B62-DRB1*0401* haplotype has been found in male RA patients and a higher frequency of the *HLA-B44-DRB1*0401* haplotype in female patients (Jaraquemada *et al.*, 1989; Ollier *et al.*, 1984). This has been confirmed in further RA cohorts and particular TNF microsatellite profiles have been established for each haplotype (Hajeer *et al.*, 1997).

In male RA patients, it is possible that *HLA-B62-DRB1*0401* is a high risk RA haplotype and, as men are usually less susceptible to RA, a higher frequency of this haplotype would be expected in males who develop the condition. Alternatively, the *HLA-B62-DRB1*0401* haplotype may contain allelic forms of other genes important for the development of RA in males, and it may be that these are present on this haplotype through linkage disequilibrium.

The latter explanation may have some credibility as previous studies have demonstrated that:

- levels of testosterone in both mice (Ivanyi *et al.*, 1972) and men (Spector *et al.*, 1988a) are strongly influenced by genes encoded within the major histocompatibility complex;
- *HLA-B62* is associated with lower testosterone levels in both healthy and RA males (Ollier *et al.*, 1989);
- low testosterone levels exist in male RA patients independent of treatment (Spector *et al.*, 1988b).

Although the mechanism for such genetic-based regulation of testosterone is not as yet understood, genes are present in the class III which influence steroid metabolism.

In females, other hormonal factors which play a role in RA etiology may be operating, and it is possible that alleles of genes which encode or regulate such factors are again in linkage disequilibrium with genes on the *HLA-B44-DRB1*0401* haplotype. The risk of RA is higher in nulliparous women than in parous women (Da Silva and Spector, 1992) and paradoxically is higher in women immediately post partum (Del Junco *et al.*, 1989; Silman *et al.*, 1993a). The association with nulliparity has been explained by underlying infertility and decreased fecundity in women destined to develop RA (Nelson *et al.*, 1993). Increased RA risk in post partum women appears to be due to breast feeding (Brennan and Silman, 1994). A biological link between these two observations is a high concentration of prolactin (Noel *et al.*, 1974; Davies, 1990). Prolactin is a pro-inflammatory hormone which has immunodulatory effects on the immune system. Hypophysectomized mice do not develop adjuvant-induced arthritis unless treated with prolactin (Jara *et al.*, 1991) and, furthermore, this process can be reversed by treatment with bromocryptine, a dopaminergic agonist which is used to control hyperprolactinaemia.

The gene for prolactin lies close to the HLA region and maps on the telomeric side of *HLA-C* (Evans *et al.*, 1988). Polymorphism of the prolactin gene has already been reported (Myal *et al.*, 1991), and it is possible that further polymorphisms will be found in either the structural or promoter regions of the gene. If such polymorphisms are associated with either the level of prolactin or its regulation, it raises the possibility that, through linkage disequilibrium, certain HLA haplotypes may be linked with a characteristic pattern of prolactin production. This suggestion is supported by the report of an association between HLA and prolactin-secreting adenomata (Farid *et al.*, 1980). Recently we have presented evidence to demonstrate that for some haplotypes including *HLA-B44-DRB1*0401*, linkage disequilibrium extends from *HLA-DRB1* to microsatellite markers close and telomeric to prolactin. Thus, in this region, linkage disequilibrium can exist for distances greater than 10 cM.

8.8 How much of the genetic contribution is due to HLA?

In addition to case–control association studies of HLA in RA, a considerable number of family studies have been performed using affected sibling pairs (Ollier *et al.*, 1986). These have reproducibly demonstrated significant linkage with the HLA complex. Families have also been used to determine the risk of RA in siblings who are HLA identical to the proband (Deighton *et al.*, 1989). From this data it has been calculated that HLA is responsible for approximately one-third of the total genetic component. This estimate can also be confirmed by taking the approach of Risch. Using a sibling recurrence risk of 3.9% (Deighton *et al.*, 1989) and an RA prevalence of 0.8% (Silman and Hochberg, 1993) a λs of 4.9 is obtained. The λs for HLA can be estimated by calculating the ratio of the expected proportion of sibling pairs sharing zero HLA haplotypes (25%) to that observed. In a collection of 100 UK affected sibling pair RA families (Worthington *et al.*, 1994) the proportion sharing zero HLA haplotypes is 14% which generates a λ_{HLA} of 1.8, approximately 40% of the total genetic contribution. Thus it would appear that HLA by itself is responsible for a major proportion of the total genetic component. However, this still leaves an appreciable target for those embarking on whole genome screening studies in RA.

Given the extensive international effort being directed at RA it is hoped that within the near future most if not all of the genetic risk factors making a significant contribution to RA will be identified. The association of HLA with RA has been known for over 20 years although we have not categorically established how it contributes to etiopathogenesis. Ironically, HLA, which probably makes the biggest contribution to RA susceptibility and was discovered first, may well be the last to be understood.

References

Aho K, Markku K, Tuominen J and Kaprio J. (1986) Occurrence of rheumatoid arthritis in a nationwide series of twins. *J. Rheumatol.* **133**, 899–902.

Albani S, Tuckwell JE, Esparza L, Carson DA and Roudier J. (1992) The susceptibility to rheumatoid arthritis is a cross-reactive B cell epitope shared by the *Escherichia coli* heat shock protein dnaJ and the histocompatibility leukocyte antigen DRB1*0401 molecule. *J. Clin. Invest.* **89**, 327–331.

Albani S, Keystone EC, Nelson JL *et al.* (1995) Positive selection in autoimmunity: abnormal immune responses to a bacterial dnaJ antigenic determinant in patients with early rheumatoid arthritis. *Nature Med.* **1**, 448–452.

Arnett FC, Edworthy SM, Bloch DA *et al.* (1988) The American Rheumatism Association 1987 revised criteria for the classification of rheumatoid arthritis. *Arth. Rheum.* **31**, 315–324.

Auger I, Escola JM, Gorval JP and Roudier J. (1996) HLA-DR4 and HLA-DR10 motifs that carry susceptibility to rheumatoid arthritis bind 70-kD heat shock proteins. *Nature Med.* **2**, 306–310.

Brautbar C, Naparstek Y, Yaron M *et al.* (1986) Immunogenetics of rheumatoid arthritis in Israel. *Tissue Antigens* **28**, 8–14.

Brennan FM, Field M, Chu CQ, Feldmann M and Maini RN. (1991) Cytokine expression in rheumatoid arthritis. *Br. J. Rheumatol.* **30**, 76–80.

Brennan P and Silman A. (1994) Breast-feeding and the onset of rheumatoid arthritis. *Arth. Rheum.* **37**, 808–813.

Cornelis F, Faure S, Martinex M *et al.* (1996) Systemic screening of the entire genome in rheumatoid arthritis families reveals 3 major susceptibility loci. *Arth. Rheum.* (Supplement) **39**, S73.

Da Silva JAP and Spector TD. (1992) The role of pregnancy in the course and aetiology of rheumatoid arthritis. *Clin. Rheumatol.* **11**, 189–194.

Davies JRE. (1990) Prolactin and related peptides in pregnancy. *J. Clin. Endocrinol. Metab.* **4**, 273–290.

Deighton CM, Walker DJ, Griffiths ID and Roberts DF. (1989) The contribution of HLA to rheumatoid arthritis. *Clin. Genet.* **36**, 178–182.

DeJongh BM, van Romunde KJ, Valkenberg HA, de Lange GG and van Rood JJ. (1984) Epidemiological study of HLA and GM in rheumatoid arthritis and related symptoms in an open Dutch population. *Ann. Rheum. Dis.* **43**, 613–619.

Del Junco DJ, Annegers JF, Coulam CB and Luthra HS. (1989) The relationship between rheumatoid arthritis and reproductive function (abstract). *Br. J. Rheumatol.* (Supplement 1) **28**, 33.

Dizier MH, Eliaou JF, Babron MC, Combe B, Sany J, Clot J and Clerget-Dapoux F. (1993) Investigation of the HLA component involved in rheumatoid arthritis (RA) by using the marker association -segregation χ^2 (MASC) method: rejection of the unifying-shared-epitope hypothesis. *Am. J. Hum. Genet.* **53**, 715–721.

Elliot MJ, Maini RN, Feldman M *et al.* (1994) Randomised double-blind comparison of climeric monoclonal antibody to tumour necrosis factor α (cA2) versus placebo in rheumatoid arthritis. *Lancet* **344**, 1105–1110.

Evans AM, Petersen JW, Sekhon GS and DeMars R. (1988) Use of human lymphoblastoid deletion mutations to map the prolactin gene on human chromosome 6p (abstract). *Am. J. Hum. Genet.* **43**, A143.

Falconer DS. (1989) *Introduction to Quantitative Genetics.* Longman Scientific, Harlow.

Farid NR, Noel EP, Sampson L and Russell NA. (1980) Prolactin secreting adenomata are possibly associated with HLA-B8. *Tissue Antigens* **15**, 333–335.

Firestein GS and Zvaifler NJ. (1990) How important are T cells in chronic rheumatoid synovitis? *Arth. Rheum.* **33**, 768–773.

Gao X, Olsen NJ, Pincus T and Stastny P. (1990) HLA-DR alleles with naturally occurring amino acid substitutions and risk for development of rheumatoid arthritis. *Arth. Rheum.* **33**, 939–946.

Gregersen PK, Silver J and Winchester RJ. (1987) The shared epitope hypothesis: an approach to understanding the molecular genetics of susceptibility to rheumatoid arthritis. *Arth. Rheum.* **30**, 1205–1213.

Hajeer A, Worthington J, Silman AJ and Ollier WER. (1996) Association of tumor necrosis factor microsatellite polymorphisms with HLA-DRB1*04-bearing haplotypes in rheumatoid arthritis patients. *Arth. Rheum.* **39**, 1109–1114.

Hajeer A, John S, Ollier W *et al.* (1997) TNF microsatellite haplotypes are different in male and female RA patients. *J. Rheumatol.* **24**, 217–219.

Hardwick LJ, Walsh S, Butcher S, Nicod A, Shatford J, Bell J, Lathrop M and Wordsworth BPW. (1996) Mapping and characterisation of the genes involved in rheumatoid arthritis. *Br. J. Rheumatol* (Supplement 2) **35,** 14.

Harvald B and Hauge M. (1965) Hereditary factors elucidated by twin studies. In: *Genetics and the Epidemiology of Chronic Disease* (eds JV Neel, MW Shaw and MJ Schull), pp. 61–76. US Public Health Service, Washington, DC.

Harvey J, Lotze M, Stevens MB, Lambert G and Jacobson D. (1981) Rheumatoid arthritis in a Chippewa band: I. Pilot screening study of disease prevalence. *Arth. Rheum.* **24**, 717–721.

Hazes JMW and Van Zeben D. (1991) Oral contraceptive and its possible protection against rheumatoid arthritis. *Ann. Rheum. Dis.* **50**, 72–74.

Ivanyi P, Hampl R, Starka L and Mickova M. (1972) Genetic association between H-2 gene and testosterone metabolism in mice. *Nature New Biol.* **238**, 280–281.

Jara LJ, Lavalle C, Fraga A, Gomez-Sanchez C, Silveira LA, Martinez-Osuna P, Germain BF and Espinoza LR. (1991) Prolactin immunoregulation and autoimmune diseases. *Sem. Arth. Rheum.* **20**, 273–284.

Jaraquemada D, Ollier W, Awad J, Young A and Festenstein H. (1986) HLA and rheumatoid arthritis: susceptibility or severity? *Dis. Markers* **4**, 43–53.

Jaraquemada D, Ollier W, Awad J *et al.* (1989) HLA and rheumatoid arthritis: a combined analysis of 440 British patients. *Ann. Rheum. Dis.* **45**, 627–636.

Klouda PT, Corbin SA, Bidwell JL, Bradley BA, Ahern MJ and Maddison PJ. (1986) Felty's syndrome and HLA-DR antigens. *Tissue Antigens* **27**, 112–113.

Lawrence JS. (1970) Heberden oration 1969. Rheumatoid arthritis nature or nuture? *Ann. Rheum. Dis.* **29**, 357–379.

Lawrence JS and Wood PHN. (1968) Genetics of rheumatoid arthritis. In: *Rheumatic Diseases* (eds JJR Duthie and RRM Alexander), pp. 19–28. William & Wilkins, Baltimore, MD.

MacGregor A. (1994) Modelling heritability: the study of twins. MSc. Faculty of Medicine, University of Manchester, UK.

MacGregor A, Ollier W, Thomson W, Jawaheer D and Silman AJ. (1995) HLA-DRB1*0404 genotype and rheumatoid arthritis: increased association in men, young age at onset, and disease severity. *J. Rheumatol.* **22**, 1032–1036.

Mehra NK, Vaidya MC, Taneja V, Agarwar A and Malaviya AN. (1982) HLA-DR antigens in rheumatoid arthritis in North India. *Tissue Antigens* **20**, 300–302.

Morgan K. (1990) What do anti-collagen antibodies mean? *Ann. Rheum. Dis.* **49**, 62–65.

Myal Y, Dimattia GE, Gregory CA, Friesen HG, Hamerton JL and Shiu RPC. (1991) A BglII RFLP at the human prolactin gene locus on chromosome 6. *Nucleic. Acids Res.* **19**, 1167.

Nelson JL, Koepsell TD, Dugowson CE, Voigt LF, Daling JR and Hansen JA. (1993) Fecundity before disease onset in women with rheumatoid arthritis. *Arth. Rheum.* **36**, 7–14.

Noel GL, Suh HK and Frantz AG. (1974) Prolactin release during nursing and breast stimulation in postpartum and nonpostpartum subjects. *J. Clin. Endocrinol. Metab.* **38**, 413–423.

Ollier WER and MacGregor A. (1995) Genetic epidemiology of rheumatoid disease. *Br. Med. Bull.* **51**, 267–285.

Ollier W and Thomson W. (1992) Population genetics of rheumatoid arthritis. *Rheum. Dis. Clin. North. Am.* **18**, 741–759.

Ollier W, Venables PJW, Mumford PA, Maini RN. Awad J, Jaraquemada D, D'Amaro J and Festenstein H. (1984) HLA antigen association with extra articular rheumatoid arthritis. *Tissue Antigens* **24**, 279–291.

Ollier W, Silman A, Gosnell N *et al.* (1986) HLA and rheumatoid arthritis: an analysis of multi-case families. *Dis. Markers* **4**, 85–98.

Ollier W, Carthy D, Cutbush S, Okoye R, Awad J, Fielder A, Silman A and Festenstein H. (1988) HLA-DR4 associated Dw types in rheumatoid arthritis. *Tissue Antigens* **33**, 30–37.

Ollier W, Spector T, Silman A, Perry L, Ord J, Thomson W and Festenstein H. (1989) Are certain HLA haplotypes responsible for low testosterone levels in males? *Dis. Markers* **7**, 139–143.

Panayi GS, Lanchbury JS and Kingsley GH. (1992) The importance of the T cell in initiating and maintaining the chronic synovitis of rheumatoid arthritis. *Arth. Rheum.* **35**, 729–735.

Pociot F, Briant L, Jongeneel CV, Molvig J, Worsaae H, Abbal M, Thomsen M, Nerup J and Cambon-Thomsen A. (1993) Association of tumor necrosis factor (TNF) and class II major histocompatibility complex alleles with the secretion of TNF-α and TNF-β by human mononuclear cells: a possible link to insulin dependent diabetes mellitus. *Eur. J. Immunol.* **23**, 224–231.

Risch N. (1987) Assessing the role of HLA-linked and unlinked determinants of disease. *Am. J. Hum. Genet.* **40** 1–14.

Roudier J, Petersen J, Rhodes G, Luka J and Carson DA. (1989) Susceptibility to rheumatoid arthritis maps to a T cell epitope shared by the HLA Dw4 DR beta 1 chain and the Epstein–Barr virus glycoprotein gp110. *Proc. Natl Acad. Sci. USA* **86**, 5104–5108.

Sanchez B, Moreno I, Magarino R, Garzan M, Gonzales HF, Garcia A and Nuñez-Rolan A. (1990) HLA-DRw10 confers the highest susceptibility to rheumatoid arthritis in a Spanish population. *Tissue Antigens* **36**, 174–176.

Scott DGI, Bacon PA and Tride CR. (1981) Systemic rheumatoid vasculitis: a clinical and laboratory study of 50 cases. *Medicine* **60**, 288–297.

Silman AJ and Hochberg MC. (1993) *Epidemiology of the Rheumatic Diseases.* Oxford University Press, Oxford.

Silman AJ, Kay A and Brennan P. (1992) Timing of pregnancy in relation to the onset of rheumatoid arthritis. *Arthr. Rheum.* **35**, 152–155.

Silman AJ, Ollier W, Holligan S, Birrell F, Adebajo A, Asuzu MC, Thomson W and Pepper L. (1993a) Absence of rheumatoid arthritis in a rural Nigerian population. *J. Rheumatol.* **20**, 618–622.

Silman AJ, MacGregor A, Thomson W, Holligan S, Carthy D and Ollier WER. (1993b) Twin concordance rates for rheumatoid arthritis: results of a nationwide study. *Br. J. Rheumatol.* **32**, 903–907.

Spector TD, Ollier WER, Perry LA and Silman AJ. (1988a) Evidence for similarity in testosterone levels in haplotype identical brothers. *Dis. Marker* **6**, 119–125.

Spector TD, Perry LA, Tubb G, Silman AJ and Huskisson EC. (1988b) Low free testosterone levels in males with rheumatoid arthritis. *Ann. Rheum. Dis.* **47**, 65–68.

Stastny P. (1978) Association of B-cell alloantigen DRw4 with rheumatoid arthritis. *N. Engl. J. Med.* **298**, 869–871.

Symmons DPM, Barrett EM, Bankhead CR, Scott DGI and Silman AJ. (1994) The incidence of rheumatoid arthritis in the United Kingdom: results from the Norfolk Arthritis Register. *Br. J. Rheumatol.* **33**, 735–739.

Thomson W, Pepper L, Payton A, Carthy D, Scott D, Ollier W, Silman A and Symmons D. (1993) Absence of an association between HLA-DRB1*04 and RA in newly diagnosed cases from the community. *Ann. Rheum. Dis.* **52**, 539–541.

Vassalli P. (1992) The pathophysiology of tumour necrosis factors. *Annu. Rev. Immunol.* **10**, 411–452.

Wasmuth AG, Veale AMO, Plamer DG and Heighton TC. (1978) Prevalence of rheumatoid arthritis in families. *Ann. Rheum. Dis.* **31**, 85–91.

Weyand CM, Hicok KC, Conn DL and Goronzy JJ. (1992a) The influence of HLA-DRB1 genes on disease severity in rheumatoid arthritis. *Ann. Intern. Med.* **117**, 801–806.

Weyand CM, Xie CP and Goronzy JJ. (1992b) Homozygosity for the HLA-DRB1 allele selected for extra articular manifestations in rheumatoid arthritis. *J. Clin. Invest.* **89**, 2033–2039.

Wilkens RF, Nepom GT, Marks CR, Nettles JW and Nepom BS. (1991) Associations of HLA-Dw16 with rheumatoid arthritis in Yakima Indians. *Arth. Rheum.* **34**, 43–47.

Wilson AG, Symons JA, McDowell TL, Di Giovine FS and Duff GW. (1994) Effects of a tumour necrosis factor (TNFa) promoter base transition on transcriptional activity. *Br. J. Rheum.* **33**, 89.

Winchester RJ. (1977) B-lymphocyte allo-antigens, cellular expression and disease significance with special reference to rheumatoid arthritis. *Arth. Rheum.* **20**, 159.

Wordsworth BP, Lachbury JSS, Sakkas LI *et al.* (1989) HLA-DR4 subtype frequencies in rheumatoid arthritis indicates that DRB1 is the major susceptibility locus within the HLA class II region. *Proc. Natl Acad. Sci. USA* **86**, 10049–10053.

Wordsworth P, Pile KD, Buckely JD, Lanchbury JSS, Ollier B, Lathrop M and Bell JI. (1992) HLA heterozygosity contributes to susceptibility to rheumatoid arthritis. *Am. J. Hum. Genet.* **51**, 585–591.

Worthington, J, Ollier WER, Leach MK *et al.* (1994) The Arthritis and Rheumatism Council's National Repository of family material: pedigrees from the first 100 rheumatoid arthritis families containing affected sibling pairs. *Br. J. Rheumatol.* **33**, 970–976.

Young A, Jaraquemada D, Awad J, Festenstein H, Corbett M, Hay FC and Roitt IM. (1984) Association of the HLA-DR4/Dw4 and DR2/Dw2 with radiological changes in a prospective study of patients with rheumatoid arthritis. *Arth. Rheum.* **26**, 20–25.

9

Practical benefits from understanding the genetics of chronic diseases

Roger R. Williams, Steven C. Hunt, Paul N. Hopkins, Lily Wu and Susan Stephenson

9.1 Introduction

In the past two decades, astonishing progress has been made regarding our understanding of the genetic factors promoting common chronic diseases. In the next two decades, similar progress should dramatically change practical approaches to diagnosing, treating and preventing common diseases like strokes, coronary atherosclerosis, hypertension, cancer, diabetes, asthma, schizophrenia, osteoporosis and Alzheimer's disease. We stand at the intersection of discovery and application!

We will examine three major dimensions of progress through genetics: better treatment, improved diagnosis and more successful practical applications. Although much of this discussion looks mostly toward the future, there is actually much we can do now to start practising the medicine of the future as illustrated by collaborators in 25 countries who 'make early diagnoses to prevent early deaths in medical pedigrees with familial hypercholesterolaemia' ('MED PED FH').

9.2 From genes to treatment and prevention

9.2.1 Pathophysiology from genetics

Most serious chronic diseases currently lack a 'cure' or even truly effective therapy. Examples include atherosclerosis, cancer, arthritis,

Genetics of Common Diseases: future therapeutic and diagnostic possibilities,
edited by I. Day and S. Humphries. © 1997 BIOS Scientific Publishers Ltd, Oxford

diabetes, schizophrenia, morbid obesity, multiple sclerosis, psoriasis, osteoporosis, cystic fibrosis, asthma and Alzheimer's disease. An online computer registry of mutations reflects rapid progress in finding some genetic contribution for some of these chronic diseases (OMIM, 1997). Finding and understanding genes that cause such disorders is helping us elucidate exact pathophysiologic mechanisms of disease causation. Comprehensive knowledge of pathophysiology in turn forms the basis for designing the most effective strategies for treatment and prevention! Even though some mutations are detected from rare subsets of such diseases, the expanded knowlege will often help conquer more common forms of the disease as well.

In 1985, Brown and Goldstein received the Nobel Prize for their discovery of the mechanism of inherited defects in low-density lipoprotein (LDL) receptors which causes familial hypercholesterolaemia (FH) (Hobbs *et al.*, 1992). Even though FH only occurs once in every 500 persons, this discovery opened the door to understanding the whole concept of receptors in molecular biology and now applies broadly to many other areas such as estrogen receptors relating to breast cancer, neurotransmitter receptors relating to neurological disorders and mental illness, and angiotensin receptors involved in blood pressure control.

Recent discoveries of genes causing cancer are now unlocking the secrets of disordered cellular and molecular physiology that allows malignant cells to emerge and metastasize (Cillo *et al.*, 1996). Studies of environmental factors interacting with defined genes are beginning to explain why some persons are very susceptible and others quite resistant to the same environmental factors (Williams, 1984). Detailed pictures of pathophysiology are being discovered including: genes and gene products; molecular and cellular processes; biochemical and hormonal functions; environmental influences; and final expressions in disordered tissues and organs. As we come to fully understand most factors contributing to a particular disease process, we can design the best strategies for treatment and prevention.

9.2.2 New and better drugs from genetics

Persons with FH are born with twice normal cholesterol levels that could never be brought down to normal until recently. Now, a new class of pharmacological agents called 'statins' make it possible for FH heterozygotes to finally achieve normal cholesterol levels (Williams *et al.*, 1995). Because an inherited disorder (FH) led Brown and Goldstein to discover LDL receptors and their physiology, pharmacologists were able to determine the beneficial effects of 'statin' medication, derived in large part from up-regulating LDL receptors. While this is especially effective therapy for FH patients who have an inherited problem with half of their LDL receptors, it turns out that statins are highly effective medications used for most patients with elevated cholesterol, even though most of them have normal LDL receptors. Thus, discovery of an uncommon gene defect has helped elucidate an effective mechanism for a common treatment for blood cholesterol elevation.

There are currently no highly effective medications for morbid obesity, Alzheimer's disease or breast cancer. However, genetic studies underway offer the promise for discovering enough pathophysiology to devise effective medications for these disorders. If statins can now normalize twice normal cholesterol levels in persons with heterozygous FH, then similar discoveries in the near future will normalize twice normal body fat indices in persons with morbid obesity, prevent dementia in persons otherwise destined to develop Alzheimer's disease and cure or prevent breast cancer in persons carrying the dominant genes *BRCA1* and *BRCA2* (Shattuck-Eidens *et al.*, 1995; Tavtigian *et al.*, 1996). Several of the world's most successful drug companies have already invested millions of dollars in collaborative genetic research because they believe this prediction to be valid.

In addition to helping us develop new medications, genetic discoveries are helping us to find out which patients are most susceptible to both the therapeutic and toxic effects of medications by testing for genetic factors that influence the metabolism of specific drugs. For example, genetically slow metabolizers of specific drugs have prolonged effects or cumulative levels that can lead to toxicity; while genetically rapid metabolizers can eliminate the pharmacologic agent so rapidly that the therapeutic effect is diminished (Idle and Smith, 1995).

9.2.3 Genetically focused disease targets for tailored medications

In the early days of antibiotics we only had penicillin and sulfa drugs, so everyone with pneumonia was treated with these primordial antibiotics. Three decades later we take it for granted that life-threatening pneumonia requires a 'culture and sensitivity' (C&S) response. We identify the exact organism responsible and indicate which antibiotic will be most effective for eradicating the infection. Patients whose pneumonia would not respond to 'pen and sulfa' died in former days and now live because we diagnose specific causes and use antibiotics tailored to their infection. Some genetic diagnoses and tailored therapies are now beginning to emerge as a genetic version of C&S for chronic diseases. Two examples are discussed below.

Glucocorticoid remediable aldosteronism (GRA) results from a dominant chimeric mutation (Lifton *et al.*, 1992) with combined fragments of two genes, the aldosterone synthase gene and the steroid 11-beta-hydroxylase gene, located next to each other on chromosome 8. During recombination, unequal crossing over produces the mutant gene with sequences and functions of both genes combined into the variant. Administration of exogenous glucocorticoids (like dexamethasone) can often suppress aldosterone, abnormal steroids and the severe blood pressure elevations. Persons with the gene for GRA often have early severe hypertension and close relatives dying of cerebral haemorrhage in their forties. GRA will usually not respond to ordinary antihypertensive medications but can respond to prednisone (which suppresses hormone production), spironolactone (which competitively inhibits aldosterone receptor) or amiloride (which inhibits the distal renal epithelial sodium channel response to mineralocorticoid action).

Liddle's syndrome results from dominant mutations (Shimkets *et al.*, 1994) of the epithelial sodium channel gene on chromosome 16p. Liddle's syndrome causes suppressed aldosterone secretion in contrast to the hyperaldosteronism seen in GRA. Excessive reabsorption of sodium and exchange for potassium in the distal nephron probably account for the hypertension and hypokalaemia. Both of these features of Liddle's syndrome are responsive to triamterine or amiloride which specifically inhibit the epithelial sodium channel. Unlike GRA, this syndrome is not responsive to spironolactone which inhibits the mineralocorticoid receptor. Kidney transplant also seems to eliminate the problem.

In persons with severe hypertension due to GRA or Liddle's syndrome, making a specific genetic diagnosis and prescribing medication tailored to the diagnosis can prevent death from strokes before age 50. Failure to make the proper genetic diagnosis and treating GRA or Liddle's syndrome with usual antihypertensive medication is like treating an undiagnosed *Pseudomonas* pneumonia with ordinary penicillin. In both cases, the patient will die from lack of effective treatment tailored to a specific diagnosis.

In three decades we have progressed from two antibiotics for all pneumonias to a host of specific organisms and tailored antibiotics. In the next three decades, genetic discoveries will help us make similar progress for the common chronic diseases.

9.2.4 From shotgun public health to genetically focused prevention

We are now entering an era of genetic guidance for more individualized hygienic recommendations for specific individuals. This should improve compliance with the lifestyle factors of greatest importance for each individual wishing to preserve their health through prudent daily actions. In the commonly communicated 'shotgun approach' to public health, everyone is advised that they should: jog daily; avoid sugar, fat and sodium; eat plenty of vitamins like folic acid and vitamin C; have regular screening exams for cancer (mammography or prostate exam, proctoscopy for colon polyps, etc.); and a host of other hygienic measures to prevent everything from atherosclerosis to arthritis. In fact, few human beings in a normal daily life are effectively complying with all of these recommendations. In some cases, an intervention such as jogging may help prevent one disorder (Syndrome X-induced atherosclerosis) but may promote progression of another (osteoarthritis of weight-bearing joints). If a person knew they were genetically susceptible to arthritis and genetically resistant to atherosclerosis they might be most prudent by choosing not to jog!

This topic has risen to the level of public debate for sodium restriction for persons with hypertension. Some respected public health authorities continue strongly to advocate compliance for all persons with hypertension or a tendency to develop it, while others challenge both the feasibility and value of such recommendations. Harriet Dustan, a prominent hypertension researcher and prudent peacemaker suggested the answer to this debate lies in the ability to identify the people who have 'salt-sensitive hypertension' (Ziporyn, 1996). Progress in genetics offers the opportunity to understand why some environmental interventions seem to be more efficacious in some persons than in others and to tailor preventive interventions for specific persons to their genetic susceptibility. Below are two examples relating to hypertension and atherosclerosis.

Recent studies provide support for a hypertension, and pre-eclampsia, promoting genetic variant at the angiotensinogen locus which is carried by 30% of Caucasians and about 90% of blacks (Caulfield *et al.*, 1994; Hata *et al.*, 1993; Jeunemaitre *et al.*, 1992; Ward *et al.*, 1993). An association of this gene with disordered pathophysiology of the renin–angiotensin–aldosterone system suggests the mechanism relates to one previously defined form of 'salt-sensitive hypertension' (Hopkins *et al.*, 1996). In studies being prepared for publication, angiotensinogen genotyping for persons with hypertension or borderline hypertension seems to identify some whose blood pressure responds little or not at all to sodium restriction and potassium supplementation, and others with a 'responsive genotype' (6–10 mmHg drop in systolic blood pressure in response to dietary electrolyte interventions). If a person were being asked to change the salt content in their diet for the rest of their life, they would surely like to have a gene test to indicate if this major change in daily life would result in a 10 mm drop or no drop at all in systolic blood pressure.

Homocysteine levels above 13–19 μmol l^{-1} seem to promote coronary atherosclerosis (Hopkins *et al.*, 1995). The level of homocysteine can be affected by a common 'heat-labile' mutation of the gene for methylene tetrahydrofolate reductase (*MTHFR*) as well as by dietary intake of a common vitamin, folic acid. A recent report indicates that persons who developed the high homocysteine levels associated with high risk for coronary atherosclerosis generally required both genetic and environmental risk exposure (Jacques *et al.*, 1996). They were homozygous for the *MTHFR* heat-labile mutation and had low folic acid intake (as reflected in serum levels). Supplementing everyone's food with enough folic acid could prevent hyperhomocystenaemic atherosclerosis. The alternative would be to find approximately 10% of persons in the general population who have homozygous *MTHFR* mutations and be sure they take enough folic acid to prevent the problem.

We are just now beginning to learn how focused lifestyle and screening recommendations can be more successful because a few of them can be strongly advised for specific persons with particular genetic susceptibility detected by genetic testing. Three decades from now such focused prevention should be more often the rule rather than the exception.

9.2.5 Gene therapy – the prescription of the future

The Wright brothers travelled a very short distance in a very crude aeroplane, but it marked a beginning, and today air travel is a key part of our society, relied upon each day by many ordinary people. Gene therapy is already beyond the historic first flight. For atherosclerotic diseases the first short flight was launched in 1992 for a person with homozygous FH (Grossman *et al.*, 1994). To replace the genetically absent LDL receptors, some of his liver cells were surgically removed, infected with a virus containing the human gene sequence for normal LDL receptors and 'returned' to his liver. It required statin drug stimulation of the few newly acquired LDL receptors to produce a perceptible drop in serum cholesterol from this procedure. The effect was much too small to have a meaningful clinical result; however, like the first flight at Kitty Hawk, the first gene therapy for FH in Michigan 'got off the ground'.

At this beginning stage, some probably view gene therapy as a last chance hope for rare FH homozygotes. However, it offers much more! For starters it offers a life long cure for 10 million FH heterozygotes in the world. Fully successful gene replacement will be a cure, not just a treatment. We have drugs that help many FH heterozygotes now, but non-compliance deprives many of them of the full benefit of therapy. Normal genes work night and day without side effects. A recipient of successful gene therapy will not have to worry about forgetting to take the next dose. For the rest of their life, gene therapy could give them normal cholesterol levels and let them take their newly acquired normal LDL receptors for granted, like the rest of us do.

Basic technical challenges of gene therapy include: (i) obtaining a gene sequence with beneficial effects when expressed in the appropriate tissue, (ii) attaching the therapeutic DNA to a vehicle like a virus; and (iii) delivering it in large numbers to cells where its effect is needed. These are not trivial tasks, but solutions to technical challenges seem just to take time until ingenious minds forge the path from first flight to jet flight.

It has been estimated that about 2000 persons are currently being tested with various forms of gene therapy for diseases ranging from CF to cancer. In the case of CF there is reason for optimism. The homozygous gene defect in CF results in total loss of a transmembrane ionic pump. Experts estimate that restoration of function to just 10% of these ion pumps would reverse the disorder. Even with our current limitations in the success of vector delivery of therapeutic genes, it seems likely that the near future should see vector technology capable of therapeutically infecting at least 10% of cells with the normal gene to ablate CF in these homozygotes.

Gene therapy has potential beyond replacing defective genes. It can override the effects of existing genes. It can carry the benefits of antiatherogenic genes from naturally protected humans to others born without this good fortune.

Citing some current gene therapy research projects illustrates progress well under way. In Rochester, Minnesota, the endothelial nitric oxide synthase gene (*eNos*) is under study because nitric oxide is a potent vasodilator which may also inhibit platelet aggregation and SMC proliferation. An adenoviral vector encoding cDNA for eNOS was generated by homologous recombination and applied *in vitro* to porcine coronary SMC achieving increased levels of nitrate and cyclic guanine monophosphate (cGMP) and diminished cell proliferation (O'Brien *et al.*, 1996).

In Milan, Italy, apoE-deficient mice were given intramuscular injections of naked supercoiled plasmids containing complete human apoE cDNA. The injected DNA is maintained in an episomal, circular form and does not replicate. However, transgene persistence has been demonstrated up to 19 months. In these apoE-deficient mice, average total cholesterol levels were significantly decreased from the first week after plasmid injection and achieved a drop from 441 mg dl^{-1} to 273 mg dl^{-1} by the eighth week (Vazio *et al.*, 1996).

In Houston, Texas, synthetic DNA complexes have been constructed that are as efficient vectors as viruses but lack the immunological limitations. Furthermore, specific high level expression of these exogenous genes was achieved by receptor-mediated delivery of synthetic DNA vectors, coated, condensed and targeted by lipophilic non-exchangeable derivatives of apoE-3 peptides, which are high affinity ligands for the low-density lipoprotein (LDL) and very low-density lipoprotein (VLDL) receptors. This method is being used to identify and modify the barriers to targeted delivery to hepatocytes *in vivo* (Smith *et al.*, 1996).

In Parma, Italy, a molecular variant called ApoA-I Milano is under study because it seems to protect against coronary disease (despite causing a low HDL cholesterol). In transgenic mice it seems to promote more efficient cholesterol efflux from cells ('reverse cholesterol transport') (Chiesa *et al.*, 1996).

Within three decades, we may well be writing prescriptions for genes rather than medications. Compliance will be complete and automatic. High risk genes causing tragedy in some families will be conquered, and protective genes that now prevent atherosclerosis in a few lucky families may be shared with everyone who needs them. The technology seems likely to succeed. The social and economic factors may be the rate-limiting steps!

9.3 Improved diagnosis

9.3.1 Earlier diagnosis through genetics

As we succeed in developing successful therapeutic and preventive options as outlined above, it becomes even more important to find those who can benefit as early as possible. Today we can perform targeted screening for genetic disorders in young relatives of persons known to have treatable inherited conditions like FH, GRA or Liddle's syndrome. In the future, batteries of genetic tests will offer diagnosis for many people at early ages comfortably before disease consequences develop, especially for the common chronic disorders with delayed onset like atherosclerosis, cancer, arthritis, diabetes, schizophrenia, Alzheimer's disease and osteoporosis.

9.3.2 More accurate diagnosis through genetics

Even our best current diagnostic tests have problems with sensitivity and specificity. Imagine a test that has 99% sensitivity and 99% specificity, and you have just described a properly functioning genetic test. Even with the challenges of genetic testing that currently exist (such as the need to sequence the entire loci for *BRCA1* and *BRCA2* for women with familial breast cancer), the main current challenge is expense not accuracy. This currently available panel for breast cancer genes already offers much better specificity and sensitivity than most standard clinical tests. As more affordable and more efficient gene tests become available, we will gain greater ability to make more accurate and greater numbers of diagnoses and to make them much earlier.

Patients are currently asked to take expensive medications for the rest of their lives with only crude descriptive diagnoses like 'high blood cholesterol', hypertension, arthritis or depression. While the severity of chronic conditions is lessened, few individuals are truly returned to fully normal health because of therapy matched to their inherited problem. There is a lot of room for improvement in the accuracy of chronic disease diagnosis to match our transition from a 'pen and sulfa' drug level of therapy for chronic diseases, to a C&S model of specific diagnosis and tailored therapy.

9.4 More successful practical applications: MED PED FH – A practical approach to early diagnosis, treatment and prevention

New opportunities are emerging for preventing the consequences of serious diseases because of the discovery of causal mutations in specific gene loci. MED PED is an international collaboration to *M*ake *E*arly *D*iagnoses and *P*revent *E*arly *D*eaths in *MED*ical *PED*igrees.

The initial target of this effort is heterozygous FH. This is one of the most common and best understood serious single gene disorders. Published clinical diagnostic criteria have been reported to have 98% specificity and 89% sensitivity in families with FH (Williams *et al.*, 1993). The clinical diagnosis of FH is reliable, relatively inexpensive and available worldwide. In some countries, even gene testing is readily available for patients suspected of FH. Potent and effective medications are available that act specifically at the site of genetically defective LDL receptors. Clinical trials have established the ability of medical therapy to normalize cholesterol levels, arrest or reverse atherosclerotic coronary artery lesions and significantly reduce morbidity and mortality. Persons with FH have a well defined problem: half the normal number of properly functioning LDL receptors in their liver and twice normal LDL cholesterol levels in their blood. If patients with FH are diagnosed early in life (perhaps by age 18), and treated with full dose medication, it is possible to achieve normal cholesterol levels. Thus, if this treatment can be maintained for the rest of their lives, could we not expect them to have a normal life expectancy? Also, if normal life is achievable with socially affordable resources surely we should be outraged if short sightedness or misplaced priorities prevent this life-saving knowledge from being applied on behalf of persons carrying the gene for FH.

Over 10 million persons in the world have FH and about 200 000 of them die with premature ischaemic heart disease each year. Unfortunately, few patients with FH are receiving the benefit of recent advances in the diagnosis and treatment of FH. Pooled estimates from MED PED countries indicate that 80% of the patients with FH are not diagnosed, 84% are taking no medication to lower their cholesterol, and only 7% have reasonably well treated cholesterol levels (Williams *et al.*, 1996). Moreover, few are adequately evaluated to detect and treat already established coronary artery disease.

The international MED PED collaborators have organized to share successful approaches and combine efforts to find and help persons with FH. The initial computer registries in 25 countries contain 16 585 FH patients together with their relatives and physicians. Four international MED PED subcommittees have been organized to focus on specific efforts:

(i) patient and physician education (sharing educational materials and support programmes)
(ii) government affairs and publicity (increasing public awareness and funding)
(iii) research and molecular genetics (promoting collaboration and research progress)
(iv) computer data management (standardizing tools for FH family registry data).

A major feature of the MED PED approach is concentration on high risk families for rapid case-finding. From each known index case in the registry, several new FH cases can be identified among close relatives, and sometimes many FH cases can also be found among distant relatives identified using pedigree expansion. Education, treatment and long-term support are also thought to be more effective in families. The MED PED approach has been tested in some locations for up to 8 years with considerable success. The MED PED model seems appropriate for other treatable dominant single gene diseases such as: familial defective apoB, dominant Type III hyperlipidaemia, GRA, Liddle's syndrome, multiple endocrine neoplasia II, long QT sudden arrhythmic death syndrome and some forms of breast and colon cancer (see *Table 9.1*). This approach should work for diseases that meet these criteria:

(i) a single dominant gene which causes preventable serious illness
(ii) availability of validated diagnostic tests (gene test or clinical test)
(iii) availability of some form of treatment or prevention which is shown to be effective.

9.5 More public awareness and financial support are needed

The group of treatable dominant genetic disorders listed above are more common than AIDS and currently more treatable than AIDS. While government agencies continue to fund important efforts for diseases like AIDS, they should also begin to invest in activities that will help prevent needless early deaths in persons with treatable genetic diseases. Achieving this balanced government funding approach will require co-ordinated efforts of a large number of patients and their physicians speaking in concert.

In the meantime, the initial efforts of MED PED to help FH families has proceeded thanks to the combined support from several sources (Merck Human Health Inc. and its international affiliates, the World Health Organization, family and national heart associations in several countries, and some government agencies like state and provincial health departments and the US Centers for Disease Control).

Can you imagine the outrage that would arise if 600 jumbo jet airplanes crashed each year due to defective engine bolts that could have been repaired but were not? Then help raise a similar outrage against the estimated 200 000 premature deaths worldwide in middle aged men and women who die early each year because their FH is not diagnosed or properly treated! Then multiply this figure by the number of other similar treatable genetic disorders with severe consequences. If we work together we can *M*ake *E*arly *D*iagnoses and *P*revent *E*arly *D*eaths in *MED*ical *PED*igrees! **The collection of detailed medical family histories, together with contacting and helping relatives in high risk pedigrees like these with FH, FDB, GRA and Liddle's syndrome, should become part of routinely supported medical care.**

Because far-sighted physicians and medical researchers in 25 collaborating countries listed in the Appendix already see this vision of future, they are working to help the medical care systems in each of their countries to adapt the new methods and technology related to advances in genetics to help citizens avoid the tragedy of preventable morbidity and mortality. Presently these dedicated MED PED collaborators report that most relatives they contact and find with FH have not been previously diagnosed. Even those who have had some diagnosis of high cholesterol have rarely received adequate therapy.

At the present time the MED PED collaborators are setting an example of case detection through relative screening. The MED PED effort is supported by humanitarian funding to demonstrate the need and feasibility of this approach to justify future support by government and health insurance agencies.

Collaborators report diverse experiences. South African Afrikaners have one of the highest population prevalences of FH in the world. However, in one large clinic, they report a dire situation. Their clinic follows about 800 FH heterozygotes who are poor and depend on the state health service for treatment. This includes an essentially free coronary bypass surgery when their coronary disease becomes manifest. However, drug budgets for prevention are viewed as excessive. Their budget allows the patients 20 mg of simvastin for about 6 months of the year (i.e. 6 months of half a reasonable dose, and no medication for the other 6 months). Furthermore, even with a recent attempt to provide free simvastatin by courtesy of Merck Human Health, the medical staff in the FH clinic do not have the resources to see this large number of patients even if free medications are provided.

In Germany, MED PED physicians are trying to get the government health service to pay for screening of relatives in FH families with support billed to the account number of the FH proband until relatives are found who can carry their own reimbursable diagnosis of FH. The German collaborators have also developed special computer programs for collecting and managing the complex family data needed to support an active MED PED programme. In Denmark, FH experts have been communicating intensively with government officials to help address some of the concerns for protecting the privacy and autonomy of relatives in a way that will still allow MED PED clinics to find and help relatives with FH. They have helped others in the European community to deal with this key issue.

In British Columbia, the provincial government has supported the concept of outreach clinics where persons with FH can be diagnosed and treated with the back up help of a central lipid clinic at the University of British Columbia. In Quebec Province, FH also occurs with much higher population frequency. A patient society (Canadian Association of FH) has elicited the help of well known physicians and patients with FH who are entertainment celebrities to help educate the public and members of the FH Association through advertisements and newsletters.

In England, one of the oldest registries of FH patients has documented the tragic outcomes of this high risk group of patients. A Lipid Society of physicians and a lay society of patients are organized to foster the needs of familial lipid patients. In Austria, physicians from diverse locations come together in the Austrian Lipid Society and talk about taking advantage of their unity to benefit their patients with FH. They have had several successful exposures in the influential media to help get their message out to the medical community, patients and government officials.

In Israel, several large news articles have highlighted the efforts of a local FH expert who specialized in testing for FH mutations common in Israel and Lebanon. Next door in Lebanon, another MED PED collaborator has established an e-mail address so he can communicate effectively with FH colleagues around the world to help the cause of FH patients in his country, which has one of the highest FH rates in the world!

In Australia, a Family Heart Association has been organized and brochures and questionnaires developed that are very user friendly for FH patients and their relatives. In Hungary, FH collaborators are consulting with government officials to arrange for free medication for FH patients diagnosed in the MED PED programme. In the USA, both MED PED and a lay organization (Inherited High Cholesterol Foundation) advertise their shared toll-free telephone number (1-888-2Hi-Chol and encourage anyone with a cholesterol level above 320 with triglyceride below 300 to call this number for free help diagnosing FH in their family. A network of 100 lipid clinics in the USA is associated with this effort.

In Iceland, about half of the entire population of FH patients are diagnosed and entered into their registry! In Switzerland, FH physicians pay particular attention to the companion disorder familial defective apoB (FDB), which seems to have a higher prevalence in Switzerland and often requires the same attention for diagnosis and treatment to avoid early heart attacks as the LDL receptor-defective FH patients. In Italy, several highly motivated FH physicians have defined FH mutations for hundreds of FH patients and mapped geographic clusters for each mutation. They have also collected a wealth of non-invasive cardiology clinic data on FH patients showing that many need tests and interventions to prevent coronary deaths.

The sheer enthusiasm of a few dedicated physicians is leading to the current initiation of MED PED collaborations in Sweden, Belgium, France, Greece, Ireland, New Zealand, Poland, Portugal and Spain. Some communication has also begun to make possible MED PED collaboration with Japan and India.

Two of the most successful MED PED countries are Norway and Holland. Each has succeeded in establishing meaningful collaborative arrangements with their government ministries of health. Each has been highly successful in large volume gene testing for FH mutations. Both have developed extensive educational brochures for patients and physicians which they have shared with other colleagues. Each is finding about a thousand new FH patients each year!

9.6 MED PED social issues: cost and confidentiality

The cost effectiveness of drug therapy for FH has been rigorously analysed and documented (Goldman, 1993). Daily treatment with a low dose of lovastatin was projected to save money as well as lives, and higher dose therapy was associated with an acceptable cost per year of life saved in this conservative analysis.

Some social scientists raise concerns about projects like MED PED causing psychological stress by contacting relatives to talk about their family history. MED PED collaborators in several countries report a large preponderance of positive reactions from relatives contacted in FH families. The dramatic occurrence of very early heart attack deaths in FH families is usually well known to relatives. Whether they know their cholesterol level or not, many relatives in FH pedigrees already worry about having an early heart attack death long before being contacted by MED PED. Because of MED PED, thousands of relatives have been screened and found to have normal cholesterol and reassured that their fear of early death was not warranted. Others found to have very high cholesterol levels have learned something very few of them knew before MED PED: their super high cholesterol levels can be dramatically reduced and early heart attack deaths can be prevented! Similar observations are likely to be found in the dominant hypertension families.

9.7 Predictions for the future

Genetic research is unlocking the mysteries of pathophysiology. It will lead to earlier and more accurate diagnoses. It will lead to the development of more effective treatments specifically tailored to the

Table 9.1. Treatable dominant chronic diseases

Genetic trait	Description	Clinical diagnosis	Genetic diagnosis	Treatment or prevention
Familial hyper-cholesterolaemia (FH)	Very high LDL cholesterol and very early heart attack deaths	LDL cholesterol and xanthoma: quite reliable	LDL receptor gene tests (>200 causal mutations)	Drugs reduce cholesterol; probably extend life 10–30 years
Familial defective apoB (FDB)	apoB and cholesterol very high with early heart attack deaths	High cholesterols can mimic FH, but some are lower	Two specific causal mutations	Same as for FH for those with very high cholesterol
Dominant type III hyperlipidaemia	Very high beta VLDL cholesterol with early heart attack deaths	High triglyceride and abnormal triglyceride/VLDL ratio	Several specific causal mutations	Specific medications and diet can often normalize levels and prolong life
Long QT syndrome	High risk for sudden arrhythmic death especially in youth	Long duration of QT interval on electrocardiogram	Linkage and mutations found	Medication can lower risk of sudden death and prolong life
GRA hypertension	Severe high blood pressure and early deaths	Abnormal steroid hormones; BP normal after dexamethasone	Several specific causal mutations	Suppress abnormal steroid with hormones like dexamethasone
Liddle's syndrome	Severe high blood pressure and early stroke deaths	BP response to ameloride and triamterene	Causal mutations	Ameloride or triamterene
Multiple endo-neoplasia II (MEN-II)	Pheochromocytoma and other endocrine neoplasia	MRI and CT scans for tumours	Causal mutations	Surgical removal of endocrine tumours
Dominant breast cancer	Fatal metastatic breast carcinoma	Self and doctor exam mammography	*BRCA1* and *BRCA2*	Frequent screening; breast tissue removal?
Dominant ovarian cancer	Fatal metastatic cancer	Pelvic laparoscopy with biopsy	*BRCA1*	Prophylactic oophorectomy
Dominant colon cancers	Fatal metastatic cancer	Colonoscopy with biopsy	Colon cancer gene?	Frequent screening; prophylactic surgery?
Prostate cancer	Disabling and fatal prostate cancer	Doctor exam; blood PSA; needle biopsy	*PRCA1*	Frequent screening; prophylactic surgery?

underlying pathophysiologic mechanisms in individual patients. It will provide tools to improve compliance with treatment. It will help us avoid making the wrong diagnosis, missing a hidden illness or giving the wrong treatment to patients who need our help. It will improve screening and case-finding methods, making public health even more of a science. If genetic technology can follow the history of computer technology, it will also reduce the costs while improving the effectiveness of medical diagnosis, treatment and prevention. It should lead to highly cost-effective medical care.

Thirty years from now we will be the age of our parents and our children will be our age. It does not seem unreasonable to predict that by this time, the fruits of genetic discoveries will enable us to master common chronic diseases then as we generally succeed very well with common bacterial diseases now. We will have specific tests and tailored treatment, and usually succeed in overcoming or preventing the serious health problems of chronic diseases. For any diseases involving genetic susceptibility or molecular components of pathophysiology, we will understand why some individuals are at high risk while others are at low risk. More importantly, we should be able to help those at high risk become like those at low risk. Imagine a world in which heart attacks and strokes, diabetes and cancer, asthma and obesity, arthritis and osteoporosis, psoriasis and cystic fibrosis, schizophrenia, Alzheimer's disease and serious depression are all largely prevented or managed! This should be our world in the year 2027!

References

Caulfield M, Lavender P, Farrall M, Munroe P, Lawson M, Turner P and Clark AJL. (1994) Linkage of the angiotensinogen gene to essential hypertension. *New Engl. J. Med.* **330**, 1629–1633.

Chiesa G, Parolini C, Canavesi M, Rubin EM, Franceschini G and Bernini F. (1996) Cholesterol efflux potential in mice expressing human apolipoprotein A-I Milano. In: *Abstract Book: 66th Congress of the European Atherosclerosis Society, Florence, Italy, July 13–17,* p. 27. Giovanni Lorenzini Medical Foundation, Milan, Italy.

Cillo C, Cantile M, Mortarini R, Barba P, Parmiani G and Anichini A. (1996) Differential patterns of HOX gene expression are associated with specific integrin and ICAM profiles in clonal populations isolated from a single human melanoma metastasis. *Int. J. Cancer* **66**, 692–607.

Goldman L, Goldman PA, Williams LW and Weinstein WC. (1993) Cost-effectiveness considerations in the treatment of heterozygous familial hypercholesterolemia with medications. *Am. J. Cardiol.* **72**, 75D–79D.

Grossman M, Raper SE, Kozarsky K, Stein EA, Englehardt JF, Muller D, Lupien PJ and Wilson JM. (1994) Successful *ex vivo* gene therapy directed to the liver in a patient with familial hypercholesterolemia. *Nature Genetics* **6**, 335–341.

Hata A, Namikawa C, Sasaki M, Nakamura T, Tamura K and Lalouel JH. (1993) Angiotensinogen as a risk factor for essential hypertension in Japan. *J. Clin. Invest.* **93**, 1285–1287.

Hobbs HH, Brown MS and Goldstein JL. (1992) Molecular genetics of the LDL receptor gene in familial hypercholesterolemia. *Hum. Mutat.* **1**, 445–466.

Hopkins PN, Wu LL, Wu J, Hunt SC, James BC, Vincent GM and Williams RR. (1995) Higher plasma homocyst(e)ine and increased susceptibility to adverse effects of low folate in early familial coronary artery disease. *Arterioscler. Thromb. Vasc. Biol.* **15**, 1314–1320.

Hopkins PN, Lifton RP, Hollenberg NK *et al.* (1996) Blunted renal vascular response to angiotensin II is associated with a common variant of angiotensinogen gene and obesity. *J. Hypertension* **14**, 199–207.

Idle JR and Smith RL. (1995) Pharmacogenetics in the new patterns of healthcare delivery. *Pharmacogenetics* **5**, 347–350.

Jacques PF, Bostom AG, Williams RR, Ellison RC, Eckfeldt JH, Rosenberg IH, Selhub J and Rozen R. (1996) Relation between folate status, a common mutation in methylenetetrahydrofolate reductase, and plasma homocysteine concentrations. *Circulation* **93**, 7–9.

Jeunemaitre X, Soubrier F, Kotelevtsev Y *et al.* (1992) Molecular basis of human hypertension: role of angiotensinogen. *Cell* **71**, 169–180.

Lifton RP, Dluhy RG, Powers M, Rich GM, Cook S, Ulick S and Lalouel JM. (1992) A chimaeric 11β-hydroxylase/aldosterone synthase gene causes glucocorticoid-remediable aldosteronism and human hypertension. *Nature* **355**, 262–265.

O'Brien TO, Kullo I, Chen A and Katusic Z. (1996) Adenoviral-mediated gene transfer of nitric oxide synthase (NOS) to the vascular wall yields functional enzymatic activity. *Abstract Book: 66th Congress of the European Atherosclerosis Society, Florence, Italy, July 13–17*, p. 28. Giovanni Lorenzini Medical Foundation, Milan, Italy.

OMIM. (1997) Online Mendelian Inheritance in Man. National Center for Biotechnology Information home page: Http://www3.ncbi.nlm.nih.gov/Omim.

Shattuck-Eidens D, McClure M, Simard J *et al.* (1995) A collaborative survey of 80 mutations in the BRCA1 breast and ovarian cancer susceptibility gene. Implications for presymptomatic testing and secreening. *J. Am. Med. Assoc.* **273**, 535–541.

Shimkets RA, Warnock DG, Bositis CM *et al.* (1994) Liddle's syndrome: heritable human hypertension caused by mutations in the β subunit of the epithelial sodium channel. *Cell* **79**, 1–8.

Smith LC, Hauer J and Sparrow JT. (1996) Gene delivery by lipophilic apo E. In: *Abstract Book: 66th Congress of the European Atherosclerosis Society, Florence, Italy, July 13–17*, p. 28. Giovanni Lorenzini Medical Foundation, Milan, Italy.

Tavtigian SV, Simard J, Rommens J *et al.* (1996) The complete BRCA2 gene and mutations in chromosome 13q-linked kindreds. *Nature Genetics* **12**, 333–337.

Vazio VM, Rinaldi M, Ciafre SA, Signori E, Seripa D, Parrella P, Vespignani I, Farace MG, Uboldi P and Catapano AL. (1996) Functional chronic correction of dyslipidemia in apo E deficient mice by direct intramuscular injection of naked plasmid DNA. In: *Abstract Book: 66th Congress of the European Atherosclerosis Society, Florence, Italy, July 13–17*, p. 28. Giovanni Lorenzini Medical Foundation, Milan, Italy.

Ward K, Hata A, Jeunemaitre X *et al*. (1993) A molecular variant of angiotensinogen associated with preeclampsia. *Nature Genet.* **4**, 59–61.

Williams RR. (1984) Understanding genetic and environmental risk factors in susceptible persons. *Western J. Med.* **141**, 799–806.

Williams RR, Hunt SC, Schumacher MC, Hegele RA, Leppert MF, Ludwig EH and Hopkins PN. (1993) Diagnosing heterozygous familial hypercholesterolemia using new practical criteria validated by molecular genetics. *Am. J. Cardiol.* **72**, 171–176.

Williams RR, Hopkins PN, Wu LL and Hunt SC. (1995) Guidelines for managing severe familial lipid disorders. *Primary Cardiol.* **21**, 47–53.

Williams RR, Hamilton-Craig I, Kostner GM *et al*., (1996) MED-PED: an integrated genetic strategy for preventing early deaths. In: *Genetic Approaches to Noncommunicable Diseases* (eds K Berg, V Boulyjenkov and Y Christen) pp. 35–45. Springer, Heidelberg.

Ziporyn T. (1996) Shaking up conventional wisdom on salt. *Harvard Health Lett.* **22**, 6–7.

Appendix I

International MED PED collaborators. Lone Andersen, Pedro Barosa, Ulrike Beisiegel, Pascale Benlian, Stefano Bertolini, D. John Betteridge, Helen Bilianon, Victor Boulyjenkov, Sebastiano Calandra, Alberico Catapano, Ian Hamilton Craig, Andrew Czeizel, Roger Darioli, Ian Day, Joep Defesche, Olivier Descamps, Gosta Eggertsen, Mats Eriksson, Ole Faergeman, Ian Graham, Vilmundur Gudnason, Michael R. Hayden, Robert Hegele, Steve Humphries, Roger Illingworth, Selim Jambart, Henrik K. Jensen, George Jerums, John Kane, John Kastelein, Christiane Keller, Ulrich Keller, G. Kolovou, Gert M. Kostner, Maritha J. Kotze, Peter Kwiterovich, Mogens L. Larsen, Robert S. Lees, Eran Leitersdorf, Trond P. Leren, James Mann, A. David Marais, Louis Massana, Pedro Mata, J. P. Miller, Andre R. Miserez, Leiv Ose, Josef Patsch, Francisco Perez-Jiminez, Simon Pimstone, Xavier Pinto, Derick Raal, Andrzej Rynkiewicz, Herbert Schuster, Russell Scott, H.C. Seftel, Manzo Sergio, Gunnar Sigurdsson, Jonathan Silberberg, Pedro Silda, Jose Silva, Evan Stein, Elizabeth Steinhagen-Thiessen, David Sullivan, Serena Tonstadt, Michael R. Turner.

Participating countries

Australia	France	Ireland	New Zealand	Spain
Austria	Germany	Israel	Norway	Sweden
Belgium	Greece	Italy	Poland	Switzerland
Canada	Hungary	Lebanon	Portugal	UK
Denmark	Iceland	Netherlands	South Africa	USA

10

Specific approaches to pulmonary emphysema and its therapy

N.A. Kalsheker

10.1 Genetics of pulmonary emphysema

Pulmonary emphysema is a significant cause of morbidity in Western countries, affecting some 3% of the population (Hay and Robin, 1991). It is likely to become an even bigger problem as the population ages. Environmental factors, such as cigarette smoke, make a significant contribution and could potentially account for some of the familial aggregation associated with the disease. In carefully controlled family case studies, correcting for smoking history, there is still strong evidence for genetic factors predisposing to disease (Cohen *et al.*, 1975). It is as yet unclear if there are several interacting genes involved or whether one or two specific genes play a major role in a subgroup of patients. In some families there is evidence to support a single gene model (Higgins and Keller, 1975). As with all chronic diseases, there are problems associated with genetic models, including variable penetrance, errors in clinical diagnosis, genetic heterogeneity and the sporadic non-inheritable occurrence of the disease.

Pulmonary emphysema comes under the general category of lung disorders referred to as chronic obstructive airways disease (COAD) and is the end result of a number of diseases which include pulmonary emphysema, chronic bronchitis and bronchiectasis. The characteristic feature of pulmonary emphysema is the destruction of the normal

Genetics of Common Diseases: future therapeutic and diagnostic possibilities,
edited by I. Day and S. Humphries.

architecture of the lung, resulting in enlargement of the air spaces with a consequent reduction in the surface area of the lungs available for air exchange. In the long term, patients have difficulty in breathing and eventually require supplementary oxygen. Some of these diseases may co-exist, for example chronic bronchitis often co-exists with pulmonary emphysema. This review will focus on the role of α_1-antitrypsin deficiency in COAD, as this genetic disease has been studied extensively. It will also explore the approaches and potential for genetherapy for α_1-antitrypsin deficiency and CF.

10.2 α_1-Antitrypsin deficiency

The best described genetic association with COAD is α_1-antitrypsin deficiency. α_1-Antitrypsin deficiency was first reported over 30 years ago (Laurell and Eriksson, 1963) in five patients who lacked the α_1-globulin band on serum electrophoresis. The major constituent of this band is α_1-antitrypsin or α_1-proteinase inhibitor (α_1-PI). Three of the five patients had pulmonary emphysema, as did nine of 14 additional patients with α_1-antitrypsin deficiency described the following year (Eriksson, 1964), thus establishing the association between α_1-antitrypsin deficiency and COAD. Deficiency results in progressive lung damage in early adult life in cigarette smokers. Longitudinal studies suggest that 25% of individuals with α_1-antitrypsin deficiency survive to the age of 50, compared with 85% of the general population (Larsson, 1978). For patients with deficiency who smoke, life expectancy is reduced by another 10 years. α_1-Antitrypsin protects the lower respiratory tract from damage by serine proteinases released by neutrophils, in particular, elastase.

Over 75 α_1-antitrypsin or PI variants have been described. Many of these variants occur rarely, and two deficiency variants associated with disease are common in white Caucasians of northern European origin. These include the *Pi Z* and *S* variants which occur in approximately 3 and 10% of the population respectively in the UK (Carrell *et al.*, 1982). The S variant *per se* does not confer increased risk of lung disease. The risk of disease is increased if the plasma concentration of α_1-antitrypsin is less than about 35% of the mean concentration found in the normal population. This occurs with homozygous *Z* and compound heterozygotes with both the *S* and *Z* alleles. Each *S* allele results in the production of about 60% of the protein produced by the normal *M* allele and the *Z* allele accounts for about 10–20% of the protein produced by the *M* allele. Consequently, only homozygous *Z* and, to a lesser extent, compound *SZ* heterozygotes are prone to developing COAD. Individuals with the *Pi null–null* genotype who do not produce any α_1-antitrypsin develop COAD at an even earlier age compared with *Pi Z*, suggesting that even the reduced plasma concentration associated with *Pi Z* may confer some protection (Cox *et al.*, 1988). The association of α_1-antitrypsin deficiency with COAD has led to the proteinase–antiproteinase imbalance theory which proposes that excess proteinase activity has the potential to cause lung destruction. The *Pi Z* allele also predisposes to the development of juvenile cirrhosis, and this is not a feature of the *Pi null* allele. The liver disease results from accumulation of abnormally folded protein in the endoplasmic reticulum.

There is variability in pulmonary function associated with classical α_1-antitrypsin deficiency, and many individuals with deficiency do not have clinically significant lung function impairment (Silverman *et al.*, 1989). It is, as yet, unknown what proportion of patients with *Pi Z* develop COAD. Consequently, there is as yet no good case for the treatment of *Pi Z* individuals who are apparently healthy. In a recent study on deaths attributed to α_1-antitrypsin deficiency, 2.7% of all deaths with COAD among patients aged 35–44, and in 1.2% of all deaths listing hepatic disease among children aged 1–14 (Browne *et al.*, 1996), the authors highlight the fact that the disease is probably under-reported. Another study has shown that patients with α_1-antitrypsin deficiency were not diagnosed until a mean age of 41 years, after a mean of 5.7 years of respiratory symptoms or dyspnoea (Stoller *et al.*, 1993). The early detection of α_1-antitrypsin deficiency may influence the outcome of pulmonary disease, as individuals who smoke should be actively counselled to refrain from it.

Irreversible damage to the architecture of the lung results in enlargement of the air spaces and pulmonary emphysema. More recently, deficiency of another closely related serine proteinase inhibitor, α_1-antichymotrypsin, has also been shown to be associated with COAD (Poller *et al.*, 1990), highlighting the importance of serine proteinase inhibitors in the pathology of the disease. However, further studies are required to assess the significance of these observations.

10.3 Structure of α_1-antitrypsin

Mature α_1-antitrypsin consists of 394 amino acids with a molecular mass of about 52 kDa. The protein is glycosylated, with the carbohydrate residue constituting about 15% of the weight of the protein. The carbohydrate side chains are attached at residues 46, 83 and 247, which result in heterogenous molecular species, and this is detected as multiple bands by isoelectric focusing of plasma proteins. The latter technique is still used widely to determine the protein type of α_1-antitrypsin.

The tertiary structures of a molecular variant of intact α_1-antitrypsin, and cleaved forms of the molecule, have been determined (Loebermann *et al.*, 1984). α_1-Antitrypsin is a globular, highly ordered, molecule. About 30% of its structure is helical and 40% consists of β-pleated sheets. There are nine α-helices and three β-pleated sheets. The reactive centre loop is exposed on the surface of the molecule. This unusually large loop arises from a β-pleated sheet (the A sheet). When α_1-antitrypsin is cleaved in its reactive centre loop it undergoes a massive conformational change that is dominated by the insertion of the loop as one of the strands of the A sheet, gaining extraordinary stability as a result. The native molecule is thus thought to be in a stressed conformation and the cleaved protein is in a relaxed form (Elliott *et al.*, 1996).

The interaction between enzyme and inhibitor is the subject of much debate. A docking process establishes interactions between residues in the exposed loop of the inhibitor and the primary and secondary binding sites on the enzyme. As yet, no crystal structures of enzyme–inhibitor complexes are available. Recent structural data on intact α_1-antitrypsin suggest that the native molecule is poised for optimal substrate binding (Elliot *et al.*, 1996). There are three possible consequences once substrate binding occurs: either the inhibitor is cleaved; or an enzyme inhibitor complex is formed; or both events occur. It has been demonstrated recently with α_1-antichymotrypsin that the inhibitor induces a conformational change in the enzyme chymotrypsin that distorts the His–Asp–Ser catalytic triad of the enzyme, such that the enzyme has reduced activity and a slow turnover rate. This conformation may thus favour docking of the enzyme to form a complex, rather than cleavage (Rubin, 1996).

10.4 α_1-Antitrypsin gene structure and mRNA species

The α1-antitrypsin gene is located on chromosome 14 at position q32.1 (Lai *et al.*, 1983). It contains about 12 000 bp of nucleotide sequence and is located in a cluster of related genes of the serine proteinase inhibitor family (Billingsley *et al.*, 1993). The gene contains seven exons, separated by six introns. Two specific promoter regions have been identified, one for hepatocytes and one for monocytes/macrophages (Perlino *et al.*, 1987) (*Figure 10.1*). α_1-Antitrypsin is produced mainly by the liver, but small amounts (about 1%) are also produced by monocytes (Rogers *et al.*, 1983).

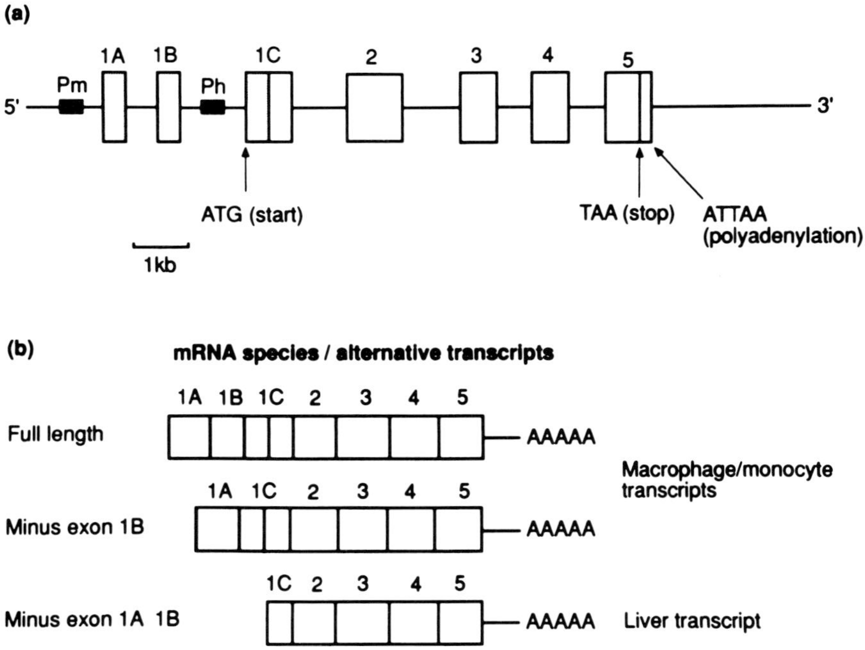

Figure 10.1. (a) The structure of the a_1-antitrypsin gene. Pm and Ph correspond to the monocyte and hepatocyte specific promoters respectively. ATG corresponds to the start of the coding sequence. (b) Alternative transcripts produced by hepatocytes and monocytes. Hepatocytes produce a single species and monocytes produce two species by alternative splicing; exon 1B may or may not be present. Reproduced from Barnes PJ and Stockley RA, *Molecular Biology of Lung Disease* with permission from Blackwell Science.

There are two tissue-specific promoters, one for hepatocytes and one for monocytes and other tissues including the gastro-intestinal tract, lung and kidneys (Carlson *et al.*, 1988; Hafeez *et al.*, 1992).

The mRNA species produced by the liver differ from that produced by the monocyte. The latter produce two distinct mRNA species (*Figure 10.1*). For convenience, the additional monocyte exons, which are essentially non-coding, are designated 1A, 1B and 1C. In the liver, transcription begins in the middle of exon C and includes 49 bp of untranslated sequence. Monocyte transcription begins 2 kbp upstream of the liver promoter and the two transcripts contain either 1A, 1B and 1C, or exons 1A and 1C. In the latter, exon 1B is excluded. There are potentially three macrophage-specific transcription initiation sites and, in hepatoma cells, there is the potential to use the macrophage-specific transcriptional initiation sites during modulation by the acute phase mediator, IL-6 (Hafeez *et al.*, 1992).

α_1-Antitrypsin is an acute phase reactant and as such concentrations increase three- to four-fold during inflammation. This is mainly mediated by the cytokine IL-6 (Kalsheker and Swanson, 1990). A number of DNA sequence elements in the α_1-antitrypsin gene have been shown to be critical for tissue-specific expression. Upstream from the transcription start site, three regulatory elements have been described (*Figure 10.2*). A dominant tissue-specific element is sited between nucleotides –137 and –37 upstream of the TATA box (Ciliberto *et al.*, 1985; de Simone *et al.*, 1987; Shen *et al.*, 1987). A second element is located between nucleotides –261 and –210, and is capable of increasing transcription approximately four- to five-fold (Shen *et al.*, 1987). The third regulatory element is located between nucleotides –488 and –356, and increases transcriptional activity three- to four-fold (Hardon *et al.*, 1988).

In addition, we have identified a 3′ enhancer, a region in which the mutation associated with COAD occurs (*Figures 10.2* and *10.3*). Wild-type sequence demonstrated an approximately 50–100% increase in activity compared with a control promoter plasmid, whereas mutant sequence demonstrated 20–40% less activity than the control promoter plasmid (Morgan *et al.*, 1993). We have recently demonstrated that the dominant enhancer is located at the 5′ end and the effects of the mutation only become manifest after stimulation by the cytokine IL-6 (unpublished observations). In effect, there is a diminished response to IL-6 in the mutant sequence in a cell transfection system, and this suggests that there may be a state of relative deficiency during the acute phase response.

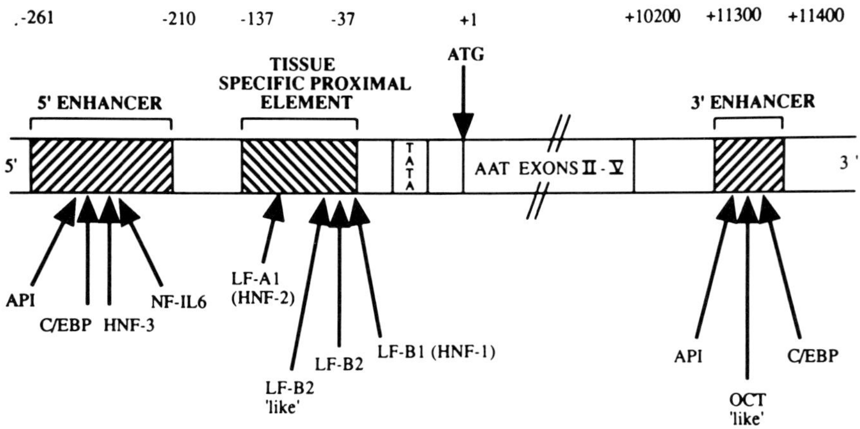

Figure 10.2. The DNA sequence elements associated with regulating expression of the a_1-antitrypsin gene and some of the transcription factors which bind to them. Reproduced from Barnes PJ and Stockley RA. *Molecular Biology of Lung Disease* with permission from Blackwell Science

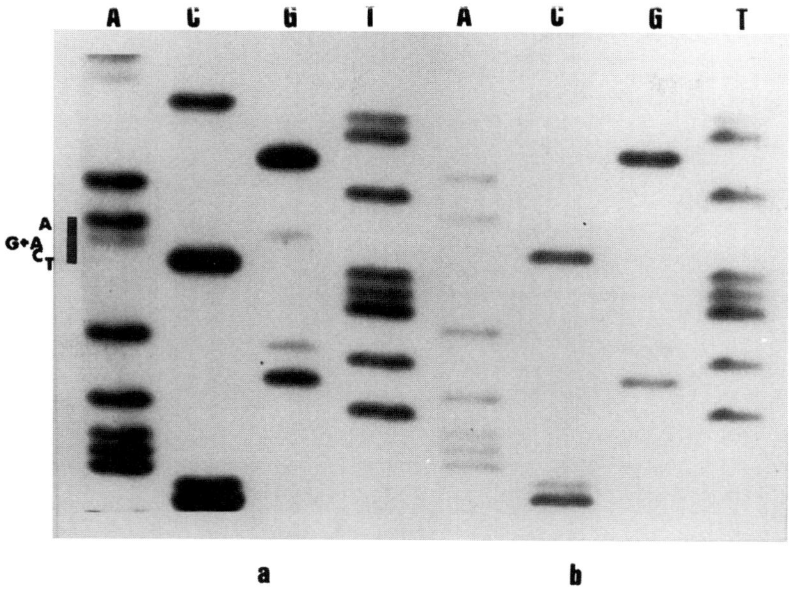

Figure 10.3. Sequence of (a) heterozygous mutant and (b) normal sequence at the Taa 1 site. Reproduced from Kalsheker NA and Morgan K with permission from the *American Journal of Respiratory and Critical Care Medicine.*

We have further demonstrated that this is due to loss of positive co-operativity between an octamer-1 binding site, where the mutation occurs, and an NF-IL-6 site, which is a key mediator of the acute phase response and highlights how the ubiquitous transcription factor octamer-1 interacts with a tissue-specific transcription factor, namely NF-IL-6, to produce a tissue-specific response.

10.5 Genetic mutations associated with disease

Given the relative inefficiency of establishing variation by analysing proteins, since only about one-third of all mutations that occur in proteins are likely to be detected, and because most mutations occur in non-coding sequences, we initiated a detailed study of the α_1-antitrypsin gene to see if we could identify additional DNA mutations and determine whether any variants were associated with disease. This hypothesis was worth testing because of the association of α_1-antitrypsin deficiency and COAD.

After an extensive investigation of the α_1-antitrypsin gene, we identified a polymorphism associated with COAD, which has been

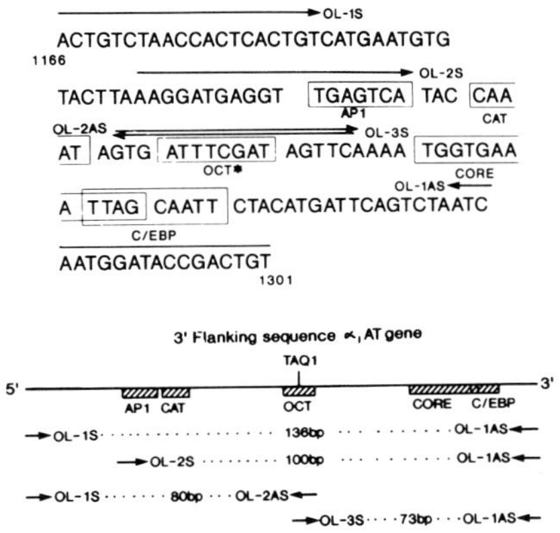

Figure 10.4. The region of the 3′ α_1-antitrypsin enhancer showing the consensus sequences for a number of transcription factors and the oligonucleotides used for amplification of selective regions containing these motifs.

confirmed by others (Morgan *et al.*, 1992; Poller *et al.*, 1990). The polymorphism arises from a single point mutation and occurred in 5% of our controls and 17% of patients with COAD (Kalsheker *et al.*, 1987). We subsequently demonstrated that the mutation occurs in an enhancer element in the 3′ flanking region of the gene (Morgan *et al.*, 1993, and *Figure 10.4*).

10.6 Therapeutic trials with α_1-antitrypsin

In assessing the potential of replacement therapy there are a number of important considerations which will help define groups of patients who may benefit. Firstly, not all patients who have α_1-antitrypsin deficiency

Table 10.1. Summary of replacement therapy approaches

		Results
■ Replace with protein, e.g. α_1-antitrypsin		Biochemical efficacy
■ Gene therapy	– viral (retrovirus, adenovirus) + tissue-specific promoters	Expression for up to 4 weeks
	– liposomes	
	– *ex vivo* into hepatocytes + transplantation	Expression for up to 47 days

develop COAD, even when taking the history of cigarette smoking into account (Silverman *et al.*, 1989). Thus, some patients appear to escape the disease. Secondly, even in families with classical α_1-antitrypsin deficiency, deficiency states may not always segregate with disease. Thirdly, another mutation which confers a risk of developing COAD has been identified in the 3′ flanking sequence of the α_1-antitrypsin gene (see below). The approaches to replacement therapy are summarized in *Table 10.1.*

It has been estimated that more than 10 million Americans are affected with COAD and about 1% of this is due to α_1-antitrypsin deficiency (Hay *et al.*, 1991). Human α_1-antitrypsin (Prolastin) has been approved by the Food and Drug Administration (FDA) in the USA for replacement therapy in patients with α_1-antitrypsin deficiency, and the therapy has been granted orphan drug status. It has been demonstrated that replacement therapy is effective in improving the levels of α_1-antitrypsin in serum and epithelial tissue. However, the long-term effects of replacement therapy in ameliorating disease are as yet unknown. This is made difficult by the chronic and insidious nature of COAD, and the number of potential beneficiaries is small. These considerations make it difficult to be certain of the outcome. Furthermore, a prospective randomized clinical trial to determine efficacy may cost over US$1 billion (Hay *et al.*, 1991).

In order to assess the costs of replacement therapy, the average medical costs to support patients with COAD have to be taken into account. An estimate of average medical costs per patient year can vary from US$587 to US$6238 (estimated in 1990) for COAD patients, depending on the number of years after onset of symptoms. Depending on the efficacy of treatment, the cost per life year saved varied from US$28 000 (70% efficacy) to US$128 000 (30% efficacy). With an efficacy of 30% or greater, this is comparable with other widely used medical interventions (Hay *et al.*, 1991). On this basis, it has been argued that α_1-antitrypsin replacement therapy should be offered under the category of an orphan drug, so that large scale clinical trials do not need to be conducted before this becomes an acceptable drug.

There are many other issues that need to be considered in relation to therapy. What are the most favoured routes for the long-term administration of α_1-antitrypsin? Should treatment be given regularly or intermittently during episodes where there may be a high neutrophil load, for example during infection? How much should be given? How does one establish long-term efficacy? An alternative approach to direct relacement of α_1-antitrypsin protein is the introduction of genes which express α_1-antitrypsin *in vivo*.

10.7 Gene therapy

Several studies have reported the successful introduction and expression of protein of α_1-antitrypsin cDNA constructs transfected into hepatocytes. The most successful of these approaches have involved the use of retroviral vectors and, although these strains are replication defective, they still constitute a potential health risk, and this approach is not likely to have widespread acceptability, particularly as there are simpler alternatives. However, the amounts of protein expressed may not be sufficient to provide protection to the lung over a long period of time and the production is not controlled properly.

10.8 Gene transfer techniques

For the adequate expression of any cDNA of interest, the gene should be linked to an appropriate promoter, and entry into the cell has to be facilitated in some way. Most studies to date have focused on ubiquitous promoters which are present in viruses such as SV40 or cytomegalovirus and, in theory, these promoters should be active in any nucleated cell. Cell-specific expression may be a better strategy to follow, as it will limit gene expression to a desired cell population. In this respect, lung-specific promoters such as surfactant protein C promoter have been investigated (Wert *et al.*, 1993).

Two agents have been used hitherto to facilitate the entry of genes into cells. These include viral vectors, amongst the most popular of which have been retro viruses, and liposomes. Retroviruses generally produce efficient gene transfer, including the integration of the DNA into the host genome. The major disadvantage of this approach is that it is possible to introduce the virus into dividing cells only, and this approach would therefore not be suitable for the airways, which contain epithelium that is terminally differentiated. There are other risks associated with retroviruses in general and, in particular, those arising from random integration into the host genome.

The recent introduction of adenoviruses has resulted in a number of attempts at gene therapy which are trophic for respiratory epithelium. The principal problem with these viral vectors is that the sequences coding for the coat proteins are retained in the vector systems used, and these are immunogenic. Consequently they result in the production of inflammation through cytotoxic lymphocytes or neutralizing antibodies, thus reducing the efficiency and duration of repeated application. The viruses used tend to be replication defective, but there is a theoretical danger that these viruses may recombine with wild-type sequence and may be rendered infectious.

Liposomes are vesicles which form spontaneously when lipid preparations of particular composition are mixed in polar solutions. Cationic liposomes form complexes with DNA (which has a net negative charge) and have been used for some years for *in vitro* gene transfer. Liposomes are relatively inefficient at gene transfer compared with viral vectors, but do have the advantage of low toxicity.

10.9 Results of administration

Direct delivery through the airways is likely to be the best route for repeated administration and can be controlled by varying droplet size. However, there may be barriers in the form of mucus and dilution by airway surface liquid. A considerable amount of work has been done in gene therapy of CF. The normal counterpart of the faulty gene, the CF transmembrane conductance regulator (CFTR) functions as a cyclic AMP-regulated chloride channel on the mucosal surface of the airway epithelium. Abnormalities in this gene result in reduced expression and are associated with the disease. Initial studies have demonstrated that it is possible to correct for the chloride defect *in vitro* (Drumm *et al.*, 1990; Rich *et al.*, 1990). It was subsequently demonstrated that the *CFTR* gene could be expressed in the airways of normal mice *in vivo* using adenovirus-and liposome-mediated gene transfer (Rosenfeld *et al.*, 1991; Yoshimura *et al.*, 1991). With these approaches it is possible to maintain expression for up to 4 weeks. It has been shown that even low levels of expression of CFTR may correct electrophysiological defects (Caplen *et al.*, 1995), and this makes gene therapy for CF a strategy worth pursuing. This is in contrast to α_1-antitrypsin deficiency where high levels of expression may be required, and this introduces problems in trying to ensure adequate expression.

10.10 Gene therapy and α_1-antitrypsin deficiency

Several different cell types have been successfully transfected *in vitro* with the human α_1-antitrypsin gene, including human fibroblasts (Garver *et al.*, 1987), canine hepatocytes (Kay *et al.*, 1992), sheep endothelium (Lemarchand *et al.*, 1993) and cotton rat airway epithelium (Rosenfeld *et al.*, 1991). Hepatocytes transfected with retrovirus *ex vivo* have been transplanted into dogs, and human α_1-antitrypsin mRNA could be detected for up to 47 days (Kay *et al.*, 1992), with low levels of production of the protein for up to 14 days (Lemarchand *et al.*, 1993). In sheep, transfection of endothelial cells with adenovirus α_1-antitrypsin cDNA resulted in α_1-antitrypsin expression only in endothelial cells, and no α_1-antitrypsin could be detected in the circulation (Lemarchand *et al.*, 1993). In the cotton rat, direct installation of adenovirus α_1-antitrypsin cDNA into the lungs was followed by the detection of the human protein in airway lining fluid, but only at 2% of the required level for function. In the lung interstitia space, the concentration of α_1-antitrypsin is approximately 3 μM, compared with a blood plasma concentration of 30 μM (Wewers *et al.*, 1987).

On the basis of these studies, one human gene therapy protocol has been submitted to the regulatory authorities in the USA. There are two initial aims of the protocol: firstly, to assess whether topical application of liposome–α_1-antitrypsin cDNA complexes into the nasal epithelium of subjects with α_1-antitrypsin deficiency results in mRNA, and secondly to see whether any histological changes occur as a result of gene transfer. A further aspect of the proposal is to deliver the complex into the lower airways of the subject scheduled for elective pneumonectomy. Three days before surgery the complex will be instilled bronchoscopically, and the site of exposure marked with a dye. At the time of surgery, the transfected area will be lavaged for assessment of α_1-antitrypsin levels, and the specimen examined for mRNA and protein.

The liver is the major site of synthesis of α_1-antitrypsin, and several studies have attempted to optimize expression in this cell type. Partial hepatectomy at the time of gene administration stimulates cell proliferation which in turn appears to enhance gene delivery and expression (Wilson *et al.*, 1992). In mice which received a 70% hepatectomy, gene expression of α_1-antitrypsin persisted for up to 6 months.

10.11 Response to therapy

Since pulmonary emphysema can take decades to manifest, it is a difficult task to assess the response to treatment. It has been suggested that, in order to demonstrate efficacy, this would require 500 subjects with a 5-year treatment period (Burrows, 1983). It will therefore be difficult to establish the efficacy of gene therapy for α_1-antitrypsin deficiency. Notwithstanding these obstacles, the task is well defined, and refinements in technology may result in efficient and stable expression of the α_1-antitrypsin gene.

10.12 Conclusions

This review has described the genetics of COAD with a particular focus on α_1-antitrypsin deficiency, an area in which a considerable amount of research has been undertaken. An understanding of the pathophysiology of the disease, combined with the prospects of replacement therapy, may help ameliorate a crippling disease in a proportion of individuals who are susceptible to COAD.

Acknowledgements

Some of the work reported here was supported by the Wellcome Trust (Grant reference numbers: 19343, 035324, 044161).

References

Billingsley GD, Walter MA, Hammond GL and Cox DW. (1993) Physical mapping of four serpin genes: α_1-antitrypsin, α_1-antichymotrypsin, corticosteroid-binding globulin and protein C inhibitor, within a 280-kb region on chromosome 14q32.1. *Am. J. Hum. Genet.* **52**, 343–353.

Browne RJ, Mannino DM and Khoury MJ. (1996) a_1-antitrypsin deficiency deaths in the United States from 1979–1991 – an analysis using multiple-cause mortality data. *Chest* **110**, 78–83.

Burrows B. (1983) Q clinical trial of efficacy of antiproteolytic therapy: can it be done? *Am. Rev. Respir. Dis.* **127**, S42–S43.

Caplen NJ, Alton EWFW, Middleton PG *et al.* (1995) Liposome mediated CFTR gene-transfer to the nasal epithelium of patients with cystic fibrosis. *Nature Med.* **1**, 272.

Carlson JA, Rogers BB, Sifers RN, Hawkins HK, Finegold MJ and Woo SLC. (1988) Multiple tissues express α_1-antitrypsin in transgenic mice and man. *J. Clin. Invest.* **82**, 26–36.

Carrell RW, Jeppsson JO, Laurell C-B, Brennan SO, Owen MC, Vaughan L and Boswell DR. (1982) Structure of human α_1-antitrypsin. *Nature* **298**, 329–334.

Ciliberto E, Dente L and Cortese R. (1985) Cell-specific expression of a transfected human α_1-antitrypsin gene. *Cell*, **41**, 531–540.

Cohen BH, Ball WC and Bias WB. (1975) A genetic epidemiologic study of chronic obstructive pulmonary disease: study design and preliminary observations. *Johns Hopkins Med. J.* **137**, 94–104.

Cox DW and Levison H. (1988) Emphysema of early onset associated with a complete deficiency of α_1-antitrypsin deficiency (null homozygotes). *Am. Rev. Respir. Dis.* **137**, 371–375.

De Simone V, Ciliberto G, Hardon E, Paonessa G, Palla F, Lundberg L and Cortese R. (1987) *Cis*- and *trans*-acting elements responsible for the cell-specific expression of the human α_1-antitrypsin gene. *EMBO J.* **6**, 2759–2766.

Drumm ML, Pope HA, Cliff WH, Rommens JM, Marvin SA, Tsui LC, Collins

FS, Frizzel RA and Wilson JM. (1990) Correction of the cystic fibrosis defect *in vitro* by retrovirus mediated gene transfer. *Cell* **62**, 1227–1233.

Elliott PR, Lomas DA, Carrell RW and Abrahams J-P. (1996) Inhibitory conformation of the reactive loop of α_1-antitrypsin. *Nature Struct. Biol.* **3**, 676–681.

Eriksson S. (1964) Pulmonary emphysema and α_1-antitrypsin deficiency. *Acta Med. Scand.* **175**, 197–205.

Garver RI, Chytil A, Courtney M and Crystal RG. (1987) Clonal gene therapy: transplanted mouse fibroblast clones express human α_1-antitrypsin gene *in vivo. Science* **237**, 762–764.

Hafeez W, Ciliberto G and Perlmutter DH. (1992) Constituitive and modulated expression of the human α_1-antitrypsin gene. *J. Clin. Invest.* **89**, 1214–1222.

Hardon EM, Frain M, Paonessa G and Cortese R. (1988) Two distinct factors interact with the promoter regions of several liver-specific genes. *EMBO J.* **7**, 1711–1719.

Hay JW and Robin ED. (1991) Cost-effectiveness of α_1-antitrypsin replacement therapy in treatment of congenital chronic obstructive pulmonary disease. *Am. J. Public Health,* **81**, 427–433.

Higgins M and Keller J. (1975) Familial occurrence of chronic respiratory disease and familial resemblance in ventilatory capacity. *J. Chronic Dis.* **28**, 239-251.

Kalsheker NA and Swanson T. (1990) Exclusion of an exon in monocyte α_1-antitrypsin mRNA after stimulation of U937 cells by interleukin-6. *Biochem. Biophys. Res. Commun.* **172**, 1116–1121.

Kalsheker NA, Hodgson IJ, Watkins GL, White JP, Morrison HM and Stockley RA. (1987) DNA polymorphism of the α_1-antitrypsin gene in chronic lung disease. *Br. Med. J.* **294**, 1511–1514.

Kay MA, Baley P, Rothenberg S *et al.* (1992) Expression of human α_1-antitrypsin in dogs after autologous transplantation of retroviral transduced hepatocytes. *Proc. Natl Acad. Sci. USA* **89**, 89–93.

Lai EC, Kao FT, Law ML and Woo SLC. (1983) Assignment of the α_1-antitrypsin gene and a sequence related gene to human chromosome 14 by molecular hybridisation. *Am. J. Hum. Genet.* **35**, 385–392.

Larsson C. (1978) Natural history and life expectancy in severe α_1-antitrypsin deficiency PiZ. *Acta Med. Scand.* **204**, 345–351.

Laurell C-B and Eriksson S. (1963) The electrophoretic alpha$_1$-globulin pattern of serum in α_1-antitrypsin deficiency. *Scand. J. Clin. Lab. Invest.* **15**, 132–140.

Lemarchand P, Jones M, Yamada I and Crystal R. (1993) *In vivo* gene transfer and expression in normal uninjured blood vessels using replication deficient recombinant adenovirus vectors. *Circ. Res.* **72**, 1132–1138.

Loebermann H, Tokuoka R, Deisenhofer J and Huber R. (1984) α_1-Proteinase inhibitor: crystal structure analysis of two crystal modifications, molecular model and preliminary analysis of the implications for function. *J. Mol. Biol.* **177**, 531–557.

Morgan K, Scobie G and Kalsheker N. (1992) The characterisation of a mutation of the 3′ flanking sequence of the α_1-antitrypsin gene commonly associated with chronic obstructive airways disease. *Eur. J. Clin. Invest.* **21**, 134–137.

Morgan K, Scobie G and Kalsheker N. (1993) Point mutation in a 3′ flanking sequence of the α_1-antitrypsin gene associated with chronic respiratory disease occurs in a regulatory sequence. *Hum. Mol. Genet.* **2**, 253–257.

Perlino E, Cortese R and Ciliberto G. (1987) The human α_1-antitrypsin gene is transcripted from two different promoters in macrophages and monocytes. *EMBO J.* **6**, 2767–2771.

Poller W, Meissen C and Olek K. (1990) DNA polymorphisms of the α_1-antitrypsin gene region in patients with chronic obstructive pulmonary disease. *Eur. J. Clin. Invest.* **20**, 1–7.

Rich DP, Anderson MP, Gregory RJ, Cheng SH, Paul S, Jefferson DM, McCann JD, Klinger KW, Smith AE and Welsh MJ. (1990) Expression of cystic fibrosis transmembrane conductance regulator corrects defective chloride channel regulation in cystic fibrosis airway epithelial cells. *Nature* **347**, 358–363.

Rogers J, Kalsheker N, Wallis S, Speer A, Coutelle CH, Woods D and Humphries SE. (1983) The isolation of a clone for human α_1-antitrypsin and the detection of α_1-antitrypsin in mRNA from liver and leukocytes. *Biochem. Biophys. Res. Commun.* **116**, 375–382.

Rosenfeld MA, Siegfried W, Yoshimura K *et al.* (1991) Adenovirus-mediated transfer of a recombinant α_1-antitrypsin gene to the lung epithelium *in vivo*. *Science* **252**, 431–434.

Rubin H. (1996) Serine protease inhibitor (SERPINS): where mechanism meets medicine. *Nature Med.* **2**, 632–633.

Shen RF, Li Y, Sifers RN, Wang H, Hardwick C, Tsai SY and Woo SLC. (1987) Tissue specific expression of the human α_1-antitrypsin gene is controlled by multiple *cis*-regulatory elements. *Nucleic Acids Res.* **15**, 8399–8415.

Silverman EK, Pierce JA, Province MA, Roa DC and Campbell EJ. (1989) Variability of pulmonary function in α_1-antitrypsin deficiency: clinical correlates. *Ann. Intern. Med.* **111**, 982–991.

Stoller JK, Spray J and Smith P. (1993) Demographics and impact of α_1-antitrypsin deficiency: results of a mail survey (Abstract). *Am. Rev. Respir. Dis.* **147**, A871.

Wert SE, Glasser SW, Korthagan TR and Whisett JA. (1993) Transcriptional elements from the human SP-C gene direct expression in the primordial respiratory epithelium of transgenic mice. *Dev. Biol.* **156**, 426–443.

Wewers MD, Casolaro MA, Sellers SE, Swayze SC, McPhaul KM, Wittes JT and Crystal RG. (1987) Replacement therapy for α_1-antitrypsin deficiency associated with emphysema. *N. Engl. J. Med.* **316**, 1055–1062.

Wilson JM, Grossman M, Cabrera JA, Wu CH and Wu GY. (1992) Hepatocyte-directed gene transfer *in vivo* leads to transient improvement of hypercholesterolemia in low density lipoprotein receptor-deficient rabbits. *J. Biol. Chem.* **267**, 963–967.

Yoshimura K, Rosenfeld MA, Nakamura H *et al.* (1991) Expression of the human CFTR gene in the mouse lung after *in vivo* intratracheal plasmid mediated gene transfer. *Nucleic Acids Res.* **20**, 3233–3240.

11

Gene therapy for neurological diseases: *quo vadis*? Achievements and expectations of, and challenges for, the brave new technology

P.R. Lowenstein, J. Jaszai and M.G. Castro

"The conceptual basis for gene therapy has been established beyond reasonable doubt, and few if any can still doubt that many human diseases will eventually be treated by complementation of the offending genetic aberrations rather than by manipulation of secondary metabolic aberrations underlying genetic defects."

(Theodore Friedmann, 1994)

11.1 Introduction

Childhood leukaemias are mostly curable in the late twentieth century. It was not so 50 years ago when clinical trials for these deadly diseases were first attempted. They did not immediately achieve long-term survival either; it took several decades to do so. Today, however, many childhood leukaemias belong to those diseases where medicine has been successful. This is our dream, and these are our goals for gene therapy of brain diseases.

Genetics of Common Diseases: future therapeutic and diagnostic possibilities,
edited by I. Day and S. Humphries.

This chapter offers a concise review on the current status of gene therapy for neurological disorders. We will identify several central, but so far unaddressed, major questions. Answers to these questions ought to indicate new directions for gene therapy for brain disorders, and further the development of experimental work aiming towards the clinical implementation of these novel therapies. Gene therapy has now entered, what Friedmann called the 'implementation phase' (Friedmann, 1994). The idea of gene therapy, even for the treatment of brain disease, is not 'new'. Gene therapy was proposed more than 20 years ago, and has been discussed ever since (Friedmann, 1983; Friedmann and Roblin, 1972; Wolff and Lederberg, 1994). There are few doubts, from a theoretical standpoint, that gene therapy ought to work. The current challenge, however, is to implement it within clinical settings (Friedmann, 1994).

11.2 Challenges towards clinical implementation: the central questions and the development of a viable clinical strategy

Even if gene therapy can be developed in theory for any disease, it is likely that certain diseases will become preferential targets, when taking into account the adequacy of current available vectors (e.g. their toxicity, longevity of expression, level of expression, etc.), in conjunction with the characteristics of individual diseases to be treated (e.g. their pathophysiological characteristics, genetic component, incidence, etc.) (Lowenstein, 1995). Such considerations led to the early gene therapy trials for severe combined immunodeficiency (SCID) due to mutations in the adenosine deaminase (ADA) gene. It is interesting to go back and examine some of the historical discussions and perspectives (Culver, 1994; Friedmann, 1983; Lyon and Gorner, 1995; Thompson, 1994; Wolff & Lederberg, 1994) to realize that it was not until relatively recently that the target disease for the initial experiments was eventually identified. The experiments on ADA deficiency began in earnest in 1990, and the first peer-reviewed results were published in late 1995 (Blaese *et al.*, 1995; Bordignon *et al.*, 1995).

We ought to remember that the current perceived excitement provided by gene therapy is underpinned by the fact that it represents a powerful new addition to existing pharmacological tools. Pharmacology and gene therapy go hand in hand, complementing each other. The theoretical and practical breakthrough provided by gene therapy is due to the fact that whereas pharmacological intervention through the use of drugs is limited to ***modifying*** the cell's functions, gene therapy, through the delivery of particular DNA sequences, can engineer new functions in diseased cells – a theoretical revolution in the pharmacological approach to disease management.

For any disease to be targeted by gene therapy, researchers must be able to address the following questions in order to develop a sound strategy that will lead to a clinically relevant reversion of the phenotypic manifestations of disease:

(i) *How* is the nucleic acid to be transferred to the target cells?
(ii) *Which cells* need to be targeted?
(iii) *What percentage* of the affected cells need to be transduced?
(iv) *What level of expression* has to be achieved for the therapeutic transgene to revert the disease phenotype to normal?
(v) For *how long* does the transgene have to be expressed?
(vi) At *what stage* of the disease does the transgene have to be expressed?

How to transfer the gene depends on the gene itself (e.g. its size, introns, regulatory sequences, etc.), the disease to be targeted (e.g. its pathophysiology, duration, etc.), and especially the interactions among the vector, transgenic construct and target cells. Importantly, all currently available vectors have advantages as well as disadvantages. The fact that supporters of one or another system tend to cast a blind eye on the failings of their own preferred system does not force the immediate dismissal of such shortcomings.

The answer to (ii) depends on the pathophysiology of the disease (i.e. which is the affected brain cell type). This is not always straightforward to determine, especially because even in the case of single-gene inherited disorders, this information is *not* provided by molecular genetics. Molecular genetics will only determine the mutation of a particular gene and determine its overall role as the cause of the disease. However, it does not provide information on the type of cells in which expression of a wild-type copy of the diseased gene is needed in order to protect the whole organ from the manifestations of disease.

This can nevertheless be investigated, and answers for many diseases have been provided. For example, the phenotype of most inherited leukodystrophies and the brain manifestations of several mucopolysaccharidoses are highly likely to be due to lack of expression of mutated genes in brain microglial cells, implying that genetic engineering and/or replacement of microglial cells could revert the disease phenotype to normal; in the gangliosidoses, however, it is neurons that accumulate pathologic lysosomes, and those cells will need to be targeted. Dopaminergic nigro-striatal neurons degenerate in

Table 11.1. Inherited 'trinucleotide repeat' diseases

Neurogenetic disease	Affected gene	Chromosomal location	Expanded triplet (intragenic location)	Transgenic model available
Huntington's disease	Huntingtin (*IT15*)[a]	4p16.3	CAG (in coding region)	yes[i]
Smith's disease (DRPLA)	Atrophin-1 (*CTG-B37*)[b]	12p12–pter	CAG (in coding region)	no
SCA1	Ataxin-1[c]	6p22-p23	CAG (in coding region)	yes[j]
Machado-Joseph disease (SCA3)	*MJD-1*[d]	14q32.1	CAG (in coding region)	yes[k]
Kennedy's disease (SMBA)	Androgen receptor[e]	Xq11–q12	CAG (in coding region)	yes[l]
Fragile X mental retardation (FRAXA)	*FMR-1*[f]	Xq27.3	CGG (in the 5'-UTR)+	yes[m]
Fragile X mild mental retardation (FRAXE)	FMR-2[g]	Xq28	CCG/CGG	no
Friedreich's ataxia	X25[h]	9q13	GAA (in intron 1)	no

Key: [a]The Huntington's Disease Collaborative Research Group, 1993; [b]Koide *et al.*, 1994; [c]Banfi *et al.*, 1994; [d]Kawaguchi *et al.*, 1994; [e]La Spada *et al.*, 1991; [f]Verkerk *et al.*, 1991; [g]Gu *et al.*, 1996; Gecz *et al.*, 1996; [h]Campuzano *et al.*, 1996; [i]Mangarini *et al.*, 1996; [j]Burright *et al.*, 1995; [k]Ikeda *et al.*, 1996; [l]Bingham *et al.*, 1995; [m]Dutch-Belgian Fragile X Consortium.

+Lack of FMR1 protein due to the transcriptional suppression of the *FMR1* gene as a result of CGG trinucleotide expansion and CpG island hypermethylation.

Parkinson's disease, and neurons throughout the basal ganglia and neocortex degenerate in Huntington's disease; however, the cell type to be targeted by gene therapy will depend on whether the primary defect occurs in neurons themselves, or whether neuronal cell death is secondary to a defect in glial cells.

Transgenic animal models could help provide some of these crucial answers. Recently, improved transgenic animal models displaying human-like pathology have been developed for Huntington`s disease (Mangiarini *et al.*, 1996), Alzheimer`s disease (Hsiao *et al.*, 1996), ataxia teleangiectasia (Barlow *et al.*, 1996) and for Charcot–Marie–Tooth disease type 1A (Huxley *et al.*, 1996; Sereda *et al.*, 1996) (see also *Table 11.1*). Importantly, these new models now exhibit clinical, biochemical and pathological characteristics similar to the human diseases.

The persistent lack of a complete understanding of the pathophysiology of many diseases limits any therapies to symptomatic treatment. It is still not known why neurons degenerate in Huntington's, Parkinson's or Alzheimer's disease. However, transplantation therapy has been proposed by many scientists worldwide, and has already proceeded to clinical stages in Parkinson's, while it is very close to clinical implementation in Huntington's disease. The basis for the human experiments rests with rodent or primate data. These experiments were done on animals *not* suffering from any of these disorders. The possibility that transplanted cells will eventually succumb to disease, in the same way as the patient's own, must be taken into account. Unfortunately, it still remains difficult to design useful and predictive clinical human experiments. This opens up the ethical argument, which also needs urgent attention. New therapies ought to proceed to clinical trials only when the scientific evidence of gene transfer, gene expression-induced phenotype correction, and lack of toxicity, have all been established (Wivel and Walters, 1993). Compassionate use in the absence of supporting scientific data should never be allowed to proceed to the clinical stage; lack of evidence against an unproven therapy does by no means represent any sort of evidence in favour of its therapeutic utility.

The identification of the diseased cell type (i.e. the target for gene therapy) can then be used to develop the therapeutic strategy (Lowenstein, 1995). The importance and difficulties associated with the correct identification of the proper target cells is also seen in the development of cystic fibrosis gene therapy; it will certainly be a different experimental challenge to target the respiratory epithelium than to target the submucous pulmonary glands. It needs to be stressed that the therapeutic gene therapy strategy to be utilized will depend critically on many characteristics of the affected cell type to be transduced (e.g. its location, division rate, etc.) (Crystal, 1995).

What percentage of transgene expression (compared to normals), and in what percentage of diseased cells, needs to be achieved by a therapeutically effective gene therapy? *Table 11.2* shows the correlation between levels of enzyme activity and clinical phenotype in Lesch-Nyhan syndrome, an inherited neurological disorder. Note that this table provides the level of enzyme activity in *all* cells; thus patients are free of neurological symptoms if 100% of their neurons express 10% of hypoxanthine-guanine phosphoribosyl transferase (HPRT) enzyme activity. Gene transfer however, might only target a percentage of the total population of cells. In that case, our question becomes: what level of transgene expression in what percentage of target cells needs to be achieved, for normal brain function to be maintained or recovered? Thus, further pathophysiological considerations become a crucial concern; if eventual neuronal death depends on the intracellular accumulation of a toxic metabolite, *all* neurons that ought to be preserved will have to be successfully transduced, since only transduced cells will be able to keep the disease causing metabolite below the toxic level.

Transgenic and chimaeric animal technology experiments could provide relevant answers to questions relating to the percentage of target cells needing to be transduced and the level of transgene expression that will have to be achieved by gene therapy. Especially in the cases where mutations causing human disease have been identified, transgenic animal models can be generated using these mutations to construct transgenic lines expressing mutations conferring a wide range of gene product activities (Dorin *et al.*, 1995, 1996). Thus, the percentage of necessary enzyme activity or gene product can be assessed, and the minimum level of gene product that prevents the appearance of the pathological phenotype in transgenic animals *in vivo* can be accurately determined. This approach has now been used in the case of the cystic fibrosis animal model, in which transgenic lines expressing different levels of the cystic fibrosis transmembrane regulator (CFTR) have been generated (Dorin *et al.*, 1995, 1996). This allows a very accurate estimate *in vivo* of the minimum amount of CFTR activity required to prevent the expression of the pathologic phenotype. By using the transgenic lines expressing various levels of gene product to generate chimaeras of CFTR transgenic and normal mice, one could actually determine not only at what level the transgene needs to be expressed, but also in what percentage of cells it ought to be expressed *in vivo* (Porteus and Hastie, personal communication). Importantly, while several groups are now

Table 11.2. Phenotype of HPRT deficiency

Disease symptoms		HPRT activity % of control
Classical Lesch–Nyhan	Mental retardation, self-mutilation, choreoathetosis, hyperuricaemia	1.4
Intelligent Lesch–Nyhan	Self mutilation, choreoathetosis hyperuricaemia	1.4–1.6
Neurological Lesch–Nyhan	Choreoathetosis, hyperuricaemia	1.6–8
Neurologically normals	Hyperuricaemia	8–60
Neurologically normals	Normal uricaemia	≥60

trying to provide these answers, they have only been performed for a transgenic model of cystic fibrosis and animal models are not available as yet to guide clinical experiments in other fields.

For implementing clinical trials, we strongly support the notion that we do need to know the answers to questions ii-vi in advance, since in a number of diseases, the phenotype will depend exquisitely on how much gene product is actually expressed. An extreme case of this is the gene dosage effect observed in peripheral neuropathies (Charcot–Marie–Tooth disease type 1A) caused by duplications or point mutations of the *PMP-22* gene, in which the neurological phenotype varies dramatically depending on the expression of one, two, three or four copies of *PMP-22*. This can now be examined in transgenic animal models (Huxley *et al.*, 1996; Sereda *et al.*, 1996). Also, overexpression of one gene, even if not deleterious by itself, can deregulate the expression of other genes, as is the case for lysosomal enzymes; for example, overexpression of one lysosomal enzyme can lead to the cell being unable to retain other lysosomal enzymes due to the saturation of the lysosomal targeting pathway (Anson *et al.*, 1992).

The longevity of transgene expression required will depend on the disease: short-term expression will be sufficient for the treatment of brain tumours, while many decades of transgene expression will be needed for continuously preventing the ongoing neuronal degeneration in Huntington's disease. The stage of disease progression at which gene expression is needed can be answered from a careful examination of clinical data for each disease and the use of clinically relevant transgenic animal models. For example, age of onset of symptoms depends in many cases on the effects on gene activity produced by individual mutations; in many lysosomal inherited diseases age of onset depends on the percentage of remaining enzyme activity and, in inherited neurological diseases due to expansion of trinucleotide repeats (see *Table 11.1*) e.g. Huntington's disease, an inverse correlation can be seen between the age of onset and the number of repeats present in the affected gene. A recent transgenic mice model for Huntington's disease also supports such a correlation (Mangiarini *et al.*, 1996). However, neuronal degeneration can precede the onset of symptoms by many years, and thus gene therapy, ideally, ought to be attempted long before symptoms make their appearance. In certain cases, when the genetic diseases manifest in the perinatal period, we will have to evaluate the necessity for prenatal (*in utero*) gene delivery to completely prevent the disease from developing (Douar *et al.*, 1996; Pergament and Fiddler, 1995).

11.3 Gene therapy tools: their status circa June 1997

Only two general methods exist by which to transfer genes into the brain: either the gene of interest is inserted into a platform cell which is then transplanted into the brain; or a vector is used to transfer the gene directly into the constituent brain cells themselves. The first method achieves the transfer of the gene of interest into the brain, but not into the brain cells, while the second does so.

The first method is exemplified by the engineering of various cell types *in vitro* followed by transplantation (e.g. fibroblasts, tumour cell lines, myoblasts, neuronal precursor cells) (Fisher and Gage, 1993), and several recent in-depth reviews describe these methods (see Bibliography). An interesting variant of this general strategy is the transplantation of normal or *in vitro* engineered and transduced bone marrow (Stewart *et al.*, 1997; Walkey *et al.*, 1994, 1996). Microglial cells ('brain macrophages') derive from the bone marrow, and enter the brain continuously throughout life. Thus, the transplantation of genetically modified bone marrow amounts to the delivery of modified microglial cells to the brain. The transplantation of engineered cells can be used with two general aims: (i) to deliver an extracellularly active molecule (e.g. dopamine in animal models of Parkinson's disease, neuronal growth factors in animal models of neurodegeneration, or lysosomal enzymes which can be taken up by brain cells in metabolic brain disease) or (ii) to replace degenerated cells (e.g. through the transplantation of neuronal precursors) (Whittemore and Snyder, 1996).

Most gene transfer vectors are either viral, or non-viral. The first generation of chimaeric viral vectors have recently been described (Fischer *et al.*, 1996; Noguiez-Hellin *et al.*, 1996; Savard *et al.*, 1997). Viral vectors are derived from retrovirus, adenovirus, herpes simplex virus 1 (HSV-1), or adeno-associated virus, while the non-viral vectors are either simple plasmids, or more complex vectors such as the adenovirus–liposome complexes or liposomes themselves (see 'bibliography'). Retroviruses derived from murine retroviruses need at least one round of cell division to integrate their provirus into the target cells and express their transgenes, and their use is thus limited to the transduction of dividing cells in the brain, especially brain tumours. However, novel retroviral vectors based on HIV can effectively transduce non-proliferating, cell cycle-arrested cell lines *in vitro* (Lever, 1996; Lewis *et al.*, 1992), as well as terminally differentiated striatal and hippocampal neurons *in vivo* (Naldini *et al.*, 1996). Adeno-associated viruses apparently do so as well (During and Leone, 1996), but it remains to be formally shown whether they can actually integrate into the host cell genome after infecting non-dividing cells. Vectors derived from adenovirus, herpes simplex virus1, or adeno-associated virus will all infect non-dividing, as well as dividing cells.

Undoubtedly all current vectors have advantages and disadvantages. The ideal vector for gene transfer into the brain has yet to be developed. Those currently used, and some of their advantages for brain gene therapy, are shown in *Table 11.3*. Several recent books and reviews describe the methods for the preparation and use of viral vectors for gene transfer into the brain, and these are referenced in the bibliography at the end of this chapter (see Bibliography).

Even if many of the viruses used for gene transfer can potentially cause encephalitis very little is known on how these viruses interact with brain cells. There is a clear knowledge gap in our understanding of how these viruses inter-relate with the different cell types present in the brain. In the case of herpes simplex virus type 1 (HSV–1), it is still unclear how the virus moves throughout neurons and the central nervous system under natural conditions. While much effort has been devoted to disabling HSV-1 in order to convert it into a safer vector, and although such vectors can enter a state phenomenologically similar to latency (Kesari *et al.*, 1996), it is unclear whether any gene functions are actually required for the virus to enter latency and how the lack of these functions will affect the new viral vector.

Out of the approximately 80 genes identified in the HSV1 genome, the most thoroughly understood ones are those involved in viral gene expression, DNA replication and the generation of new viral particles

Table 11.3. A comparison between gene transfer methods of use in neuroscience

Vectors	Adenovirus	HSV-1/r	HSV-1/a	Adeno-associated virus	Retrovirus (murine and human derived)	Vaccinia virus	Semliki-Forest Virus	Micro-injection	Transfection
Size (kb)	36	152	5–30	4.68	7–10	185	11,51	10–30	10–30
Cloning capacity	7.5	10–30	5–25	4.5	7–8	30	9–20	10–30	10–30
Neuronal transduction									
In vivo?	yes	yes	yes	yes	yes	yes	yes	no	no
In vitro?	yes	yes	yes	yes	yes	yes	yes	yes	yes
Very long-term gene expression	?	in DRG hippo-campus?	yes?	yes	yes	no	no	yes	yes
Gene therapy?	yes	yes	yes	yes	yes	no	no	no	yes
Vaccination	yes	?	?	?	?	yes	yes	no	no

(Ward and Roizmann, 1994). However, the details of HSV-1 neuron interactions are still relatively poorly understood and the following questions still unanswered:

- How is the virion transported to the nucleus?
- How do virions move from cell body to dendrites when moving from one neuron to another?
- What aspects of the neuronal machinery are exploited by HSV-1 during neuronal infection?
- Is there any cell death during natural infection of target neuronal structures or viral reactivation?
- What can we learn from varicella zoster virus, a herpesvirus which expresses many genes during latency – as opposed to HSV-1 – even in the absence of reactivation?
- What determines the apparent selectivity for HSV-1 to infect neurons *in vivo*, while *in vitro* it has a preference for glial cells?
- What happens to cellular promoters inserted into HSV-1 vectors?
- Can either intraneuronal anterograde or retrograde axonal transport be exploited to achieve a distribution of the virus throughout the CNS?

The toxicity of these vectors towards neurons still needs to be fully addressed and understood at the molecular level. It becomes clear that furthering our understanding of the changes induced by herpes infection of neurons could greatly contribute to our exploitation of these viruses as vectors for gene therapy. For example, we have recently shown that infection of secretory cells with an adenoviral or HSV-1 derived vector can modify regulated neuropeptide secretion (Castro *et al.*, 1997; Tomasec *et al.*, 1997). Thus, HSV-1 infection can profoundly alter neuronal physiology. Furthering the understanding of such interactions will lead to better and safer vectors in the future (Efstathiou and Minson, 1995; Fink *et al.*, 1996; Lachmann and Efstathiou, 1997).

Similar challenges await to be met concerning the use of adenovirus vectors. Wild-type adenovirus is replication defective in rodents; thus, the behaviour of such vectors in rodents does not predict their behaviour in human cells. Although cotton rats are permissive to adenovirus infections, they are unfortunately only rarely used in work characterizing the use of adenovirus as gene transfer vectors. Also, adenovirus administration generates an important immune response even after its administration to the brain (Byrnes *et al.*, 1995, 1996; Yang *et al.*, 1996a,b). Which adenoviral antigens induce the cytotoxic T-cell response, still remains to be clarified.

Minimal transgene expression can still be detected two months after adenovirus administration to the brain in spite of the immune response (Byrnes *et al.*, 1995). Downregulation of the immune response in experimental animals allows for multiple succesive administrations of adenovirus vectors (Yang *et al.*, 1995). Otherwise, no transgene expression is seen upon a second administration of adenovirus vectors. It is still unclear whether any of these methods could be used in humans, in some of whom transgene expression will have to be achieved over many decades. It has also been proposed that the use of adenovirus in neonate animals could render animals tolerant to these viruses, and thus allow for very long-term transgene expression (Vincent *et al.*, 1993). The modification of co-stimulatory molecules can also increase the longevity of transgene expression (Kay *et al.*, 1995). The full potential of these techniques and the full extent of transgene expression remain to be determined. These experiments have yet to be performed after the administration of adenovirus into the nervous system.

11.4 How to achieve cell type-specific expression

The targeting of viral vectors to predetermined cell-types constitutes an active area of research (Cosset and Russel, 1996; Harris and Lemoine, 1996; Hart and Vile, 1995). Many of the normal processes involved in viral entry, expression, intracellular transport, etc., could potentially be exploited to target expression to certain cell types only. Several groups have reported cell type-specific expression in transgenic animals using specific neuronal promoters, and cell type-specific expression using neuronal or viral promoters has now been achieved using viral vectors (Lowenstein *et al.*, 1995; Oh *et al.*, 1996; Robert and Mallet, personal communication; Shering *et al.*, 1997). However, recent progress in this area indicates cell type-specific expression can be achieved in various tissues with different viral vectors (Brown *et al.*, 1997; Cosset and Russel, 1996; Douglas *et al.*, 1996; Hart and Vile, 1995; Lan *et al.*, 1996; Oh *et al.*, 1996; Rothmann *et al.*, 1996; Shering *et al.*, 1997; Vile *et al.*, 1996). In particular, HSV-1 contains a number of very strong and promiscuous transcriptional activators, and these could modify the pattern of transcription obtained from a cell-specific promoter (Shering *et al.*, 1997). Our group has recently shown that the pattern of expression from a 'short' major immediate early human cytomegalovirus promoter (MIEhCMV) is dependent on the viral background in which it is inserted (Lowenstein *et al.*, 1995; Shering *et al.*, 1997). We found that such a promoter had a cell type-specific pattern of activity when inserted into an adenovirus vector (it is active in glial cells, but not in neocortical neurons), which was not apparent when the same promoter-transgene cassette was inserted into a variety of herpes simplex 1 defective mutants (it became active in both, glial cells and neocortical neurons). Similar lack of expression from a recombinant adenovirus vector containing the MIEhCMV promoter was recently confirmed by Di Polo *et al.*, 1996. Following retinal injection of adenoviral vectors, reporter gene expression in retinal neurons was found from the Rous sarcoma virus long terminal repeat but not from the MIEhCMV promoter (Di Polo *et al.*, 1996).

Peripheral administration of viral vectors to achieve transgene expression in the brain also remains a rather distant goal due to the existence of a very effective blood–brain barrier. It has been reported that osmotic shock opening of the blood–brain barrier allows for virions to infect brain tumours (Doran *et al.*, 1995; Nilaver *et al.*, 1995); also an alternative approach of transferring genes through the blood–brain barrier or the peripheral blood–nerve barrier has recently been demonstrated by intravenous administration of retrovirally transduced nerve growth factor-expressing nerve-specific autoimmune T-lymphocytes in an animal model of Guillain-Barre polyneuritis (Kramer *et al.*, 1995). How efficient these techniques will be to deliver vectors and transgenes to human brain remains to be determined.

Interactions between the viral vectors and the extracellular matrix in the brain is another area where more work is needed. The large area of brain tissue to be potentially transduced in any disease remains an important limitation for the development of clinically relevant neurological gene therapy. Understanding how viruses interact with the extracellular matrix might lead to ways in which to improve the diffusion of viral vectors throughout the brain after direct intra-parenchymal injections.

We envisage that the two clinical areas that will most rapidly grow within the field of gene therapy for neurological disorders within the next 5 years comprise: (i) the treatment of brain tumours, and (ii) all those diseases which would be amenable to improvements through the use of bone marrow transplantation (Krivit *et al.*, 1995) [e.g. leukodystrophies (see *Tables 11.4* and *11.5*)]. The attraction afforded by the treatment of brain tumours by gene therapy are the relatively small size of the tumour, the obviousness of the therapeutic objective (i.e. to destroy all tumour cells), and the clearcut therapeutic endpoint (tumour ablation and patient survival), which are both easy to assess clinically. Approaches for gene therapy of brain tumours, include the targeting of

Table 11.4. Total number of current gene therapeutic clinical trials for neurological diseases

Neurological disease	Therapeutic approach	Number of clinical trials
Brain tumors	Growth factor antisense	2
	Immunostimulators	4
	Conditionally cytotoxic genes	11
Metabolic brain disorders	Gene replacement using viral vectors	1
	Bone marrow transplantation	
	-autologous (genetically engineered)	In preparation
	-heterologous	In clinical use for > 10 years
Neurodegenerative disorders	Amyotrophic lateral sclerosis (using ciliary neurotrophic factor delivery)	1

endothelial cells of the tumour neovasculature by interruption of the paracrine growth factor pathways responsible for the neovascularization of solid tumours (Saleh *et al.*, 1996), the use viral vectors expressing cytotoxic or conditionally cytotoxic genes (Kramm *et al.*, 1995) or the manipulation of the cell-cycle regulatory cyclin-dependent kinase inhibitors (Chen *et al.*,1996).

11.5 Conclusions

As the molecular understanding of many brain disorders continues to advance, so does the knowledge concerning the transfer of genes into the post-mitotic cells which constitute most of the brain. While the field is moving towards the implementation phase, clinically acceptable strategies continue to be proposed. Viable and ethically acceptable proposals for the treatment of childhood inherited metabolic brain disorders, brain tumours and neurodegenerative diseases in adults have already been started (see *Tables 11.4* and *11.5*). As vectors continue to be developed, it is difficult to restrain the feeling of excitement when imagining the realisation of what only 5 years ago appeared as a dream. It is very clear that the war is not yet over but many battles have already been won.

Table 11.5. Sporadic and inherited CNS disorders which display major neuropsychiatric symptoms, and their possible treatment by gene therapy

Disease	Non-inherited		Inherited			
	Tumours	PD	HD	Frax-A	Leukodystrophies	CMT-1A
Outcome	Fatal	Fatal	Fatal	Non-fatal	Fatal	Non-fatal
Distribution	Focal	Focal[a]	Diffuse	Diffuse	Diffuse	Focal
Age at manifestation	Variable	Late	Variable	Prenatal	Early	Variable
Presymptomatic diagnosis	No	No	Yes[e]	Yes[b]	Yes	Yes
Genotype predicts disease severity	Not determined	Not determined	Yes	No	Yes / No[f]	Yes
Gene cloned	Yes[c]	Yes[d]	Yes	Yes	Yes	Yes
Expression needed	Short-term	Long-term	Long-term	Long-term	Long-term	Long-term
Experimental gene therapy strategy	Kill tumour cells	Dopamine replacement	Reduce expression of expanded allele	Express FMR1 in brain	Bone marrow (BM) transplantation; BM expressing mutated genes	Normalizing *PMP-22* expression
Clinically relevant available therapeutic protocols	Yes	Yes	No	No	Yes	No

Key: CMT-1A, Charcot–Marie–Tooth disease (the commonest genetic alteration is duplication of the region of chromosome 17p11.2 containing the *PMP-22* gene or point mutation of the *PMP-22*); Frax-A, Fragile X syndrome type A; HD, Huntington's disease; PD, Parkinson's disease (idiopathic).

[a]The primary pathology is loss of dopaminergic neurons from the *pars compacta* of the *substantia nigra*, but other lesions within PD are distributed throughout the brain.

[b]Molecular genetic diagnosis is possible in the fetus, but it is arguable whether this would be prior to pathological changes having occurred.

[c]The genes mutated in inherited brain tumours have been cloned. However, therapeutic gene therapy is more likely to utilize the transduction of cytotoxic genes to eliminate tumour cells.

[d]The genes for tyrosine hydroxylase and DOPA decarboxylase have been cloned. These are not however mutated in either familial or sporadic PD.

[e]Even if presymptomatic detection is available, it is not done routinely on every pregnancy; it is performed only if there is any suspicion due to a positive family history

[f]Genotype predicts disease severity in metachromatic and Krabbe's leukodystrophy, but not yet in adrenoleukodystrophy.

Acknowledgements

Tables 11.2 and 11.5 have been modified from MacMillan and Lowenstein (1996), and Table 11.3 has been modified from Lowenstein *et al.* (1996), with permission. Work on gene therapy in our laboratory is funded by The Parkinson's Disease Society, The Cancer Research Campaign, UK, The Lister Institute of Preventive Medicine, The Wellcome Trust, The North West Regional Health Authority, The Muscular Dystrophy Group (UK), The Royal Society, The Sandoz Foundation for Gerontological Research, and The Sir Halley Stewart Trust. PRL is a Research Fellow of The Lister Institute of Preventive Medicine.

References

Anson DS, Bielicki J and Hopwood JJ. (1992) Correction of mucopolysaccharidosis type I fibroblasts by retroviral mediated transfer of the human α-L-iduronidase gene. *Hum. Gene Ther.* **3**, 371–379.

Banfi S, Servadio A, Chung MY *et al.* (1994) Identification and characterization of the gene causing type 1 spinocerebellar ataxia. *Nature Genet.* **7**, 513–520.

Barlow C, Hirotsune S, Paylor R *et al.* (1996) *Atm*-deficient mice: a paradigm of ataxia teleangiectasia. *Cell,* **86**, 159–171.

Bingham PM, Scott MO, Wang S, McPhaul MJ, Wilson EM, Garbern JY, Merry DE and Fischbeck KH. (1995) Stability of an expanded trinucleotide repeat in the androgen receptor gene in transgenic mice. *Nature Genet* **9**, 191–196.

Blaese RM, Culver KW, Miller AD *et al.* (1995) T lymphocyte-directed gene therapy for ADA-SCID: initial trial results after 4 years. *Science* **270**, 475–480.

Bordignon C, Notarangelo LD, Nobili N *et al.* (1995) Gene therapy in peripheral blood lymphocytes and bone marrow for ADA-immunodeficient patients. *Science* **270**, 470–475.

Brown AB, Santer RM, Shering A, Larregina AT, Morelli AE, Southgate TD, Castro MG and Lowenstein PR. (1997) Gene transfer into enteric neurons of the rat small intestine in organ culture using a replication defective recombinant herpes simplex virus type 1 (HSV-1) vector, but not recombinant adenovirus vectors. *Gene Ther.* **4**, 331–338.

Burright EN, Clark HB, Servadio A, Matilla T, Feddersen RM, Yunis WS, Duvick LA, Zoghbi HY and Orr HT. (1995) SCA1 transgenic mice: a model for neurodegeneration caused by an expanded CAG trinucleotide repeat. *Cell* **82**, 937–948.

Byrnes AP, Rusby JE, Wood MJA and Charlton HM. (1995) Adenovirus gene-transfer causes inflammation in the brain. *Neuroscience,* **66**, 1015–1024.

Byrnes AP, MacLaren RE and Charlton HM. (1996) Immunological instability of persistent adenovirus vectors in the brain: peripheral exposure to vector leads to renewed inflammation, reduced gene expression and demyelination. *J. Neurosci.* **16**, 3045–3055.

Campuzano V, Montermini L, Molto MD *et al.* (1996) Friedreich's ataxia: autosomal recessive disease caused by an intronic GAA triplet repeat expansion. *Science,* **271**, 1423–1427.

Castro MG, Goya RG, Sosa YE, Rowe J, Larregina A, Morelli A and Lowenstein PR. (1997) Expression of transgenes in normal and neoplastic anterior pituitary cells using recombinant adenoviruses: long term expression, cell cycle dependency and effects on hormone secretion. *Endocrinology* **138**, 2184–2194.

Chen J, Willingham T Shuford M and Nisen PD. (1996) Tumor suppression and inhibition of aneuploid cell accumulation in human brain tumor cells by ectopic overexpression of the cyclin-dependent kinase inhibitor p27KIP1. *J. Clin. Invest.* **97**, 1983–1988.

Cosset FL and Russel SJ. (1996) Targeting retrovirus entry. *Gene Ther.* **3**, 946–956.

Crystal RG. (1995) Transfer of genes to humans: early lessons and obstacles to success. *Science* **270**, 404–410.

Culver KW. (1994) *Gene Therapy.* Mary Ann Liebert Inc., New York.

Di Polo A. Bray GM and Aguayo AJ. (1996) Gene expression in the adult rat retina mediated by different adenovirus vectors. *Soc. Neurosci. Abstr.* **22**, 1724.

Doran SE, Ren XD, Betz AL, Pagel MA, Neuwelt EA, Roessler BJ and Davidson BL. (1995) Gene expression from recombinant viral vectors in the central nervous system after blood brain barrier disruption. *Neurosurgery,* **36**, 965–970.

Dorin JR, Webb S, Farini E, Delaney S, Wainwright B, Smith S, Farley R, Alton WSW and Porteus DJ. (1995) Phenotypic consequence of CFTR modulation in mutant mice; implications for somatic gene therapy. *Pediat. Pulmonol.* (Supplement) **12**, 231.

Dorin JR, Farley R, Webb S, Smith SN, Farini E, Delaney SJ, Wainwright BJ, Alton EWFW and Porteous DJ. (1996) A demonstration using mouse models that successful gene therapy for cystic fibrosis requires only partial gene correction. *Gene Ther.* **3**, 797–801.

Douar A-M, Themis M and Coutelle C. (1996) Fetal somatic gene therapy. *Mol. Hum. Reprod.* **2**, 633–641.

Douglas JT, Rogers BE, Rosenfeld ME, Michael SI, Meizhen F and Curial DT. (1996) Targeted gene delivery by tropism-modified adenoviral vectors. *Nature Biotechnol.* **14**, 1574–1578.

During MJ and Leone P. (1996) Adeno-associated virus vectors for gene therapy of neurodegenerative disorders. *Clin. Neurosci.* **3**, 292–300.

Dutch–Belgian Fragile X Consortium (1994) Fmr1 knockout mice: a model to study fragile X mental retardation. *Cell* **78**, 23–33.

Efstathiou S and Minson AC. (1995) Herpes virus based vectors. *Br. Med. Bull.* **51**, 45–55.

Fink DJ, DeLuca N, Goins WF and Glorioso JC. (1996) Gene transfer to neurons using herpes simplex virus based vectors. *Annu. Rev. Neurosci.* **19**, 265–287.

Fisher KJ, Kelley WM, Burda JF and Wilson JM. (1996) A novel adenovirus–adeno-associated virus hybrid vector that displays efficient rescue and delivery of the AAV genome. *Hum. Gene Ther.* **7**, 2079–2087.

Fisher LJ and Gage FH. (1993) Grafting in the mammalian central nervous system. *Physiol. Rev.* **73**, 583–616.

Friedmann T. (1983) *Gene Therapy, Fact and Fiction.* Cold Spring Harbor Laboratory Press, New York.

Friedmann T. (1994) Editorial: the promise and overpromise of human gene therapy. *Gene Ther.* **1,** 217–218.

Friedmann T and Roblin R. (1972) Gene therapy for human genetic disease? *Science* **175**, 949–955.

Gecz J, Gedeon AK, Sutherland GR and Mulley JC. (1996) Identification of the gene FMR2, associated with FRAXE mental retardation. *Nat. Genet.* **13**, 105–108.

Gu Y, Shen Y, Gibbs RA and Nelson DL. (1996) Identification of FMR2, a novel gene associated with the FRAXE CCG repeat and CpG island. *Nature Genet.* **13**, 109–113.

Harris JD and Lemoine NR. (1996) Strategies for targeted gene therapy. *Trends Genet.* **12**, 400–405.

Hart IR and Vile RG. (1995) Targeted gene therapy. *Br. Med. Bull.* **51**, 647–655.

Hsiao K, Chapman P, Nilsen S, Eckman C, Harigaya Y, Younkin S, Yang FS and Cole G. (1996) Correlative memory deficits, Aβ elevation, and amyloid plaques in transgenic mice. *Science,* **274**, 99–102.

Huntington's Disease Collaborative Research Group (1993) A novel gene containing a trinucleotide repeat that is expanded and unstable on Huntington's disease chromosomes. *Cell* **72**, 971–983.

Huxley C, Passage E, Manson A, Putzu G, Figarella-Branger D, Pellissier JF and Fontes M. (1996) Construction of a mouse model of Charcot–Marie–Tooth disease type 1A by pronuclear injection of human YAC DNA. *Hum. Mol. Genet.* **5**, 563–569.

Ikeda H, Yamaguchi M, Sugai S, Aze Y, Narumiya S and Kakizuka A. (1996) Expanded polyglutamine in the Machado–Joseph disease protein induces cell death *in vitro* and *in vivo*. *Nature Genet.* **13**, 196–202.

Kawaguchi Y, Okamoto T, Tanikawi M *et al.* (1994) CAG expansions in a novel gene for Machado-Joseph disease at chromosome 14q32.1. *Nature Genet.* **8**, 221–228.

Kay MA, Holterman AX, Meuse L, Gown A, Ochs HD, Linsley PS and Wilson CB. (1995) Long-term hepatic adenovirus mediated gene expression in mice following CTLA4LG administration. *Nature Genet.* **11**, 191–197.

Kesari S, Lee VM, Brown SM, Trojanowski JQ and Fraser NW. (1996) Selective vulnerability of mouse CNS neurons to latent infection with a neuroattenuated herpes simplex virus 1. *J. Neurosci.* **16**, 5644–5653.

Koide R, Ikeuchi T, Onodera O *et al.* (1994) Unstable expansion of CAG repeat in hereditary dentatorubral-pallidoluysian atrophy (DRPLA). *Nature Genet.* **6**, 9–13.

Kramer R, Zhang Y, Gehrmann J, Gold R, Thoenen H and Wekerle H. (1995) Gene transfer through the blood–nerve barrier: NGF-engineered neuritogenic T lymphocytes attenuate experimental autoimmune neuritis. *Nature Med.* **1**, 1162–1166.

Kramm CM, Senaesteves M, Barnett FH *et al.* (1995) Gene therapy for brain tumors. *Brain Pathol.* **5**, 345–381.

Krivit W, Lockman LA, Watkins PA, Hirsch J and Shaprio EG. (1995) The future for treatment by bone marrow transplantation for adrenoleukodystrophy, metachromatic leukodystrophy, globoid cell leukodystrophy and Hurler syndrome. *J. Inherited Metab. Dis.* **18**, 398–412.

Lachmann RH and Efstathiou S. (1997) Utilization of the herpes simplex virus type 1 latency-associated regulatory region to drive stable reporter gene expression in the nervous system. *J. Virol.* **71**, 3197–3207.

Lan KH, Kanai F, Shiratori Y *et al.* (1996) Tumor specific gene expression in carcinoembryonic antigen producing gastric cancer cells using adenovirus vectors. *Gastroenterology* **111**, 1241–1251.

La Spada AR, Wilson EM, Lubahn DB, Harding AE and Fischbeck KH. (1991) Androgen receptor gene mutations in X-linked spinal and bulbar muscular atrophy. *Nature* **352**, 77–79.

Lever AML. (1996) HIV and other lentivirus-based vectors. *Gene Ther.* **3**, 470–471.

Lewis PF, Hensel M and Emerman M. (1992) Human immunodeficiency virus infection of cells arrested in the cell cycle. *EMBO J.* **11**, 3053–3058.

Lowenstein PR. (1995) Degenerative and inherited neurological disorders. In: *Molecular and Cell Biology of Human Gene Therapeutics* (ed. G Dickson), pp. 300–349. Chapman and Hall, London.

Lowenstein PR, Bain D, Shering AF, Wilkinson GWG and Castro M. (1995) Cell type specific expression from viral promoters within replication-deficient adenovirus recombinants in primary neocortical cultures. *Rest. Neurol. Neurosci.* **8**, 37–39.

Lowenstein PR, Wilkinson GWG, Castro MG, Shering AF, Fooks AR and Bain D. (1996) Non-neurotropic adenovirus: a vector for gene transfer to the brain and possible gene therapy of neurological disorders. In: *Genetic Manipulation of the Nervous System* (ed. DS Latchman), pp. 11–39. Academic Press, London.

Lyon J and Gorner P. (1995) *Altered Fates: The Re-engineering of Human Life.* W.W. Norton, New York.

MacMillan JC and Lowenstein PR. (1996) The strategic development of gene therapy approaches to the treatment of human neurological disease. The future of neurological therapies? In: *Protocols for Gene Transfer in Neuroscience. Towards Gene Therapy of Neurological Disorders* (eds PR Lowenstein and LW Enquist), p. 456. John Wiley and Sons, Chichester.

Mangiarini L, Sathasivam K, Seller M *et al.* (1996) Exon-1 of the HD gene with an expanded CAG repeat is sufficient to cause a progressive neurological phenotype in transgenic mice. *Cell* **87**, 493–506.

Naldini L, Blomer U, Gallay P, Ory D, Mulligan R, Gage FH, Verma IM and Trono D. (1996) *In vivo* gene delivery and stable transduction of nondividing cells by a lentiviral vector. *Science* **272**, 263–267.

Nilaver G, Muldoon LL, Kroll RA, Pagel MA, Breakefield XO, Davidson BL and Neuwelt EA. (1995) Delivery of herpesvirus and adenovirus to nude rat intracerebral tumors after osmotic blood brain barrier disruption. *Proc. Natl Acad. Sci. USA* **92**, 9829–9833.

Noguiez-Hellin P, Robertlemmeur M, Salzmann JL and Klatzmann D. (1996) Plasmoviruses – nonviral vectors for gene therapy. *Proc. Natl Acad. Sci. USA* **93**, 4175–4180.

Oh YJ, Moffat M, Wong S, Ullrey D, Geller AI and O'Malley KL. (1996) A herpes simplex virus-1 vector containing the rat tyrosine hydroxylase promoter directs cell type-specific expression of beta-galactosidase in cultured rat peripheral neurons. *Mol. Brain Res.* **35**, 227–236.

Pergament E and Fiddler M. (1995) Prenatal gene therapy: prospects and issues. *Prenat. Diagn.* **15**, 1303–1311.

Rothmann T, Katus HA, Hartong R, Perricaudet M and Franz WM. (1996) Heart muscle specific gene expression using replication defective recombinant adenovirus. *Gene Ther.* **3**, 919–926.

Saleh M, Stacker SA and Wilks AF. (1996) Inhibition of growth of C6 glioma cells *in vivo* by expression of antisense vascular endothelial growth factor sequence. *Cancer Res.* **56**, 393–401.

Savard N, Cosset F-L and Epstein A. (1997) Use of defective HSV-1 vectors harbouring GAG, POL and ENV genes to rescue defective retrovirus vectors. *J. Virol.* **71**, 4111–4117.

Sereda M, Griffiths I, Puhlhofer A *et al.* (1996) A transgenic rat model for Charcot–Marie–Tooth disease. *Neuron* **16**, 1049–1060.

Shering AF, Bain D, Castro MG, Wilkinson GWG and Lowenstein PR. (1997) Cell-type specific expression in brain cell cultures from a short human cytomegalovirus major immediate early promoter depends on whether it is inserted into herpesvirus or adenovirus vectors. *J. Gen. Virol.* **78**, 445–459.

Stewart K, Brown O, Morelli A *et al.* (1997) Uptake of α-(L)-iduronidase produced by retrovirally transduced fibroblasts into neuronal and glial cells *in vitro*. *Gene Ther.* **4**, 63–75.

Thompson L. (1994) *Correcting the Code: Inventing the Genetic Cure for the Human Body.* Lyon and Gorner, New York.

Tomasec P, Rowe J, Preston CM *et al.* (1997) Herpes simplex virus type 1 infection of secretory cells blocks regulated release of peptide hormones. *J. Neurosci.* (In press).

Verkerk AJ, Pieretti M, Sutcliffe JS *et al.* (1991) Identification of a gene (FMR-1) containing a CGG repeat coincident with a breakpoint cluster region exhibiting length variation in fragile X syndrome. *Cell* **65**, 905–914.

Vile RG, Diaz RM, Miller N, Mitchell S, Tuszanski A and Russel SJ. (1996) Tissue specific gene expression from MoMLV vectors with hybrid LTRs containing the murine tyrosinase enhancer promoter. *Virology* **214**, 307–313.

Vincent N, Ragot T, Gilgenkrantz H *et al.* (1993) Long term correction of mouse dystrophic degeneration by adenovirus mediated transfer of a minidystrophin gene. *Nature Genet.* **5**, 130–134.

Vincent MC, Trapnell BC, Baugham RP, Wert SE, Whitsett JA and Iwamoto HS. (1995) Adenovirus-mediated gene transfer to the respiratory tract of fetal sheep in utero. *Hum. Gene Ther* **6**, 1019–1028.

Walkley SU, Thrall MA, Dobrenis K et al. (1994) Bone marrow transplantation corrects the enzyme defect in neurons of the central nervous system in a lysosomal storage disease. *Proc. Natl Acad. Sci. USA* **91**, 2970–2974.

Walkley SU, Thrall MA and Dobrenis K. (1996) Targeting gene products to the brain and neurons using bone marrow transplantation: a cell-mediated delivery system for therapy of inherited metabolic human disease. In: *Protocols for Gene Transfer in Neuroscience. Towards Gene Therapy of Neurological Disorders* (eds PR Lowenstein and LW Enquist), pp. 275–302. John Wiley and Sons, Chichester.

Ward PL and Roizman B. (1994) Herpes simplex genes. The blueprint of a successful human pathogen. *Trends Genet.* **10**, 380–385.

Whittemore SR and Snyder EY. (1996) Physiological relevance and functional potential of central nervous system-derived cell lines. *Mol. Neurobiol.* **12**, 13–38.

Wivel NA and Walters L. (1993) Germ line specific modification and disease prevention: some medical and ethical perspectives. *Science* **262**, 533–538.

Wolff JA and Lederberg J. (1994) An early history of gene transfer and therapy. *Hum. Gene Ther.* **5**, 469–480.

Yang Y, Trinchieri G and Wilson JM. (1995) Recombinant IL-12 prevents formation of blocking IgA antibodies to recombinant adenovirus and allows repeated gene therapy to mouse lung. *Nature Med.* **1**, 890–893.

Yang YP, Haecker SE, Su Q and Wilson JM. (1996a) Immunology of gene therapy with adenoviral vectors in mouse skeletal muscle. *Hum. Mol. Ther.* **5**, 1703–1712.

Yang YP, Su Q and Wilson JM. (1996b) Role of viral antigens in destructive cellular immune responses to adenovirus vector-transduced cells in mouse lungs. *J. Virol.* **70**, 7209–7212.

Bibliography

Detailed in-depth reviews from the theory to practice. Please also check the books quoted in the references.

Dickson G. (ed.) (1995) *Molecular and Cell Biology of Human Gene Therapeutics.* Chapman and Hall, London.

Dunnet SB and Bjorklund AT. (1994) *Functional Neural Transplantation.* Raven Press, New York.

Fields BN. (1996) *Fundamental Virology,* 3rd Edn. Raven Press, New York.

Gage F and Christen Y. (eds) (1992) *Gene Transfer and Therapy in the Nervous System.* Springer Verlag, Berlin.

Kaplitt MG and Loewy AD. (1995) *Viral Vectors: Gene Therapy and Neuroscience Applications.* Academic Press, San Diego.

Latchman D. (ed.) (1996) *Genetic Manipulation of the Nervous System.* Academic Press, London.

Lowenstein PR and Enquist LW. (1996) *Protocols for Gene Transfer in Neuroscience: Towards Gene Therapy of Neurological Disorders.* John Wiley and Sons, Chichester.

Roth MG. (1994) *Protein Expression in Animal Cells. Methods in Cell Biology,* Vol. 43. Academic Press, San Diego.

Rothwell NJ. (ed.) (1995) *Immune Responses in the Nervous System.* BIOS Scientific Publishers, Oxford.

Vos JMH. (ed.) (1995) *Viruses in Human Gene Therapy.* Chapman and Hall, London.

Wolff JA. (ed.) (1994) *Gene Therapeutics: Methods and Applications of Direct Gene Transfer.* Birkhauser, Boston.

12

Impact of genomics on the discovery and development of modern medicines

David S. Bailey and Geoffrey I. Johnston

12.1 Introduction

The deployment of genomics within pharmaceutical research and development is markedly influencing both the efficiency and focus of the drug discovery and development process. The molecular biology of target genes, including studies of their structure, sequence variation and their patterns of expression in specific tissues, diseases, patient populations and drug treatment regimes, provides radically new benchmarks both for the discovery of new drugs and for optimizing their efficacy, quality and potential utility.

Above all, genomics provides a new opportunity for diversifying treatment of important chronic and acute disease through the identification at the molecular level of multiple new disease targets, the use of which, in drug discovery, promises a much broader portfolio of drugs than is evident in clinical practice today, where a relatively few drugs dominate prescriptions: currently, drugs targeting only six different molecular mechanisms account for nearly 50% of the sales of the leading prescription drugs worldwide [*Table 12.1,* summarized from *Pharma Business* (July–August), 1996].

In this chapter, we describe the collision between these new genomic methodologies and the more conventional ways of discovering,

Genetics of Common Diseases: future therapeutic and diagnostic possibilities,
edited by I. Day and S. Humphries.

developing and testing drugs, and point to ways in which both drug discovery and prescription are likely to be enhanced.

Table 12.1. Target mechanisms of the top-selling drugs in 1995

Therapy area	Target mechanism	Gene family	Drugs	Sales (million $)
Gastric ulcers	Histamine H2 antagonism	G-protein coupled receptor	Zantac, Tagamet Pepcid, Axid	5464
Hypertension	ACE inhibition	Metalloproteinase	Vasotec, Capoten, Zestril	4771
Lipid-lowering	HMG CoA reductase inhibition	Reductase	Zocor, Mevacor, Pravachol	3985
Depression	5HT transport inhibition	Ion channel	Prozac, Zoloft, Paxil	3837
Hypertension/ angina	Ca channel blockade	Ion channel	Norvasc, Adalat, Procardia	3658
Gastric ulcers	H^+/K^+ ATPase inhibition	Phosphatase	Losec	3009

12.2 The traditional route to pharmaceutical discovery

Pharmaceutical discovery ultimately relies on establishing a set of pharmacological screens with which to profile novel chemical 'leads'. Such leads are early-stage agents which then become 'candidates' for clinical development upon optimization through medicinal chemistry. In this context, we will reserve the term 'drug' for chemical agents which are in clinical trials.

Traditional pharmacological screens have ranged from isolated preparations measuring, for example, muscle contraction, to *in vivo* studies of the regulation of mammalian blood pressure. The emphasis has been placed at a physiological – or whole tissue – level, enabling multiple drug features such as tissue bioavailability and even some elements of pharmacokinetic behaviour, to be assessed, in addition to functional potency at the drug target. This process of drug discovery, based on observation and measurement of functional responses, summarized in *Figure 12.1*, has been very successful, with highly effective drugs such as beta-blockers, H2-antagonists and calcium channel blockers all emanating from a similar discovery paradigm (for examples see *Table 12.1* and Sneader, 1985).

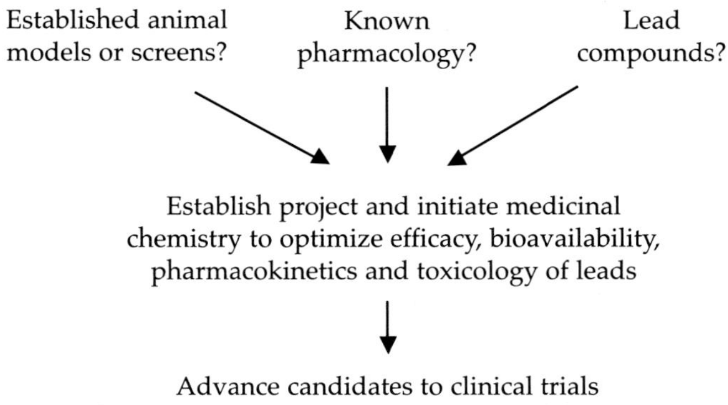

Figure 12.1 The traditional drug discovery process.

The process is chemistry-driven: projects depend upon the availability of leads with clear functional effects before medicinal chemistry can be initiated. Although each drug discovery campaign is unique, lead / candidate-seeking periods of up to 10 years (or even longer) are not uncommon, which, coupled to similarly lengthy drug development times, indicates just how time-consuming drug discovery can be. A typical time-frame under which these programmes operate is illustrated with reference to the Pfizer drug, Fluconazole (Richardson *et al.*, 1985), in *Figure 12.2.*

Not surprisingly, considerable efforts are being expended by the pharmaceutical industry to reduce these time-frames, both through the deployment of new combinatorial chemical methodologies to improve

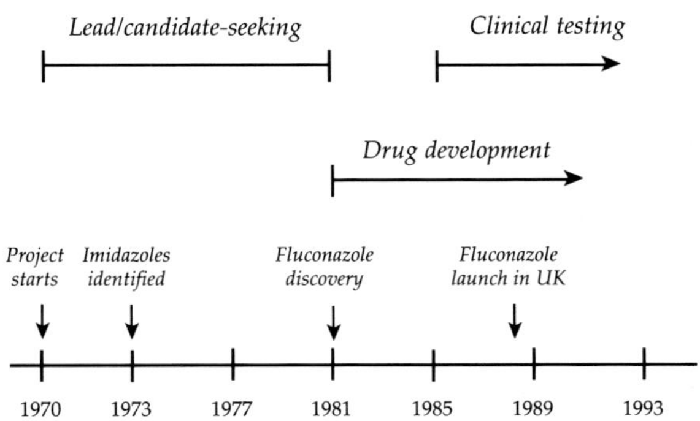

Figure 12.2. The discovery and development of Fluconazole, an antifungal azole.

lead discovery and optimization, and, more relevant for this review, the use of genomics to support both target discovery and candidate characterization.

12.3 Supporting the discovery process through genomics

There are many ways of defining the drug discovery process: one way is shown in *Figure 12.3*, where it is represented as a series of inter-linked cycles, each of which can contribute to the process as a whole.

The whole process of drug discovery and development, as we have seen, takes decades, and costs hundreds of millions of dollars, factors which have governed the rapid development of any technology which promises to optimize decision-making during the process. As we shall see below, the deployment of genomics can be highly influential in the decision-making process, from the earliest stages of target discovery through to the final trials of drugs in the clinic, providing as it does new techniques for identifying and selecting drug mechanisms, and measuring the comparative efficacy and quality of the ultimate products.

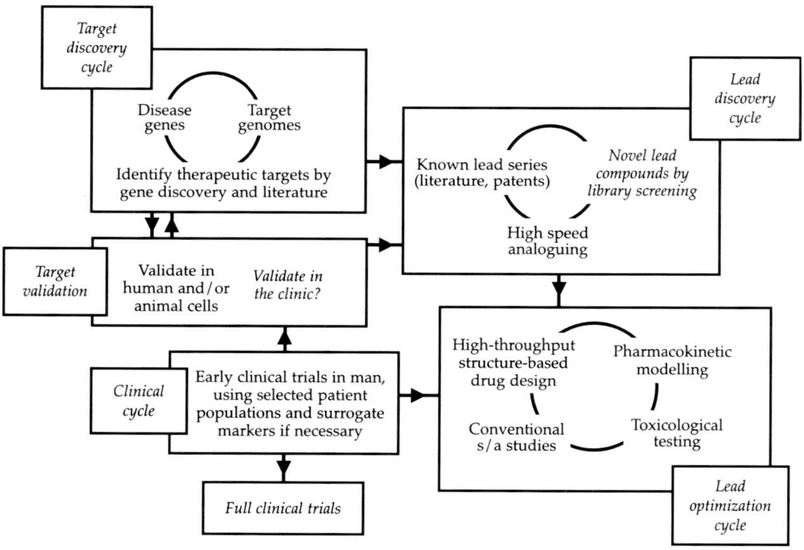

Figure 12.3. Drug discovery cycles.

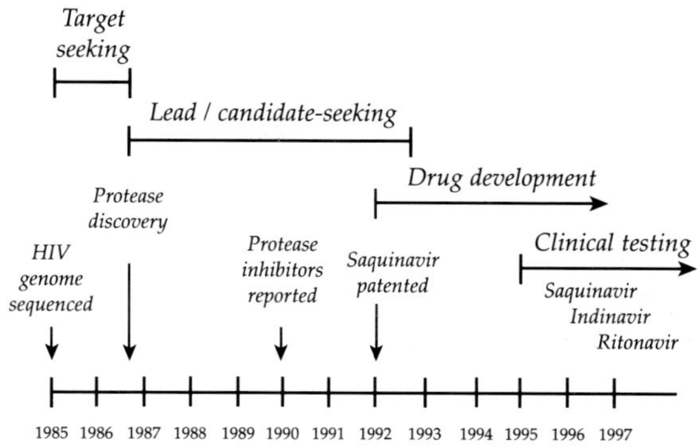

Figure 12.4. The discovery and development of HIV protease inhibitors.

12.3.1 The target discovery cycle

Before the advent of genomics, the identification of suitable biological targets from which to mount a drug discovery project was largely based on pharmacological feasibility. There has been a radical change in the last decade during which more emphasis has been placed on biochemical and molecular mechanisms for target discovery.

This is nowhere more powerfully illustrated than in the search for new therapeutic approaches to AIDS, where a particular class of agent, the HIV protease inhibitors, are proving particularly exciting. Chronological progress in the identification and progression of HIV protease as a drug target in AIDS is shown in *Figure 12.4*. For the first time, the availability of a complete pathogen sequence, in this case that of the HIV virus, has allowed the rational identification of new therapeutic approaches to HIV control.

In the case of HIV, the time taken for this process of genomics-based target discovery was dramatically short, with the first reports of an assay for the novel target HIV protease (Debouck *et al*., 1987; Kramer *et al*., 1986) appearing less than 18 months after the publication of the HIV-1 genomic sequence (Ratner *et al*., 1985). The availability of screens for protease inhibitors has resulted in the clinical development of a number of protease inhibitors, the first of these, Roche's product, Saquinavir, being patented in 1989 (Craig *et al*., 1991; Martin and Redshaw, 1989; Roberts *et al*., 1990).

With only nine open reading frames, potential drug targets for HIV were relatively easy to identify, although few of these gene products have

Table 12.2. A selection of genomes

(a) Viruses		
Bacteriophage ΦX174	5.4 kb (9 ORFS)	Published: Sanger, 1977
HIV-1	9.7 kb (9 ORFS)	Published: Ratner *et al.*, 1985; Miller and Sarver, 1997
HSV-1	152 kb (75 ORFS)	Published: McGeoch *et al.*, 1988; Ward and Roizman, 1994
EBV	172 kb (77 ORFS)	Published: Baer *et al.*, 1984
Cytomegalovirus	229 kb (200 ORFS)	Published: Chee *et al.*, 1990; Bankier *et al.*, 1991
(b) Bacteria		
E. coli	4.7 Mb	Completed: (University of Wisconsin-Madison)
Haemophilus influenzae	1.8 Mb (1750 ORFS)	Published: Fleischmann *et al.*, 1995
Helicobacter pylori	1.7 Mb	Completed: [The Institute of Genome Research (TIGR) Genome Therapeutics Corporation]
Legionella pneumophila	4.1 Mb	Underway: (TIGR)
Mycobacterium tuberculosis	4.4 Mb	Underway: (TIGR/Sanger Centre/Genome Therapeutics Corporation/Pasteur)
Mycoplasma genitalium	0.6 Mb (470 ORFS)	Published: Fraser *et al.*, 1995
Mycoplasma pneumoniae	0.8 Mb (677 ORFS)	Published: Himmelreich *et al.*, 1996
Neisseria gonorrhoeae	2.2 Mb	Underway: (Oklahoma)
Salmonella typhimurium	4.5 Mb	Underway: (TIGR)
Staphylococcus aureus	2.8 Mb	Underway: (Genome Therapeutics Corporation/Incyte Pharmaceuticals)
Streptococcus pneumoniae	2.5 Mb	Underway: (TIGR/Medimmune/Genome Therapeutics Corporation/Incyte Pharmaceuticals)
Streptococcus pyogenes	2.0 Mb	Underway: (Oklahoma/Incyte Pharmaceuticals)
Vibrio cholerae	2.5 Mb	Underway: (TIGR)
(c) Fungi		
Saccharomyces cerevisiae	13 Mb (6200 ORFS)	Published: Dujon, 1996
Schizosaccharomyces pombe	16 Mb	Underway: (Sanger Centre/Cold Spring Harbor Laboratories)
Aspergillus fumigatus	Unknown	Underway: (Incyte Pharmaceuticals)
Aspergillus nidulans	31 Mb	Underway: (Oklahoma)
Candida albicans	17 Mb	Underway: (Incyte Pharmaceuticals)
(d) Other eukaryotes		
Caenorhabditis elegans	90 Mb	Underway: (Sanger Centre/Washington University)
Drosophila melanogaster	170 Mb	Underway: (National Center for Human Genome Resources/LBNL/HHMI)
Homo sapiens	3000 Mb (~100 000 ORFS)	Underway: (Multiple sites)

proven tractable to drug discovery (Miller and Sarver, 1997). Unfortunately, the functional dissection of larger genomes, such as the human genome, promises to take much longer, simply due to their considerably greater complexity.

Shown in *Table 12.2* for comparison with HIV are a further selection of infectious agents, ranging from relatively simple viruses such as the herpes viruses, to more complex pathogens such as *Aspergillus fumigatus*, together with the size of their genomes, current status of sequence availability and projected coding potential, alongside similar data for the human genome.

Generic approaches towards the identification of new targets for human drug discovery are now routinely practised within pharmaceutical companies. Early in the process is the 'mining' of the currently available DNA sequence databases (e.g. the proprietary Incyte and Human Genome Sciences databases, or the public-domain IMAGE and GenBank databases) to identify novel members of gene families with existing 'drug target' pedigrees which may be involved in either the development of specific diseases or new strategies for their treatment. Simply trawling these databases for potential targets expressed in diseased tissue has already yielded novel homologues of key enzymes and receptors, many of which have been patented as drug discovery targets [e.g. cathepsin O (Adams *et al.*, 1995) and interleukin-1β converting enzyme-like proteases (Craig *et al.*, 1996; He *et al.*, 1996); a novel phospholipase C (Hawkins and Seilhamer, 1995); a novel amine receptor (Li and Ruben, 1996); novel potassium channels (Adams *et al.*, 1996); and novel amine and neurotransmitter transporters [(Erickson *et al.*, 1996; Fleischmann and Li, 1995; Li *et al.*, 1996)].

A more pharmaceutically oriented approach is to search for novel members of certain key receptor families which are already known from pharmacological studies to be present in a target tissue. This combination of pharmacology and molecular biology is proving particularly interesting, identifying far greater heterogeneity amongst targets than had previously been thought, with both receptor subtypes and the differential splicing of individual genes contributing to this complexity.

A good example of combining pharmacology and molecular biology in target discovery is the search for the elusive functional α1-adrenergic receptor in human prostatic smooth muscle, inhibition of which is a therapeutic strategy for managing benign prostatic hypertrophy (BPH) (reviewed in Andersson *et al.*, 1997; Hieble and Ruffolo, 1997). Several α1-adrenergic receptor antagonists are already available for the management of BPH [e.g. the Pfizer drug, Cardura (Andersson *et al.*, 1997)]. However, traditional agents show relatively little discrimination between α1-subtypes, some of 1which are not found in the prostate, and several of which can control pharmacologies which may complicate treatment. The initial cloning of the hamster α1-adrenergic receptor in 1988 was closely followed by the definition of a further two different mammalian subtypes (Cotecchia *et al.*, 1988; Lomasney *et al.*, 1991; Schwinn *et al.*, 1990), and has subsequently been broadened to the identification by standard gene cloning approaches of at least three human subtypes (Bruno *et al.*, 1991; Hirasawa *et al.*, 1993; Ramarao *et al.*, 1992; Schwinn *et al.*, 1995). As an illustration of the power of database mining, all three human subtypes are present in at least one of the main sequence databases, indicating how efficient pharmacologically driven database searching can be with at least some corresponding DNA sequence in hand. A further indication of the potential utility of the database mining approach is the number of major pharmaceutical companies who now subscribe to proprietary databases – 15 companies subscribe to the Incyte database alone.

12.3.2 The target validation cycle

Unlike traditional pharmacological targets, for which 'functional' leads are always available by definition, *in silico* genomic approaches generate only a wealth of hypothetical molecular mechanisms which ideally should be validated as real drug discovery targets. There are two primary ways of validating such targets before initiating large-scale projects. The first, based on observing the effects of deleting or up-regulating individual genes within target genomes, relies on methods of genome manipulation for success, and has led to the development of the comparatively new area of 'functional genomics' (for a review of this emerging area, see Fields, 1997). The other, more mature, approach involves ablation of individual gene products, *in vitro* or *in vivo*, and uses mainly post-transcriptional methods such as anti-sense oligonucleotide and intracellular antibody intervention to alter the expression of particular proteins and observe phenotypic sequelae.

The context of these experiments is important. One setting for target validation is to employ easily manipulable model systems (e.g. the nematode, *C. elegans*, the fruit fly, *Drosophila*, or the yeast, *S. cerevisiae*) through which to probe and exploit the function of specific gene products and, by comparison, define the functions of their mammalian equivalents. Another approach is to use more complex mammalian systems, such as the transgenic mouse or rat. Coupled with the development of automated systems for phenotype monitoring, such as fluorescence-activated cell scanning, the connection between genotype and phenotype, and the modulation of this connection through drugs, can also be examined.

Another, independent route to target validation in humans lies in the use of 'disease gene hunts' to identify the key gene products whose mutation is directly implicated in the disease process. Although such studies primarily involve human patient populations, they can often be mirrored in animal models of disease. Their major advantage lies in bringing the target validation process closer to its main objective – target identification in humans.

Specific abnormalities in key genes are contributing factors in the development and pathogenesis of many human diseases [see Bassett *et al.*, 1997, and the Internet site, On-line Mendelian Inheritance in Man (OMIM) for detailed coverage of this vast area (http://www.ncbi.nlm.nih.gov/omim/)]. Often, the value of this for drug discovery lies not so much in identifying the genetic defects associated with pathophysiologies *per se*, but rather in the insight it provides into the development of disease at a molecular and biochemical level. The identification of genetic abnormalities associated with previously untreatable chronic diseases exposes the pharmacologist to new ways of thinking about intervention points in the disease process. The familial context in which these gene hunts are carried out often dictates extensive clinical involvement, adding an important dimension to these studies.

Common diseases such as type 2 diabetes (Harris *et al.*, 1996), asthma (Holgate, 1997; Marsh *et al.*, 1997), multiple sclerosis (Ebers *et al.*, 1996; Haines *et al.*, 1996; Sawcer *et al.*, 1996; reviewed in Bell and Lathrop, 1996), stroke (Rubattu *et al.*, 1996; reviewed in Gunel and Lifton, 1996) and certain types of cancer (see, e.g., the involvement of *BRCA-1* in breast cancer in Boyd, 1995), diseases for which there are few effective drug therapies, have strong genetic linkage, and disease gene hunts are very likely to pinpoint novel targets/pathways for intervention.

A third and rather separate route to defining novel molecular mechanisms of disease is by using photo-affinity labelling to identify specific effectors of disease, piggy-backing drug leads of unknown mechanism which have already shown efficacy. The conversion of appropriate 'mystery mechanism' leads to fast-second chemistry-led projects is a major objective in many pharmaceutical companies. [For examples of such studies, see Miller *et al.* (1990) and Lee *et al.* (1994).] In such approaches, the final identity of the protein target is often revealed by searching genome databases for the sequence of the labelled protein, reinforcing the utility of comprehensive genome databases.

Looking to the future, a particularly efficient way of validating targets for drug programmes might be to use rapidly generated small molecule probes for approach validation directly in animal models, or even humans. A therapeutic target, selected with attention to specific criteria from the 'target discovery cycle', might be established as a high-throughput screen and, using rapid combinatorial library techniques linked to 'smart' library design, could be used to generate lead compounds with acceptable pharmacokinetics for animal studies. If the probes had appropriate safety characteristics, they might also be used for acute human studies. Demonstrated efficacy in either animal models of disease, or the clinic, would directly validate the mechanism and lead on to a full chemistry programme to obtain an optimized candidate.

Clearly, our growing understanding of genome structure and function presents the pharmaceutical scientist with powerful new systems for target validation before costly and potentially fruitless drug discovery projects are initiated.

12.3.3 The lead discovery cycle

At the heart of the discovery process lies the lead discovery cycle. Sourcing leads is a primary concern of most pharmaceutical companies. Originally the serendipitous product of a pharmacologist's curiosity, lead discovery is now rarely left to chance. Pharmaceutical companies, when taken together, have many millions of well characterized, structurally diverse compounds in their collections, each of which can be tested in high-throughput assays against specific biological targets. The logistics of matching informative pharmacological assays to such large numbers of potential leads is a major challenge to the industry – and has been approached in a variety of different ways.

It is evident that the traditional pharmacological 'preparation' is not well suited to screening many thousands of compounds. Over the last decade increasing emphasis has turned to biochemical assays, often using recombinant receptors and enzymes, as platforms for lead discovery. An increasing understanding of the molecular mechanisms underlying pharmacological responses has allowed the configuration of highly automated 96-well plate assays for many isolated targets, aided in large part by molecular biological approaches to the cloning and expression of specific targets. The advent of genomics has for the first time given a clear appreciation of the heterogeneity in most pharmacological targets which usually exist as specific members of extended 'gene families'. The incorporation of other members of key target families in primary or secondary screening programmes promises to improve the specificity of the lead discovery process, giving molecular pharmacologists new tools with which to dissect disease processes.

A key development in the process of lead discovery has been the use of combinatorial chemistry to improve the structural diversity of leads (reviewed in Terrett *et al.*, 1995). The close matching of combinatorial library sets with specific gene family targets promises to increase the effectiveness of the lead discovery process. Just how far lead discovery has been influenced by automation and robotics can be gauged by the fact that some companies are predicting routine screening of up to a million structurally defined compounds in high-throughput screens in less than a month by the year 2000. In these circumstances, the major challenge will be for the molecular biologist to resource the increasing number of rapidly changing screens with primary and secondary screening targets.

12.3.4 The lead optimization cycle

The progression of a lead to candidate status depends primarily on optimizing its potency against specific pharmacological targets. Building potency through traditional medicinal chemistry approaches is now, in many cases, being enhanced through the use of protein structural information, largely derived from crystal structures of proteins, both alone and co-crystallized with specific chemical compounds. Using genetic engineering technologies, it has become possible to purify to homogeneity sufficient quantities of certain recombinant proteins to mount effective X-ray structure determination programmes. Two examples of drug discovery programmes where crystal structures have helped, or are currently helping, in the lead optimization cycle are HIV protease (Lapatto *et al.*, 1989) and interleukin-1β converting enzyme (Margolin *et al.*, 1997; Wilson *et al.*, 1994).

Using existing X-ray structures, computational chemists are now able to model 3D structures of related gene family members based upon the translated DNA sequence alone (May *et al*, 1994), bringing the prospect of *ab initio* structure-based drug design (Whittle and Blundell, 1994) closer to practice. In the case of HIV-1 protease, whose structural homology to other members of the aspartyl proteinase family was recognized at an early stage (Blundell *et al.*, 1990), this has led to the rapid synthesis and testing of several series of inhibitors (reviewed in Gait and Karn, 1995), and drugs optimized using such structural information are already showing promise in clinical trials, exemplified by the compounds Viracept (Dressman *et al.*, 1995) and VX478 (Tung *et al.*, 1994). The impact which these approaches are having on drug discovery is currently limited, however, by the lack of methodologies for structural determination of membrane-bound proteins, a class of target to which many drug discovery targets belong.

However, the progress of a lead towards candidate status depends not only on its potency against its specific pharmacological or molecular target, but also on a range of other characteristics, such as its oral bioavailability, its pharmacokinetics and, importantly, its toxicological profile. In practice, studies in surrogate mammalian systems are usually used to optimize these parameters, with extensive pharmacokinetic and toxicological profiling being performed in species such as the rat and dog. The impact of genomics on these processes has yet to be fully felt, although considerable effort is currently being devoted to mirroring whole animal studies with isolated cell systems, in both primary cultures and genetically engineered cells *in vitro*. The availability from genomic projects of both human and mammalian drug-metabolizing enzymes, which often occur in highly extended gene families [for a detailed review of one of these families, the cytochrome P450s, see Gonzalez (1989)] has permitted the assembly of suites of reagents with which to profile promising compounds *in vitro*. The development of genetically engineered tools (cell lines, isolated enzymes) to complement more traditional approaches, such as rat liver slices, for studying the *in vitro* metabolism of potential candidates promises to speed up the process and improve the quality of nominated candidates. However, the key challenge in reducing such information to drug discovery practice will be to interpret and attach significance to *in vitro* observations in the wider context of drug performance *in vivo*. Much work needs to be done before such *in vitro* studies become accurate predictors of *in vivo* disposition.

Nowhere in the drug discovery process is the development of these methods of assessing candidate potential more important than in toxicology. Almost half of nominated candidates 'die' in whole animal toxicology studies – and yet we have few *in vitro* systems, apart from the Ames test measuring genetic toxicology, capable of reliably predicting such adverse toxicological effects. Results from biochemical profiling of molecular events, such as cytochrome P450 induction, liver enzyme induction and immune suppression, already influence preclinical assessment of candidate potential, and the advent of new transcript imaging techniques (see below) promises to add further dimensions to such studies.

In all these areas of lead optimization, genomics is providing significant new opportunities for improving the quality and efficiency with which safe and effective drugs are discovered.

12.3.5 The clinical discovery cycle

In just the same way that genomics can support preclinical drug discovery, it can also be applied to optimizing the clinical discovery process, pinpointing genetic abnormalities in patients before treatment, as well as providing sensitive systems with which to monitor drug-related molecular changes accompanying drug administration, aspects of the clinical development process for most drugs.

As an example, infection genomics, ranging from the detection of pathogens in the blood to their treatment with highly specific drugs, is already making a major impact on the treatment of infectious disease. A particularly worrying aspect of the latter is the emergence of resistance to widely prescribed drugs, such as Rifampicin and Isoniazid for tuberculosis and reverse transcriptase inhibitors for AIDS. The availability of techniques for following the emergence of drug-resistant organisms during clinical trials of new antimicrobials (Kozal *et al.*, 1996, see below) heralds a new era of clinical discovery – one in which the clinical performance of a drug can be objectively assessed in strict mechanistic terms as well as through the more qualitative measures of symptomatic relief. While effective disease management will remain the paramount goal for all clinical programmes, the criteria for a programme's success (and possibly failure) will increasingly include studies of related mechanistic parameters. Such information will have a major influence on the deployment of alternative strategies for the discovery and development of follow-on agents, and will be eagerly sought by preclinical discovery teams.

Wider application of 'phenotypic' genomics is to be expected in the diagnosis of not only inherited but also acquired disease, including sporadic cancers and syndromes such as immunodeficiency and other haematopoietic abnormalities. The use of the phenotypic markers provided by genomics promises to revolutionize diagnostic and preventative medicine.

12.4 Realizing new paradigms for drug discovery and development through genomics

Amongst the wealth of new technologies of relevance to drug discovery emerging from genomics, two deserve special mention: genome-wide expression analysis and pharmacogenetics.

12.4.1 Emerging techniques for genome-wide expression analysis

In practice, genomics involves not only defining individual coding elements within a particular genome, but also implementing expression analysis to define in which biological setting a particular gene may be functioning. While there are many approaches to gene expression analysis, the development of gene 'chips', based on the genome-wide assembly of unique sets of target gene sequences (Chee *et al.*, 1996; Lockhart *et al.*, 1996), opens the way to very efficient expression analysis. There are multiple applications for this technology in the drug discovery process.

Target discovery, in contrast to more general gene discovery, is critically dependent on knowing which receptor subtype, enzyme isoform or transcription factor is functionally involved in a physiological or disease process. The prospect of scanning all possible combinations of the 100 000 or so human genes to identify subsets of proteins involved in important biological events (say, for example, the smooth muscle contractile response, the process of virus infection or other vital humoral responses such as immune modulation) offers the possibility of dramatic new insights into disease biology. Gene chips are one way of efficiently accomplishing such scans.

In the context of infection genomics, the development of oligonucleotide-based PCR and gene chip technologies for the identification of resistant microbial strains is already underway. Using microarrays of oligonucleotides, Kozal *et al.* (1996) demonstrated sequence polymorphisms/mutations in the predicted amino acid sequence of the HIV-1 protease gene using 167 clinical HIV isolates from 102 patients. The data showed that the translated amino acid sequence of the protease was highly variable, with a total of 47 of the 99 residues being variant amongst the viral clinical isolates. In addition, several of the amino acid changes that were found as naturally occurring polymorphisms in patients who had never received protease inhibitors have been shown to contribute to drug resistance.

In the context of drug metabolism, the individual contributions which specific tissues make to a drug's metabolic fate are yet to be fully defined. The development of these powerful new techniques for the measurement of gene expression is allowing detailed molecular 'mapping' of key tissues, and with it the hope of more accurate prediction of the exposure of selected vascular beds to effective drug concentrations.

Such chip-based molecular diagnostics can also be applied to the diagnosis of inherited diseases such as cystic fibrosis (CF) (Cronin *et al.*, 1996) and are under development for the diagnosis of certain cancers (such as breast and colon cancer). Recently DNA chip technology was used to show that 14 out of 15 patients with known heterozygous mutations in the 3.45 kb exon 11 of the *BRCA-1* gene linked to the development of breast cancer were readily detected and that there were no false positives in 20 control samples (Hacia *et al.*, 1996). Early detection of malignant change may herald more effective therapeutic intervention for cancer, possibly indicating specific therapies attuned to the patient's genotype. The current applications and future promise of DNA chip technology seem to ensure its commercial success.

12.4.2 Pharmacogenetics

Genetic diversity amongst patient populations not only influences predisposition to disease, it also influences the relative effectiveness of many candidate drugs. For example, mutations in drug-metabolizing enzymes such as cytochrome 2D6 can radically change the rate of metabolism of certain drugs, leading to unpredictable pharmacokinetic behaviour (Eichelbaum and Evert, 1996; Linder *et al.*, 1997; Nelson *et al.*, 1996). This genetic heterogeneity amongst the patient population is known to extend to subtle mutations in key therapeutic targets. It is therefore likely that many individuals in a treatment group will respond slightly differently to drug therapies as a consequence of subtle individual genetic variations. The interface between population genetics and pharmacology has been termed 'pharmacogenetics' and is an area of considerable interest to both clinicians and pharmaceutical companies.

With tools such as gene chips available rapidly to profile individual

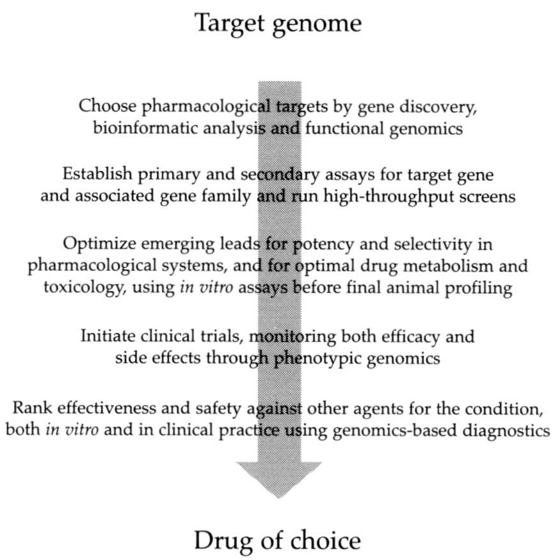

Figure 12.5. Using genomics to support drug discovery.

genetic variations, the possibility of matching specific patients to highly effective therapies, or the converse, specific avoidance of less effective or potentially hazardous treatments amongst defined patient subsets, becomes feasible. The implications of this for both drug discovery and prescription are currently being explored (Linder *et al.*, 1997).

Integrating these new genomic technologies into the drug discovery process will not be easy. It takes at least 10 years to develop a new drug, and many more to realize fully its clinical success. The time-frames over which new strategies for drug discovery need to be judged are therefore long. However, recent examples of genome-driven discovery, such as in the case of HIV protease, show that the leverage to the area of target discovery which genomics brings is realistic and can be successfully translated into major new drugs. Maybe some of the other approaches indicated in *Figure 12.5* will make equally important contributions to the overall process of drug discovery.

In this brief discussion, we have only been able to touch on some of the more obvious ways in which the drug discovery and clinical development processes are being enhanced by the application of genomics and its associated technologies. Together with the hoped for implementation of new clinical diagnostics and gene therapies, the face of both drug discovery and clinical practice is likely to be altered beyond all

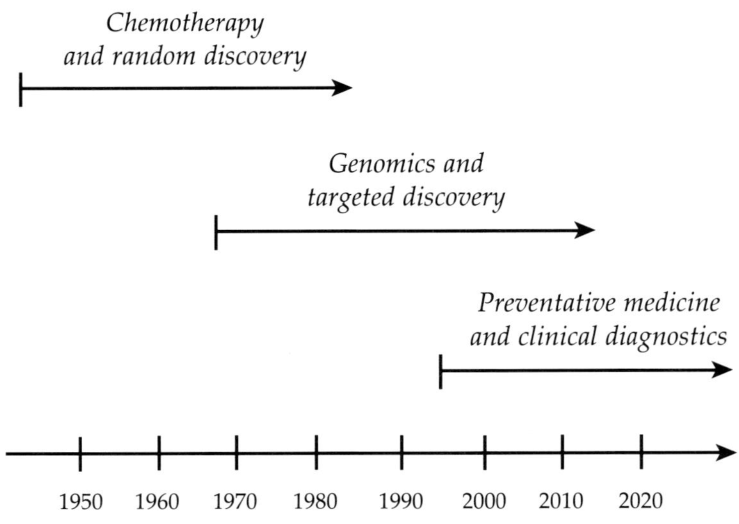

Figure 12.6. Eras of drug discovery.

recognition over the next decade. We are now moving through the era of genomics and targeted discovery to one of preventative medicine and clinical diagnostics (*Figure 12.6*), areas in large part born of genomics. The pace of such change is incredible when one remembers that DNA itself was discovered only in 1953, with the first full genome sequence appearing in 1977 (Sanger *et al.*, 1977)!

References

Adams MD, Blake JA, Drake FH, Fitzgerald LM, Fraser CM, Gowan M, Hastings GA, Kirkness EF and Lee NH. (1995) Human osteoclast-derived cathepsin. Patent WO 9524182 (Human Genome Sciences and SmithKline Beecham).

Adams MD, Li Y and White OR. (1996) Human potassium channel 1 and 2 proteins. Patent WO 9603415 (Human Genome Sciences).

Andersson KE, Lepor H and Wyllie MG. (1997) Prostatic α1-adrenoceptors and uroselectivity. *The Prostate* **30**, 202–215.

Baer R, Bankier AT, Biggin MD *et al.* (1984) DNA sequence and expression of the B95-8 Epstein–Barr virus genome. *Nature* **310**, 207–211.

Bankier AT, Beck S, Bohni R *et al.* (1991) The DNA sequence of the human cytomegalovirus genome. *DNA Sequence* **2**, 1–12.

Bassett DE, Boguski MS, Spencer F, Reeves R, Kim S, Weaver T and Hieter P. (1997) Genome cross-referencing and XREFdb: implications for the identification and analysis of genes mutated in human disease. *Nature Genet.* **15**, 339–344.

Bell JI and Lathrop GM. (1996) Multiple loci for multiple sclerosis. *Nature Genet.* **13**, 377–378.

Blundell TL, Lapatto R, Wilderspin AF, Hemmings AM, Hobart PM, Danley DE and Whittle PJ. (1990) The 3-D structure of HIV-1 proteinase and the design of antiviral agents for the treatment of AIDS. *Trends Biochem. Sci.* **15**, 425–430.

Boyd J. (1995) BRCA1: more than a hereditary breast cancer gene? *Nature Genet.* **9**, 335–336.

Bruno JF, Whittaker J, Song J and Berelowitz M. (1991) Molecular cloning and sequencing of a cDNA for a human alpha 1A-adrenergic receptor. *Biochem. Biophys. Res. Commun.* **179**, 1485–1490.

Chee MS, Bankier AT, Beck S *et al.* (1990) Analysis of the protein-coding content of the sequence of human cytomegalovirus strain AD169. *Curr. Top. Microbiol. Immunol.* **154**, 125–169.

Chee M, Yang R, Hubbell E *et al.* (1996) Accessing genetic information with high-density DNA arrays. *Science* **274**, 610–614.

Cohen LJ and De Vane CL. (1996) Clinical implications of antidepressant pharmacokinetics and pharmacogenetics. *Ann. Pharmacother.* **30**, 1471–1480.

Cotecchia S, Schwinn DA, Randall RR, Lefkowitz RJ, Carob MG and Kobilka BK. (1988) Molecular cloning and expression of the cDNA for the hamster α1-adrenergic receptor. *Proc. Natl Acad. Sci. USA.* **85**, 7159–7163.

Craig AR, Hastings GA, Hudson PL, Kirkness EF and He WW. (1996)

Interleukin-1 beta converting enzyme-like apoptosis protease-1 and -2. Patent WO 9600297 (Human Genome Sciences).
Craig, JC, Duncan IB, Hockley D, Grief C, Roberts NA and Mills JS. (1991) Antiviral properties of Ro 31-8959, an inhibitor of human immunodeficiency virus (HIV) proteinase. *Antiviral Res.* **16**, 295–305.
Cronin, MT, Fucini RV, Kim SM, Masino RS, Wespi RM and Miyada CG. (1996) Cystic fibrosis mutation detection by hybridization to light-generated DNA probe arrays. *Hum. Mutat.* **7**, 244–255.
Debouck C, Gorniak JG, Strickler JE, Meek TD, Metcalf BW and Rosenberg M. (1987) Human immunodeficiency virus protease expressed in *Escherichia coli* exhibits autoprocessing and specific maturation of the gag precursor. *Proc. Natl Acad. Sci. USA* **84**, 8903–8906.
Dressman BA, Fritz JE, Hammond M *et al.* (1995) HIV protease inhibitors. Patent WO 95/09843 (Agouron Pharmaceuticals).
Dujon B. (1996) The yeast genome: what did we learn? *Trends Genet.* **12**, 263–269.
Ebers GC, Kukay K, Bulman DE *et al.* (1996) A full genome search in multiple sclerosis. *Nature Genet.* **13**, 472–476.
Eichelbaum M and Evert B. (1996) Influence of pharmacogenetics on drug disposition and response. *Clin. Exp. Pharmacol. Physiol.* **23**, 983–985.
Erickson JD, Schafer MK, Bonner TI, Eiden LE and Weihe E. (1996) Distinct pharmacological properties and distribution in neurons and endocrine cells of two isoforms of the human vesicular monoamine transporter. *Proc. Natl Acad. Sci. USA* **93**, 5166–5171.
Fields S. (1997) The future is function. *Nature Genet.* **15**, 325–327.
Fleischmann RD and Li Y. (1995) Neurotransmitter transporter. Patent WO 9531539 (Human Genome Sciences).
Fleischmann RD, Adams MD, White O, *et al.* (1995) Whole-genome random sequencing and assembly of *Haemophilus influenzae* Rd. *Science* **269**, 496–512.
Fraser CM, Gocayne JD, White O, *et al.* (1995) The minimal gene complement of *Mycoplasma genitalium. Science* **270**, 397–403.
Gait MJ and Karn J. (1995) Progress in anti-HIV structure-based drug design. *Trends Biotechnol.* **13**, 430–438.
Gonzalez FJ. (1989) The molecular biology of cytochrome P450s. *Pharmacol. Rev.* **40**, 243–288.
Gunel M and Lifton RP. (1996) Counting strokes. *Nature Genet.* **13**, 384–385.
Hacia JG, Brody LC, Chee MS, Fodor SPA and Collins FS. (1996) Detections of heterozygous mutations in BRCA1 using high density oligonucleotide arrays and two-colour fluorescence analysis. *Nature Genet.* **14**, 441–447.
Haines JL, Ter-Minassian M, Bazyk A *et al.* (1996) A complete genomic screen for multiple sclerosis underscores a role for the major histocompatibility complex. *Nature Genet.* **13**, 469–471.
Harris CL, Boerwinkle E, Chakraborty R *et al.* (1996) A genome-wide search for human non-insulin-dependent (type 2) diabetes genes reveals a major susceptibility locus on chromosome 2. *Nature Genet.* **13**, 161–166.
Hawkins PR and Seilhamer JJ. (1995) Phospholipase C homolog. Patent US 5587306 (Incyte Pharmaceuticals).
He WW, Rosen CA, Hudson PL and Hastings GA. (1996) Interleukin-1 beta

converting enzyme-like apoptosis protease-3 and -4. Patent WO 9613603 (Human Genome Sciences).

Hieble JP and Ruffolo RR. (1997) Recent advances in the identification of α1- and α2- adrenoceptor subtypes: therapeutic implications. *Exp. Opin. Invest. Drugs* **6**, 367–387

Himmelreich R, Hilbert H, Plagens H, Pirkl E, Li B-C and Herrmann R. (1996) Complete sequence analysis of the genome of the bacterium *Mycoplasma pneumoniae. Nucleic Acids Res.* **24**, 4420–4449.

Hirasawa A, Horie K, Tanaka T, Takagaki K, Murai M, Yano J and Tsujimoto G. (1993) Cloning, functional expression and tissue distribution of human cDNA for the α1C-adrenergic receptor. *Biochem. Biophys. Res. Commun.* **195**, 902–909.

Holgate ST. (1997) Asthma genetics: waiting to exhale. *Nature Genet.* **15**, 227–229.

Kozal MJ, Shah N, Shen N *et al.* (1996) Extensive polymorphisms observed in HIV-1 clade B protease gene using high-density oligonucleotide arrays. *Nature Med.* **2**, 753–759.

Kramer RA, Schaber MD, Skalka AM, Ganguly K, Wong-Staal F and Reddy EP. (1986) HTLV-III gag protein is processed in yeast cells by the virus pol-protease. *Science* **231**, 1580–1584.

Lapatto R, Blundell T, Hemmings A *et al.* (1989) X-ray analysis of HIV-1 proteinase at 2.7 angstrom resolution confirms structural homology among retroviral enzymes. *Nature* **342**, 299–302.

Lee JC, Laydon JT, McDonnell PC *et al.* (1994) A protein kinase involved in the regulation of inflammatory cytokine biosynthesis. *Nature* **372**, 739–746.

Li Y and Ruben SM. (1996) Human amine receptor. Patent WO 9639440 (Human Genome Sciences).

Li Y, Cao L and Rosen CA. (1996) Human amine transporter. Patent WO 9627009 (Human Genome Sciences).

Linder MW, Prough RA and Valdes R. (1997) Pharmacogenetics: a laboratory tool for optimizing therapeutic efficiency. *Clin. Chem.* **43**, 254–266.

Lockhart DJ, Dong H, Byrne MC *et al.* (1996) Expression monitoring by hybridization to high-density oligonucleotide arrays. *Nature Biotechnol.* **14**, 1675–1680.

Lomasney JW, Cotecchia S, Lorenz W, Leung W, Schwinn DA, Yang-Feng T, Brownstein M, Lefkowitz RJ and Caron MG. (1991) Molecular cloning and expression of the cDNA for the α1A-adrenergic receptor. *J. Biol. Chem.* **266**, 6365–6369.

Margolin N, Raybuck SA, Wilson KP, Chen W, Fox T, Gu Y and Livingston DJ. (1997) Substrate and inhibitor specificity of interleukin-1-converting enzyme and related caspases *J. Biol. Chem.* **272**, 7223–7228.

Marsh DG, Maestri NE, Freidhoff LR *et al.* (1997) A genome-wide search for asthma susceptibility loci in ethnically diverse populations. *Nature Genet.* **15**, 389–392.

Martin JA and Redshaw S. (1989) Amino acid derivatives: compounds of the general formula X wherein R represents benzyloxycarbonyl or 2-quinolyl-carbonyl and their pharmaceutically acceptable acid addition salts inhibit proteases of viral origin and can be used as medicaments for the treatment or prophylaxis of viral infections. Patent EP432695 (Hoffman-La Roche).

May AC, Johnson MS, Rufino SD, Wako H, Zhu ZY, Sowdhamini R, Srinivasan N, Rodionov MA and Blundell TL. (1994) The recognition of protein structure and function from sequence: adding value to genome data. *Philos. Trans. R. Soc. Lond. B Biol. Sci.* **344**, 373–381.

McGeoch DJ, Dalrymple MA, Davison AJ, Dolan A, Frame MC, McNab D, Perry LJ, Scott JE and Taylor P. (1988) The complete DNA sequence of the long unique region in the genome of herpes simplex virus type 1. *J. Gen. Virol.* **69**, 1531–1574.

Miller DK, Gillard JW, Vickers PJ *et al.* (1990) Identification and isolation of a membrane protein necessary for leukotriene production. *Nature* **343**, 278–281.

Miller RH and Sarver N. (1997) HIV accessory proteins as therapeutic targets: viral regulatory proteins represent new targets for therapeutic and prevention strategies against HIV. *Nature Med.* **3**, 389–394.

Nelson DR, Koymans L, Kamataki T *et al.* (1996) P450 superfamily: update on new sequences, gene mapping, accession numbers and nomenclature. *Pharmacogenetics* **6**, 1–42.

Ramarao CS, Kincade-Denker JM, Perez DM, Gaivin RJ, Riek RP and Graham RM. (1992) Genomic organisation and expression of the human α1B-adrenergic receptor. *J. Biol. Chem.* **267**, 21936–21945.

Ratner L, Haseltine W, Patarca R *et al.* (1985) Complete nucleotide sequence of the AIDS virus, HTLV-III. *Nature* **313**, 277–284.

Richardson K, Brammer K, Marriott MS and Troke PF. (1985) Activity of UK-49,858, a bis-triazole derivative, against experimental infections with *Candida albicans* and *Trichophyton mentaggrophytes. Antimicrob. Agents Chemother.* **27**, 832–835.

Roberts NA, Martin JA, Kinchington D *et al.* (1990) Rational design of peptide-based HIV proteinase inhibitors. *Science* **248**, 358–361.

Rubattu S, Volpe M, Kreutz R, Ganten U, Ganten D and Lindpaintner K. (1996) Chromosomal mapping of quantitative trait loci contributing to stroke in a rat model of complex human disease. *Nature Genet.* **13**, 429–434.

Sanger F. (1977) Nucleotide sequence of bacteriophage ΦX174 DNA. *Nature* **265**, 687–695.

Sawcer S, Jones HB, Feakes R, Gray J, Smaldon N, Chataway J, Robertson N, Clayton D, Goodfellow PN and Compton A. (1996) A genome screen in multiple sclerosis reveals susceptibility loci on chromosome 6p21 and 17q22. *Nature Genet.* **13**, 464–468.

Schwinn DA, Lomasney J, Lorenz W, Szklut PJ, Fremeau RT, Yang-Feng TL, Caron MG, Lefkowitz RJ and Cotecchia S. (1990) Molecular cloning and expression of the cDNA for a novel α1-adrenergic receptor subtype. *J. Biol. Chem.* **265**, 8183–8189.

Schwinn DA, Johnston GI, Page SO *et al.* (1995) Cloning and pharmacological characterisation of human alpha-1 adrenergic receptors: sequence corrections and direct comparison with other species homologues. *J. Pharmacol Exp. Ther.* **272**, 134–142.

Sneader W. (1985) *Drug Discovery: The Evolution of Modern Medicines.* John Wiley and Sons, Chichester, UK.

Terrett NK, Gardner M, Gordon DW, Kobylecki RJ and Steele J. (1995)

Combinatorial synthesis – the design of compound libraries and their application to drug discovery. *Tetrahedron* **51**, 8135–8173.

Tung RD, Murcko MA and Bhisetti GR. (1994) Sulfonamide inhibitors of HIV aspartyl protease. Patent WO 94/05639 (Vertex Pharmaceuticals).

Ward PL and Roizman B. (1994) Herpes simplex genes: the blueprint of a successful human pathogen. *Trends Genet.* **10**, 267–274.

Whittle PJ and Blundell TL. (1994) Protein structure-based drug design. *Annu. Rev. Biophys. Biomol. Struct.* **23**, 349–375.

Wilson KP, Black J-A F, Thomson JA *et al.* (1994) Structure and mechanism of interleukin-1 beta converting enzyme. *Nature* **370**, 270–275.

13

Oligonucleotides and their future potential as therapeutic agents

Keith R. Fox

13.1 Introduction

Central to molecular biology is the well known paradigm that DNA makes RNA makes protein. However, despite the fact that this is an amplification system, with one copy of each gene generating many copies of RNA and protein, most therapeutic and pharmacological agents are targeted at proteins, which lie at the bottom of the pathway. There is therefore considerable interest in designing compounds which act higher up the cascade, targeting RNA and/or DNA, since this should provide a much more efficient switch of gene activity.

The potential applications for such a gene-specific intervention are almost endless. Obvious uses include antiviral and cancer therapy, as well as any disease state which results from inappropriate gene activity. We now possess a large amount of DNA sequence information on pathogenic viruses, and understand the biochemistry and molecular biology of many essential viral genes, several of which have no human homologues. Agents which can selectively inhibit the activity of viral genes (e.g. those for reverse transcriptase, integrase or protease in HIV) may have greater efficacy than conventional chemotherapeutic approaches which aim to inhibit the activity of the gene products. Such a gene-specific intervention in cancer will be less easy since, although many oncogenes have been implicated in the origin and progression of cancer,

Genetics of Common Diseases: future therapeutic and diagnostic possibilities,
edited by I. Day and S. Humphries.

these often differ from their normal cellular counterparts by only one base pair. At first sight it therefore appears that gene-specific anticancer agents will need to possess exquisite sequence recognition properties. However, some cellular oncogenes may be normally active only during development so that inhibition of the normal gene as well as the mutated form during antigene therapy may not be a problem.

Inappropriate gene activity may also be implicated in allergic and immunological diseases. In each of these cases an immense amount of research has been directed towards designing agents targeted at the (protein) gene products. Although this has produced some limited success, there are no clear design rules; compounds with disparate structures are required for inhibiting each particular protein. By contrast DNA active gene-specific agents could have much simpler design rules, since in each case the target is double-stranded DNA.

Although combinations of various synthetic small molecules have been examined in this regard for many years by several groups, this has met with limited success. Within the last 20 years oligonucleotides themselves have been shown to be useful sequence-specific binding agents since they can interact with single-stranded nucleic acids forming short duplex DNAs. In this regard antisense (and ribozyme) technologies are extremely important, but are beyond the scope of this chapter. More recently it has been realized that, in some circumstances, short oligonucleotides can bind to duplex DNA in a sequence-specific fashion, forming intermolecular triple helical structures. This provides a means for targeting unique DNA sequences, thereby affecting gene activity. Since the biological receptor for these compounds is duplex DNA, there are only two copies of each target site per diploid cell. In principle, it may therefore be possible to use very low doses of such a therapeutic agent. However, before oligonucleotides can be successfully employed as antigene agents, it is necessary that we have a thorough understanding of the rules for forming triplex structures with high affinity and stringency. This strategy will then provide a simple method for gene-specific drug design and synthesis. This chapter will describe some of the properties of triplex DNA and the rules governing its use in recognizing specific sequence in duplex DNA. For a recent comprehensive review of triple helical nucleic acids, see Soyfer and Potaman (1996).

13.1.1 Triplexes

The formation of triple helical DNA was first demonstrated in 1957 by mixing the synthetic polynucleotides poly(U) and poly(A) in the ratio 2:1 (Felsenfeld *et al.*, 1957). These formed a specific complex which produced a biphasic melting profile (Felsenfeld *et al.*, 1957) and a distinct X-ray fibre diffraction pattern (Arnott and Selsing, 1974). Although the formation of several other triplexes was demonstrated in the following years, using both ribo- and deoxyribonucleotides, these remained an interesting artefact of physical chemistry until 1987, when it was suggested that they offered a means for selectively recognizing duplex DNA (Le Doan *et al.*, 1987; Moser and Dervan, 1987). Two different triplex motifs have been characterized, which vary in the orientation of the third strand oligonucleotide. In each case, the third strand binds in the major groove of the duplex DNA, making specific contacts with substituents on the purine bases (Hélène, 1993; Thuong and Hélène, 1993). Since triplex formation requires the interaction between three polyanions, these structures are stabilized by divalent metal ions, in particular magnesium (Malkov *et al.*, 1993), and polyamines such as spermine (Hampel *et al.*, 1991).

13.2 Triplex recognition motifs

13.2.1 Parallel triplexes

The most widely studied triplexes are those containing pyrimidine-rich third strands, generating T·AT and C^{+}·GC triplets (see *Figure 13.1*). In these structures the third strand is oriented parallel to the duplex purine strand; that is the two pyrimidine-containing strands run in opposite directions (Moser and Dervan, 1987). In each base triplet, the third strand is held in place by two Hoogsteen hydrogen bonds with the purine base (i.e. to *N7* and *O6* of guanine, and *N7* and the exocyclic *6-NH$_2$* of adenine). Formation of the C^{+}·GC triplet requires protonation of the third strand cytosine. Since this base has a p*K* of 4.5, these structures are only formed

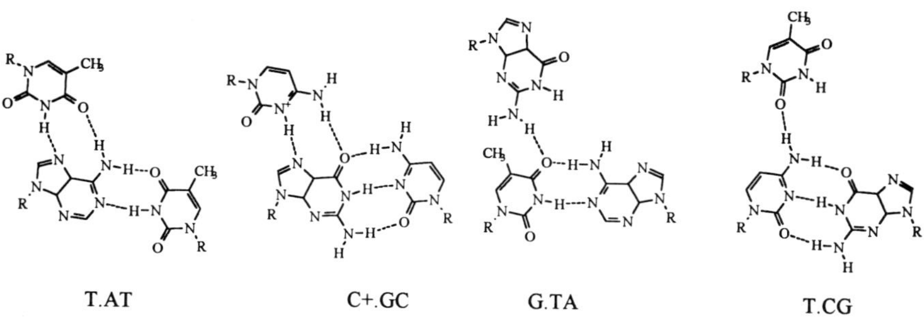

Figure 13.1. The structure of stable parallel triplets T·AT and C^{+}·GC together with the weaker triplets G·TA and T·CG.

at low pH. Various cytosine analogues have been examined in an attempt to overcome this requirement for low pH (see below). Using natural bases at physiological pH, the stability decreases with increasing numbers of C^{+}·GC triplets; this effect is more pronounced with contiguous C^{+}·GC triplets.

Since C^{+}·GC and T·AT triplets are isomorphous (i.e. the phosphodiester backbones are located in the same position relative to the base triplets) the order of the triplets is not critical, except that adjacent protonated cytosines reduce the stability. In both triplets, recognition is achieved by the formation of hydrogen bonds to the purine, rather than the pyrimidine base. Consequently triplex formation is restricted to homopurine tracts; recognition of a purine on the opposite duplex strand would require a change in the orientation of the third strand, which would also need to cross the DNA major groove. Within this motif several weaker triplets have also been described for recognition of pyrimidines, including G·TA and T·CG (Chandler and Fox, 1993; Griffin and Dervan, 1989; Radhakrishnan and Patel, 1994; Radhakrishnan *et al.*, 1994; Yoon *et al.*, 1992). In these triplets the third strand is held in place by one hydrogen bond; as a result they are much weaker and are of limited use. If these weaker triplets are included, then within this motif all four DNA bases can be recognized, generating T·AT, C^{+}·GC, G·TA and T·CG triplets.

13.2.2 Antiparallel triplexes

In the second triplex motif the third strand binds in an antiparallel orientation with respect to the duplex purine strand (Beal and Dervan, 1991). Stabilization is achieved by the formation of G·GC, A·AT and T·AT triplets (see *Figure 13.2*). Unlike the parallel triplex, these complexes are stable at physiological pH, though divalent metal ions (particularly magnesium) are still essential. Since an AT base pair can be recognized by T or A, antiparallel triplex-forming oligonucleotides are either GA- or GT-rich. In contrast to the parallel motif, these triplets are not isomorphous; that is the position of the phosphodiester backbone is different for each triplet. Consequently complex stability varies with the number of GpA and ApG steps in the duplex target.

The formation of these complexes appears to be dominated by the G·GC triplets which have a greater influence on stability than A·AT or T·AT (Fox, 1994); most work with this motif has therefore used oligonucleotides which are very G-rich, interspersed with a few A or T residues. Nonetheless, antiparallel (purine-rich) triplexes can form very stable complexes (Svinarchuk *et al.*, 1994). One problem with this motif is that the

GT-rich triplex-forming oligonucleotides can adopt stable structures which hinder interaction with their target sites (Svinarchuk *et al.*, 1996). There have been fewer studies on the properties of other (weaker) triplets within this motif, though T·CG, C·AT and A·GC have been described (Beal and Dervan, 1992a; Chandler and Fox, 1996; Dittrich *et al.*, 1994; Durland *et al.*, 1994; Ji *et al.*, 1996). It should be noted that, even if these weaker triplets are included, there is no means of recognizing a TA base pair, though C·TA is the least destabilizing mismatch (Chandler and Fox, 1996).

13.2.3 Alternate strand recognition

In both triplex motifs the most stable triplets involve hydrogen bond recognition of the purine bases. Triplex formation is therefore usually restricted to homopurine sequences. There have been several attempts to overcome this limitation, including the design of novel base analogues, third strand oligonucleotides containing 3′–3′- or 5′–5′-linkers (Horne and Dervan, 1990; Ono *et al.*, 1991a), and the use of oligonucleotides designed to utilize both binding motifs (Beal and Dervan, 1992b; Jayasena

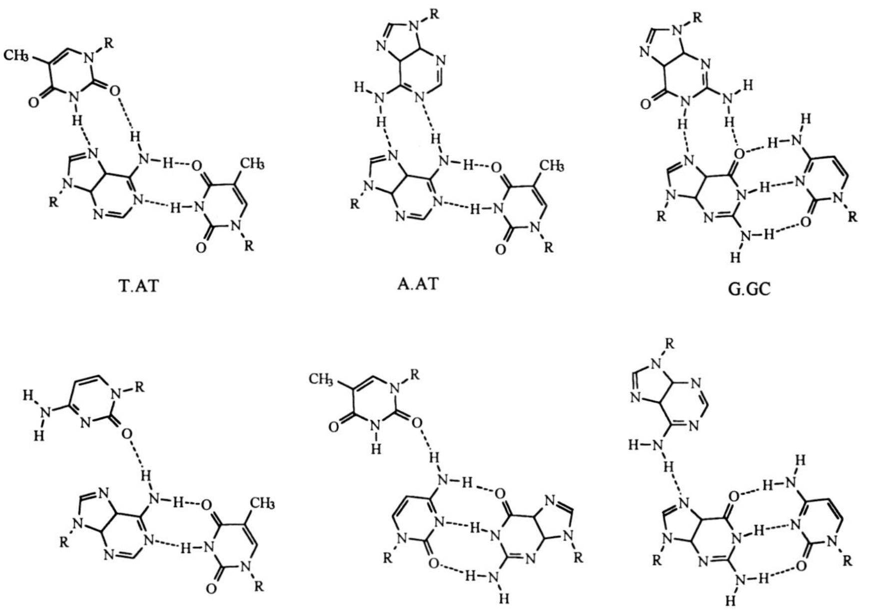

Figure 13.2. The structure of the stable antiparallel triplets G·GC, A·AT and T·AT together with the weaker triplets T·CG, C·AT and A·GC.

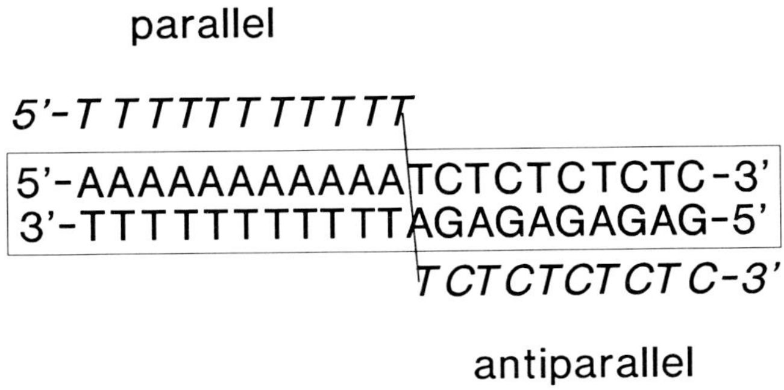

Figure 13.3. Schematic representation of an alternate strand triple helix across an RY junction. The duplex target site $A_{11}(TC)_6{\cdot}(GA)_6T_{11}$ is boxed, with the third strand oligonucleotide positioned above and below. In the left hand portion the A_{11} tract is recognized using parallel T·AT triplets, while the right hand portion forms a triplex containing antiparallel G·GC and T·AT triplets. This schematic model ignores the fact that recognition across an RY junction probably skips one or two bases in the centre.

and Johnston, 1992a,b; Washbrook and Fox, 1994a,b). The latter strategy, termed alternate strand recognition, has been used to target junctions between blocks of purine and pyrimidines. Purine bases on opposing strands are recognized by the different motifs, thereby enabling the use of an oligonucleotide with standard 5′–3′-linkages (for an example, see *Figure 13.3*). However, since the third strand has to cross the duplex major groove as it switches from one motif to the other, these complexes are not as stable as those formed at simple homopurine target sites. In general, recognition across an R_mY_n junction is easier than across Y_mR_n (Beal and Dervan, 1992b; Washbrook and Fox, 1994a,b).

In this chapter I will focus on attempts to overcome some of the current limitations in the use of triplex-forming oligonucleotides for recognition of mixed sequence DNAs under physiological conditions. In particular I will describe strategies aimed at targeting pyrimidine residues with polypurine tracts, removing the pH dependency of guanine recognition in the parallel motif, the properties of agents which stabilize triple helical DNA, and the interaction of triplex-forming oligonucleotides with nucleosome-bound DNA. For further information on other aspects of triplex formation and activity, the reader should consult other recent reviews (Soyfer and Potaman, 1996).

13.3 Current strategies

13.3.1 Stringency and mixed sequence recognition

Parallel triplexes. Although alternate strand recognition, described above, provides a means for forming triplexes across purine–pyrimidine junctions, it is not capable of recognizing mixed sequence DNAs. One approach to this problem is to make greater use of some of the weaker DNA triplets such as G·TA and T·CG. Within the parallel motif, we have examined the formation of triplexes at target sites of the type $A_8XA_8 \cdot T_8YT_8$ (X·Y = each base pair in turn) using the oligonucleotides T_8NT_8 (Chandler and Fox, 1993). The results of these experiments are summarized in *Table 13.1.*

It can be seen that even single-base changes, in the centre of either the duplex or third strand, have pronounced effects on the stability of the complex, demonstrating the stringency of triplex formation. The most stable complexes are produced with central T·AT, C^+·GC, G·TA and T·CG triplets. We have studied the stability of G·TA triplets by investigating triple helix formation at a target site within a natural DNA fragment containing a block of 11 contiguous purines interrupted by a single thymine (Brown and Fox, 1996). The triplex formed with an oligonucleotide designed to generate a G·TA triplet at this position was about 100 times more stable than one with a mismatched T·TA triplet. This demonstrates that the G·TA triplet is indeed a specific structure, rather than the best tolerated mismatch. Surprisingly we found that on

Table 13.1. Relative stability of parallel and antiparallel triplets

		Third strand base	
Base pair	Parallel	Antiparallel-GT	Antiparallel-GA
AT	T	$A > T = C > G$	$T > A = C = T$
GC	$C > T = G \gg A$	$G \geq A > C = T$	$G > A = C > T$
TA	$G > T > A = C$	$C = A \gg T = G$	$C > A > T > G$
CG	$T > C \gg G = A$	$A = T > C \gg G$	$T = C > A \gg G$

The stability of parallel triplets was determined from DNase I footprinting experiments with the sequences $A_8XA_8 \cdot T_8YT_8$ (X·Y = each base pair in turn) using the oligonucleotides T_8NT_8 (Chandler and Fox, 1993). The stability of antiparallel triplets was determined from similar experiments with the sequences $(GGA)_2GGX(GGA)_2GG \cdot (CCT)_2CCY(CCT)_2CC$ (X·Y = each base pair in turn) examining the binding of $(GCA)_2GGN(GGA)_2GG$ or $(GGT)_2GGN(GGT)_2GG$ (Chandler and Fox, 1993). All experiments were performed at pH 7.5 in the presence of 10 mM $MgCl_2$.

mutating this TA base pair to AT, generating an uninterrupted stretch of 12 purines, the perfect matched oligonucleotide generated a complex that was marginally less stable than the original one with the G·TA triplet.

In further experiments, we have explored the use of G·TA triplets for recognizing multiple TA steps within oligopurine target sites (Chandler and Fox, 1995). In theory, it might be possible to recognize alternating AT tracts with GT-containing third strands, generating blocks of alternating T·AT and G·TA triplets. We find that this structure is not stable, even in the presence of triplex-binding ligand. This could be because of the lower stability of the G·TA triplet, or because T·AT and G·TA triplets are not isomorphous. However, when this alternating AT sequence is placed adjacent to an oligoadenine tract, specific triplex formation across this mixed sequence is achieved by tethering the GT-containing third strand to a block of thymines. This forms a block of T·AT triplets which act as an anchor stabilizing the alternating T·AT and G·TA triplets. In similar experiments, we have examined the formation of adjacent G·TA triplets at thymines in the centre of an oligopurine tract. We find that a stable complex is formed only in the presence of a triplex-binding ligand, though in this instance precise stringency is lost and complexes of very similar stability are generated by placing any base opposite the TA base pairs. These weaker triplets must therefore be used with caution.

An alternative strategy for avoiding mismatches within an otherwise perfect triple helix uses abasic sites (ϕ) in the third strand oligonucleotide opposite unrecognizable bases pairs (Horne and Dervan, 1991). However, all abasic triplets are significantly less stable than T·AT or C^+·GC triplets and have a similar destabilizing effect to many triplex mismatches. A likely reason for this destabilization is that the abasic site interrupts the base stacking in the third strand.

Antiparallel triplexes. There have been fewer studies on the stringency of triplex formation with the antiparallel motif, examining triplex formation across pyrimidine residues within polypurine target sites, though T·CG, C·AT and A·GC have been reported (Beal and Dervan, 1992a; Dittrich *et al.*, 1994; Durland *et al.*, 1994).

In similar experiments to those described above, we have examined the formation of antiparallel triplexes at the target sites $(GGA)_2$-$GGX(GGA)_2GG$·$(CCT)_2CCY(CCT)_2CC$ (X·Y = each base pair in turn) using oligonucleotides of the type $(GGA)_2GGN(GGA)_2GG$ or $(GGT)_2$-$GGN(GGT)_2GG$ (Chandler and Fox, 1996), generating a single N·XY triplet in the centre of a structure containing G·GC and A·AT or T·AT triplets. The results are summarized in *Table 13.1*. We find that almost all combinations form stable complexes at low micromolar concentrations. For a central GC base pair, the most stable complex is seen with a central G·GC triplet, though A·GC is also stable. With a central AT, all four bases from stable complexes, though T·AT and A·AT are marginally more stable. For a central CG base pair, T·CG is the most stable, though A·CG and C·CG are formed with GA- and GT-containing oligonucleotides respectively. Complexes across a central TA base pair form less readily, though C·TA appears to be the most stable combination. G·YR triplets appear to be particularly unstable. It remains to be seen how and if these weaker antiparallel triplets may be used for recognition of mixed sequence DNAs.

13.3.2 pH dependency

One severe limitation to the formation of parallel triplexes is that the C^{+}·GC triplet requires conditions of low pH, necessary for protonation of the cytosine residue. The free base has a p*K* of about 4.5, though this may be elevated within triplex-forming oligonucleotides, which are often stable at pHs up to 6.0, depending on the number and location of the cytosine residues. Several cytosine analogues have been synthesised in attempts to overcome this restriction, some of which are shown in

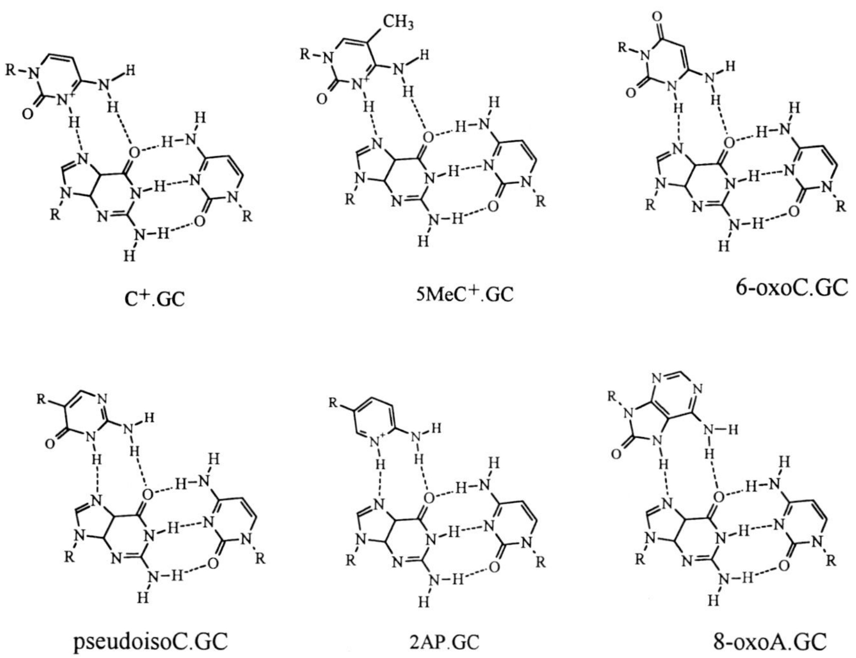

Figure 13.4 Base triplets using cytosine analogues for recognition of GC pairs.

Figure 13.4. It should be noted that the use of base analogues in a therapeutic strategy may present further problems, since the oligonucleotide degradation products may be cytotoxic, in contrast to the natural DNA bases. Recognition of guanine requires a structure presenting two hydrogen bond donors to interact with O^6 and N^7. The first analogue to be tested was 5-methylcytosine, a naturally occurring base, since this has a higher p*K* value than cytosine (Lee *et al.*, 1984; Povsic and Dervan, 1989). Triplexes containing this base are indeed more stable at slightly higher pHs, but are still not formed under physiological conditions (Koh and Dervan, 1992). It has also been suggested that the increased stability of ^{5Me}C results from the extra spine of methyl groups within the DNA major groove (Singleton and Dervan, 1992), in a similar location to the thymine methyl group, rather than from any effect on p*K*. 6-Oxocytosine (Huang *et al.*, 1996) and pseudoisocytosine (Ono *et al.* 1991b) have significantly higher p*K* values, but are not widely used in triplex strategies, though pseudoisocytosine is often used as a cytosine analogue in peptide nucleic acid (PNA) containing structures (Egholm *et al.*, 1995). 8-oxoadenine (Jetter and Hobbs, 1993; Miller *et al.*, 1992) also presents the correct arrangement of hydrogen-bonding donors for recognition of GC in a pH-independent fashion. Another promising cytosine analogue is 2-aminopyridine (2AP), which has a p*K* closer to physiological pH. Psoralen-linked oligonucleotides containing this base have been successfully targeted against a portion of the aromatase gene (Bates *et al.*, 1996). We find that $(2AP)_6T_6$ can form a stable triplex with $G_6A_6{\cdot}T_6C_6$ at pHs as high as 7.0, even though this contains a run of six contiguous GC base pairs, in contrast to $^{5Me}C_6T_6$ and C_6T_6, for which complexes are barely stable at pH 5.0.

An alternative possibility for recognition of GC base pairs at physiological pHs is to retain the use of third strand cytosines, generating a triplet which is stabilized by only one hydrogen bond (Chandler and Fox, 1993). Since T·AT is isomorphous with C·GC, the third strand stacking interactions and phosphodiester backbone configurations may allow stable complex formation without the need for two hydrogen bonds in every C·GC triplet.

13.3.3 Triplex-binding ligands

Although triplex-forming oligonucleotides are endowed with considerable selectivity, their binding may not be strong. As a result, several methods have been explored to enhance their stability. One way of achieving this is by tethering a non-specific DNA-binding agent such as acridine or psoralen to either end of the third strand oligonucleotide (Sun *et al.*, 1989, 1991). An alternative strategy is to develop compounds which bind specifically to triplex, but not duplex, DNA. Several such agents have now been described; the structures of some of these are shown in *Figure 13.5* (numbers refer to *Figure 13.5*). BePI (**1**) and BgPI are benzo-pyridoindole derivatives which selectively stabilize triple helical DNA (Escudé *et al.*, 1996; Mergny *et al.*, 1992; Pilch *et al.*, 1993a,b). Similarly coralyne (**2**) binds to several synthetic triple helices about two orders of magnitude better than to the underlying duplexes (Lee *et al.*, 1993).

Although these agents appear to bind by an intercalative mechanism, this alone is not sufficient to explain the selectivity for triplex DNA, since ethidium binds to both duplex and triplex structures, but does not selectively stabilize triple helical DNA. The naphthylquinoline derivatives (**3**) also appears to bind to triplex DNA by intercalation (Cassidy *et al.*, 1994, 1996; Chandler *et al.*, 1995; Wilson *et al.*, 1993). These naphthylquinoline compounds possess a large aromatic area to stack with the three bases in the triplex, yet, since the aromatic portions are not fused, they possess torsional flexibility and can accommodate the propeller twist of the triplets in which the three bases may not be co-planar.

We have used quantitative DNase-I footprinting to assess the ability of several of these ligands to stabilize intermolecular triplexes (Cassidy *et al.*, 1996; Chandler *et al.*, 1995) and have shown that in several instances they can reduce the oligonucleotide concentration required to generate a clear footprint (which is related to the dissociation constant) by as much as

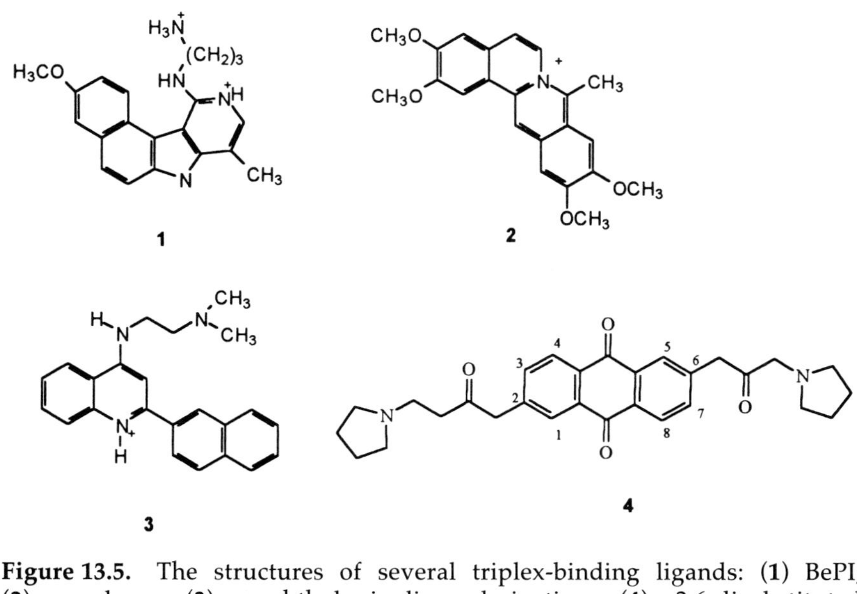

Figure 13.5. The structures of several triplex-binding ligands: (**1**) BePI; (**2**) coralyne; (**3**) naphthylquinoline derivative; (**4**) 2,6-disubstituted amidoanthraquinone.

200-fold. A further series of triplex-binding ligands are the 2,6-disubstituted amidoanthraquinones (**4**) (Fox *et al.*, 1995), which are thought to bind by spearing the DNA triplex, leaving the alkylamino side groups in both DNA grooves. Interestingly, the 1,4-disubstituted compounds do not stabilize triplex DNA, but show a greater affinity for duplex structures, in contrast to the 2,6 compounds which bind better to triplex than duplex DNA.

It appears that an essential feature of these triplex-intercalating ligands is the presence of a cationic group. In general, these compounds preferentially stabilize T·AT, rather than C^{+}·GC, presumably because of charge repulsion with the protonated cytosine in the latter (Cassidy *et al.*, 1996). In addition, these compounds have a greater effect on parallel than antiparallel complexes.

As well as stabilizing canonical triplexes, these compounds can also stabilize structures which contain triplex mismatches. By lowering the stringency of triplex formation, these ligands may promote oligonucleotide binding to sequences other than the desired target site. Although this reduced stringency suggests that these compounds must be used with care, it may be used as a means to broaden the range of potential triplex target sites, by stabilizing oligonucleotide binding at sequences for which there are no good triplet matches.

13.3.4 Nucleosomes

Although DNA triple helix formation provides a means of achieving sequence-specific recognition of duplex DNA, and has been shown to be capable of controlling gene expression, the majority of *in vitro* studies to date have examined the formation of triplex structures on naked DNA. In contrast, in eukaryotic cells DNA is complexed with a large number of proteins in the form of chromatin. These proteins may affect the formation of intermolecular triple helices. The first level of organization of cellular DNA involves the formation of nucleosomes. Each nucleosome contains about 150 bp of DNA wrapped 1.8 times around a protein octamer which consists of two copies of histones H2A, H2B, H3 and H4 (Arents *et al.*, 1991; Richmond *et al.*, 1984). Although nucleosomes are associated with many different DNA sequences, there is considerable evidence that they adopt well defined positions on DNA sequences, both *in vivo* and *in vitro*, determined by the anisotropic flexibility of DNA. Since the double helix must bend as it wraps around the protein, sequences which facilitate bending have been implicated in directing nucleosome formation. In general, GC-rich regions are positioned with their wider than average minor grooves facing away from the protein core, while the narrow minor grooves of AT sequences face towards the protein (Drew and Travers, 1985).

Westin *et al.* (1995) have shown that triplexes do not form on DNA fragments which are wrapped around nucleosome core particles. In addition, nucleosomal positioning of longer DNA fragments is altered by triplex formation, so that the triple helical regions are excluded from the nucleosome cores. This may therefore provide a means of affecting nucleosome location, and thereby affecting gene activity. In addition, we have examined the formation of intermolecular triplexes on DNA fragments which have been reconstituted on to nucleosome core particles by salt exchange. Six different DNA fragments containing polypurine triplex target sites at various positions along the 160-bp *tyr*T fragment have been generated by site-directed mutagenesis and the formation of DNA triplexes has been examined by DNase I footprinting. The positioning of this fragment on nucleosome core particles has been well documented. The sequences of these target sites are shown in *Figure 13.6* (boxed) and the third strand oligonucleotides are shown above or below the sequences. GT-containing third strands were designed to bind in an

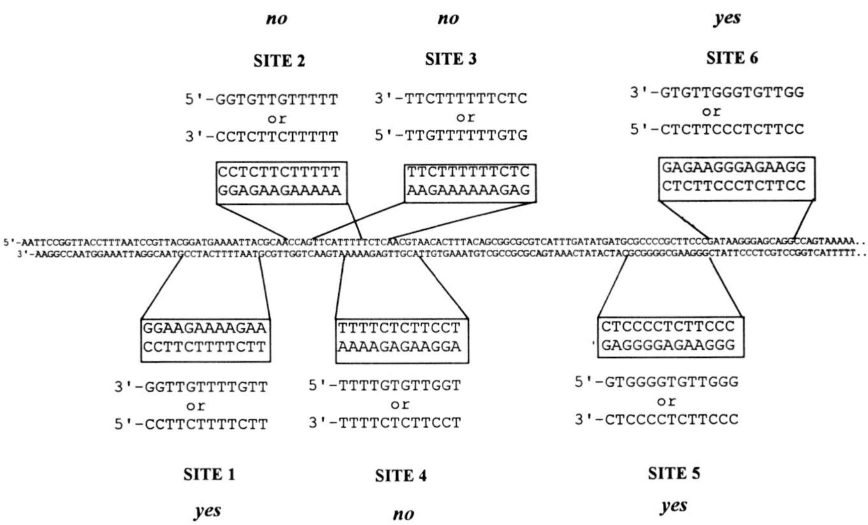

Figure 13.6. Sequences of the *tyr*T DNA derivatives used to examine triplex formation on nucleosome core particles. The novel triplex target sites (sites 1–6) are boxed, and are shown together with their respective triplex-forming oligonucleotides. In each case the upper (GT-containing) oligonucleotide is designed to bind in an antiparallel orientation, while the lower (CT-containing) is designed to bind parallel to the duplex purine strand. 'Yes' or 'no' indicate whether a triplex could be formed on reconstituted nucleosomes.

antiparallel orientation with respect to the duplex purine strand, while CT-containing oligonucleotides bind in a parallel orientation.

We have shown that nucleosome core particles can affect the formation of intermolecular DNA triplexes on DNA fragments, depending on the location of the polypurine target site. The position of the target site relative to the nucleosome core is important. Triplex formation is not affected at sites close (30 bp) to the end of the nucleosomal DNA, but is strongly inhibited at sites closer to the centre of the nucleosome.

13.4 Applications

Triplex technology has found many applications in molecular biology, some of which are described below.

13.4.1 Specific gene inactivation

The formation of intermolecular triple helices enables the design of compounds endowed with precise sequence recognition properties which have been shown to possess antigene activity in several systems.

In vitro *studies.* A number of studies have shown that triplex formation upstream of a transcriptional start site can inhibit the binding of transcription factors. Cooney *et al*. (1988) and Kim and Miller (1995) have shown the binding of oligonucleotides to the P1 and P2 promoters of the c-*myc* gene. Other cancer-related genes which have been targeted by triplex formation within their promoter regions include Her-2/*neu* (Ebbinghaus *et al.*, 1993), Ki-*ras* (Mayfield *et al.*, 1994a) and Ha-*ras* (Mayfield *et al.*, 1994b). Triplex formation can suppress binding of transcription factor Sp1 (Maher *et al.*, 1992) as too can transcription from the adenovirus major late promoter (Young *et* al., 1991) and the T7 promoter (Alunni-Fabbroni *et al.*, 1994).

In vivo s*tudies.* One of the most promising aspects of triple helix development is the demonstration that triplex-forming oligonucleotides can inhibit gene expression *in vivo*. Some of the published examples are as follows. Purine-rich oligonucleotides have been shown to inhibit c-*myc* transcription in HeLa cells (Postel *et al.*, 1991). An oligonucleotide targeted to the IL2Rα receptor promoter region inhibited its expression in T lymphocytes (Orson *et al.*, 1991). A separate group targeted the NF-κB transcription factor-binding site in the IL2Rα promoter with an acridine-linked oligonucleotide, and demonstrated inhibition of transcription on transfecting an IL2Rα–CAT reporter construct into human T cells

(Grigoriev *et al.*, 1992). Using this co-transfection approach, Macaulay *et al.* (1995) inhibited expression of P450 aromatase using psoralen-linked oligonucleotides, while Roy (1993, 1994) targeted the interferon response element. An oligonucleotide targeted to the Sp1-binding site of HIV-1 has been shown to inhibit viral transcription (McShan *et al.*, 1992), as too has transcription of a progesterone-responsive element (Ing *et al.*, 1993).

13.4.2 Site-specific cleavage

Triplex-forming oligonucleotides can be used as artificial endonucleases by linking them to a cleavage moiety. This can be achieved by covalently attaching the oligonucleotide to chemical groups such as EDTA (Moser and Dervan, 1987), copper phenanthroline (Shimizu *et al.*, 1994) or alkylating agents (Grant and Dervan, 1996; Povsic *et al.*, 1992). Photochemical agents such as psoralen (Takasugi *et al.*, 1991) or ellipticine can be attached, as can enzymes such as staphylococcal nuclease (Pei *et al.*, 1990). In an alternative 'Achilles' heel' strategy, site-specific cleavage of yeast chromosome III (340 kb) has been achieved by using the triplex to protect from *Eco*RI restriction methylase (Strobel and Dervan, 1991; Strobel *et al.*, 1991). On removing the third strand oligonucleotide, all the *Eco*RI sites are protected from cleavage, except the one which overlaps with the triplex-binding site.

13.4.3 Triplex affinity capture

Triplex formation has also been used as a method for the purification of DNA fragments containing specific polypurine tracts. The first reports of this method captured DNA fragments containing specific polypurine sequences on agarose-linked polyribonucleotides (Fox, 1992). Duplex DNA fragments containing the correct polypurine sequences bound to the polymer under triplex-forming conditions (low pH and the presence of magnesium). Bound DNA was subsequently recovered under triplex-destabilizing conditions (high pH and EDTA). A more recent development of this strategy uses biotinylated oligonucleotides, attached to streptavidin-coated magnetic beads, to enrich DNA fragments containing specific triplex-binding sites (Kiyama *et al.*, 1994; Nishikawa *et al.*, 1995).

13.4.4 Site-specific mutagenesis

Triplex-mediated mutagenesis has been achieved using oligonucleotides linked to psoralen. On ultraviolet irradiation, the psoralen moiety forms stable cross-links at TpA steps adjacent to the triplex. Site-specific mutations arise from defective excision–repair across the psoralen cross-link (Havre *et al.*, 1993; Raha *et al.*, 1996). This approach is limited by the unpredictable nature of the mutations, and the small fraction of psoralen damage that results in mutations.

13.5 Outstanding problems

The above sections have demonstrated the principles for designing oligonucleotides for recognizing unique DNA sequences. However, despite these promising developments in triplex formation, there are many obstacles to overcome and questions that need to be addressed before this becomes a realistic therapeutic strategy. One major problem is that of stability, since simple unmodified oligonucleotides are rapidly degraded by serum nucleases. The development of oligonucleotides with modified backbones and/or terminal protecting groups may be important in this regard. Toxicity is not a problem with natural oligonucleotides since the degradation products are natural cellular components. However, the metabolism of modified oligonucleotides may produce nucleoside derivatives which may become incorporated into the host genome with genotoxic consequences.

A second problem is that of delivery; although there do appear to be specific uptake mechanisms for oligonucleotides, they do not readily enter cells and do not cross the blood–brain barrier. Specific cellular mechanisms for delivering therapeutic oligonucleotides, such as liposomes, are needed. The cost of therapeutic oligonucleotides must also be considered since, although the design is straightforward, the synthesis is much more expensive than conventional medicines.

One further question to be addressed concerns which sequences should be targeted so as to produce the most effective gene inhibition. In theory, there are several ways in which these agents might act. They could be targeted to promoter regions, inhibiting the binding of transcription factors and other accessory proteins, or to coding regions, inhibiting the progression of RNA polymerase. Alternatively triplex-forming oligonucleotides could be designed to inhibit DNA replication by interacting with replication origins. In practice, this question is largely dictated by the location of polypurine tracts, which are often found in promoter regions.

Acknowledgements

Work in the author's laboratory is supported by grants from the Cancer Research Campaign and the Medical Research Council.

References

Alunni-Fabbroni MG, Manfioletti G, Manzini G and Xodo LE. (1994) Inhibition of T7 RNA polymerase transcription by phosphate and phosphorothioate triplex-forming oligonucleotides targeted to a RY site downstream from the promoter. *Eur. J. Biochem.* **226**, 831–839.

Arents G, Burlinghame RW, Wang BC, Love WE and Moudrianakis WN. (1991) The nucleosome core histone octamer at 3.1Å resolution – a tripartite protein assembly and a left handed superhelix. *Proc. Natl Acad. Sci. USA* **88**, 10148–10152.

Arnott S and Selsing E. (1974) Structures of polynucleotide complexes poly(dA)poly(dT) and poly(dT)poly(dA)poly(dT). *J. Mol. Biol.* **88**, 509–521.

Bates PJ, Laughton CA, Jenkins TC, Capaldi DC, Roselt PD, Reese CB and Neidle S. (1996) Efficient triple helix formation by oligodeoxyribonucleotides containing α- or β-2-amino-5-(2-deoxy-D-ribofuranosyl) pyridine residues. *Nucleic Acids Res.* **24**, 4176–4184.

Beal PA and Dervan PB. (1991) Second structural motif for recognition of DNA by oligonucleotide-directed triple-helix formation. *Science* **251**, 1360–1363.

Beal PA and Dervan PB. (1992a) The influence of single base triplet changes on the stability of a Pur·Pur·Pyr triple helix determined by affinity cleaving. *Nucleic Acids Res.* **20**, 2773–2776.

Beal PA and Dervan PB. (1992b) Recognition of double helical DNA by alternate strand triple helix formation. *J. Am. Chem. Soc.* **114**, 4976–4982.

Brown PM, Drabble A and Fox KR. (1996) Effect of a triplex-binding ligand on triple helix formation at a site within a natural DNA fragment. *Biochem. J.* **314**, 427–432.

Brown PM and Fox KR. (1996) Nucleosome core particles inhibit DNA triple helix formation. *Biochem. J.* **319**, 607–611.

Cassidy SA, Strekowski L, Wilson WD and Fox KR. (1994) Effect of a triplex binding ligand on parallel and antiparallel triple helices using short unmodified and acridine-linked oligonucleotides. *Biochemistry* **33**, 15338–15347.

Cassidy SA, Strekowski L and Fox KR (1996) DNA sequence specificity of a naphthylquinoline triple-helix binding ligand. *Nucleic Acids Res.* **24**, 4133–4138.

Chandler SP and Fox KR. (1993) Triple helix formation at $A_8XA_8 \cdot T_8YT_8$. *FEBS Lett.* **332**, 189–192.

Chandler SP and Fox KR. (1995) Extension of DNA triple helix formation to a neighbouring $(AT)_n$ site. *FEBS Lett.* **360,** 21–25.

Chandler SP and Fox KR. (1996) Specificity of antiparallel DNA triple helix formation *Biochemistry* **35**, 15038–15048.

Chandler SP, Strekowski L, Wilson WD and Fox KR. (1995) Footprinting studies on ligands which stabilise DNA triplexes: effects of stringency within a parallel triple helix. *Biochemistry* **34**, 7234–7242.

Cooney M, Czernuszewicz G, Postel EH, Flint SJ and Hogan ME. (1988) Site-specific oligonucleotide binding represses transcription of the human c-*myc* gene *in vitro*. *Science* **241**, 456–459.

Dittrich K, GuJ Tinder, R Hogan, M and Gao X. (1994) T·C·G triplet in an antiparallel purine·purine·pyrimidine DNA triplex. Conformational studies by NMR. *Biochemistry* **33**, 4111–4120.

Drew HR and Travers AA. (1985) DNA bending and its relation to nucleosome positioning. *J. Mol. Biol.* **186**, 773–790.

Durland RH, Rao TS, Revankar GR, Tinsley JH, Myrick MA, Seth DM, Rayford J, Singh P and Jayaraman K. (1994) Binding of T and T analogs to CG base pairs in antiparallel triplexes. *Nucleic Acids Res.* **22,** 3233–3240.

Ebbinghaus SW, Ge, JE, Rodu B, Mayfield CA, Sanders G and Miller DM. (1993) Triplex formation inhibits HER-2/*neu* transcription *in vitro*. *J. Clin. Invest.* **92**, 2433–2439.

Egholm M, Christensen L, Dueholm KL, Buchardt O, Coull J and Nielsen PE. (1995) Efficient pH-independent sequence-specific DNA binding by pseudoisocytosine-containing bis-PNA. *Nucleic Acids Res.* **23**, 217–222.

Escudé C, Sun J-S, Nguyen CH, Bisagni E, Garestier T and Hélène C. (1996) Ligand-induced formation of triple helices with antiparallel third strands containing G and T. *Biochemistry* **35**, 5735–5740.

Felsenfeld G, Davies DR and Rich A. (1957) Formation of a three-stranded polynucleotide molecule. *J. Am. Chem. Soc.* **79**, 2023–2024.

Fox KR. (1992) Wrapping of genomic polydA·polydT around nucleosome core particles. *Nucleic Acids Res.* **20**, 1235–1242.

Fox KR. (1994) Formation of DNA triple helices incorporating blocks of G·GC and T·AT triplets using short acridine-linked oligonucleotides. *Nucleic Acids Res.* **22**, 2016–2021.

Fox KR, Polucci P, Jenkins TC and Neidle S. (1995) A molecular anchor for stabilising triple-helical DNA. *Proc. Natl Acad. Sci. USA* **92**, 7887–7891.

Grant KB and Dervan PB. (1996) Sequence-specific alkylation and cleavage of DNA mediated by purine motif triple helix formation. *Biochemistry* **35**, 12313–12319.

Griffin LC and Dervan PB. (1989) Recognition of thymine·adenine base pairs by guanine in a pyrimidine triple helix motif. *Science* **245**, 967–971.

Grigoriev M, Praseuth D, Robin P, Hemar A, Saison-Behmoaras T, Dautry-Varsat A, Thuong NT, Helene C and Harel-Bellan A. (1992) A triple helix forming oligonucleotide–intercalator conjugate acts as a transcriptional repressor via inhibition of NF-κB binding to interleukin-2 receptor α-regulatory sequence. *J. Biol. Chem.* **267**, 3389–3395.

Hampel KJ, Crosson P and Lee JS. (1991) Polyamines favour DNA triple helix formation at neutral pH.

Havre PA, Gunther EJ, Gasparro FP and Glazer PM. (1993) Targeted mutagenesis of DNA using triple helix-forming oligonucleotides linked to psoralen. *Proc. Natl Acad. Sci. USA* **90**, 7879–7883.

Hélène C. (1993) Sequence-selective recognition and cleavage of double-helical DNA. *Curr. Opin. Biotechnol.* **4**, 29–36.

Horne DA and Dervan PB. (1990) Recognition of mixed-sequence duplex DNA by alternate strand triple-helix formation. *J. Am. Chem. Soc.* **112**, 2435–2437.

Horne DA and Dervan PB. (1991) Effects of an abasic site on triple helix formation characterised by affinity cleavage. *Nucleic Acids Res.* **19**, 4963–4965.

Huang C-Y, Bi G and Miller PS. (1996) Triplex formation by oligonucleotides containing novel deoxycytidine derivatives. *Nucleic Acids Res.* **24**, 2606–2613.

Ing NH, Beekman JM, Kessler DJ, Murphy M, Jayaraman K, Zendegui JG, Hogan ME, O'Malley BW and Tsai M-J. (1993) *In vivo* transcription of a progesterone-responsive gene is specifically inhibited by a triplex-forming oligonucleotide. *Nucleic Acids Res.* **21**, 2789–2796.

Jayasena S and Johnston BH. (1992a) Intramolecular triple-helix formation at $(Pu_nPy_n)(Pu_nPy_n)$ tracts: recognition of alternate strands via Pu·PuPy and Py·PyPu base triplets. *Biochemistry* **31**, 320–327.

Jayasena SD and Johnston BH. (1992b) Oligonucleotide-directed triple helix formation at adjacent oligopurine DNA tracts by alternate strand recognition. *Nucleic Acids Res.* **20**, 5279–5288.

Jetter MC and Hobbs FW. (1993) 7,8-Dihydro-8-adenine as a replacement for cytosine in the third strand of triple helices. Triplex formation without hypochromicity. *Biochemistry* **32**, 3249–3254.

Ji J, Hogan, ME and Gao XL. (1996) Solution structure of an antiparallel purine motif triplex containing a T·CG pyrimidine base triple. *Structure* **4**, 425–435.

Kim H-G and Miller DM. (1995) Inhibition of *in vitro* transcription by a triplex-forming oligonucleotide targeted to human c-*myc* P2 promoter. *Biochemistry* **34**, 8165–8171.

Kiyama R, Nishikawa N and Oishi M. (1994) Enrichment of human DNAs that flank poly(dA)poly(dT) tracts by triplex DNA formation. *J. Mol. Biol.* **237**, 193–200.

Koh JS and Dervan PB. (1992) Design of a nonnatural deoxyribonucleoside for recognition of GC base pairs by oligonucleotide-directed triple helix formation. *J. Am. Chem. Soc.* **114**, 1470–1478.

Le Doan T, Perrouault L, Praseuth D, Habhoub N, Decout JL, Thuong NT, Lhomme J and Hélène C. (1987) Sequence specific recognition, photo-crosslinking and cleavage of the DNA double helix by an α-oligothymidylate covalently linked to an azidoproflavine derivative. *Nucleic Acids. Res.* **15**, 7749–7760.

Lee JS, Wordsworth ML, Latimer LJP and Morgan AR. (1984) Poly(pyrimidine) poly(purine) synthetic DNAs containing 5-methylcytosine from stable triplexes at neutral pH. *Nucleic Acids. Res.* **12**, 6603–6614.

Lee JS, Latimer LJP and Hampel KJ. (1993) Coralyne binds tightly to both T·A·T- and $C{\cdot}G{\cdot}C^+$-containing triplexes. *Biochemistry* **32**, 5591–5597.

Macaulay VM, Bates BJ, McLean MJ, Rowlands MG, Jenkins TC, Ashworth A and Neidle S. (1995) Inhibition of aromatase expression by a psoralen-linked triplex-forming oligonucleotide targeted to a coding sequence. *FEBS Lett.* **372**, 222–228.

Maher LJ, Dervan PB and Wold B. (1992) Analysis of promoter-specific repression by triple-helical DNA complexes in a eukaryotic cell-free transcription system. *Biochemistry* **31**, 70–81.

Malkov VA, Voloshin ON, Soyfer VN and Frank-Kamenetskii MD. (1993) Cation and sequence effects on stability of intermolecular pyrimidine–purine–purine triplex. *Nucleic Acids Res.* **21**, 585–591.

Mayfield C, Squibb M and Miller D. (1994a) Inhibition of nuclear protein binding to the human ki-ras promoter by triplex-forming oligonucleotides. *Biochemistry* **33**, 3358–3363.

Mayfield C, Ebbinghaus S, Gee J, Jones D, Rodu B, Squibb M and Miller DM. (1994b) Triplex formation by the Ha-ras promoter inhibits Sp1 binding and *in vitro* transcription. *J. Biol. Chem.* **269**, 18232–18238.

McShan WM, Rossen RD, Laughter AH, Trial J, Kessler DJ, Zendegui JG, Hogan ME and Orson FM. (1992) Inhibition of transcription of HIV-1 in infected human cells by oligodeoxynucleotides designed to form DNA triple helices. *J. Biol. Chem.* **267**, 5712–5721.

Mergny JL, Duval-Valentin G, Nguyen CH, Perrouault L, Faucon B, Rougee M, Montenay-Garestier T, Bisagni E and Hélène C. (1992) Triple-helix specific ligands. *Science* **256**, 1681–1684.

Miller PS, Bhan P, Cushman CD and Trapane TL. (1992) Recognition of guanine–cytosine base pair by 8-oxoadenine. *Biochemistry* **31**, 2999–3004.

Moser HE and Dervan PB. (1987) Sequence-specific cleavage of double-helical DNA by triple helix formation. *Science* **238**, 645–650.

Nishikawa N, Oishi M and Kiyama R. (1995) Construction of a human genomic library of clones containing poly(dG-dA)·poly(dT-dC) tracts by Mg^{2+}-dependent triplex affinity capture. *J. Biol. Chem.* **270**, 9258–9264.

Ono A, Chen C-N and Kan L. (1991a) DNA triplex formation of oligonucleotide analogues consisting of linker groups and octamer segments that have opposite sugar–phosphate backbone polarities. *Biochemistry* **30**, 9914–9921.

Ono A, Ts'o POP and Kan L. (1991b) Triplex formation of oligonucleotides containing 2′-*O*-methylpseudoisocytidine in substitution for 2′-deoxycytidine. *J. Am. Chem. Soc.* **113**, 4032–4033.

Orson FM, Thomas DW, McShan WM, Kessler DJ and Hogan ME. (1991) Oligonucleotide inhibition of IL2Rα mRNA transcription by promoter regions collinear triplex formation in lymphocytes. *Nucleic Acids Res.* **19**, 3435–3441.

Pei D, Corey DR and Schultz PG. (1990) Site-specific cleavage of duplex DNA by a semisynthetic nuclease via triple-helix formation. *Proc. Natl Acad. Sci. USA* **87**, 9858–9862.

Pilch DS, Waring MJ, Sun J-S, Rougee M, Nguyen C-H, Bisagni E, Garestier T and Hélène C. (1993a) Characterization of a triple helix specific ligand. BePI {3-methoxy-7H-8-methyl-11-[(3′-amino)propylamino]-benzo[e]pyrido[4,3-b] indole} intercalates onto both double-helical and triple-helical DNA. *J. Mol. Biol.* **232**, 926–946.

Pilch DS, Martin M-T, Nguyen CH, Sun J-S, Bisagni E, Garestier T and Hélène C. (1993b) Self-association and DNA binding properties of two triple helix specific ligands: comparison of a benzo[e]pyridoindole and a benzo[g]pyridoindole. *J. Am. Chem. Soc.* **115**, 9942–9951.

Postel EH, Flint SJ, Kessler DJ and Hogan ME. (1991) Evidence that a triplex-forming oligodeoxyribonucleotide binds to the c-*myc* promoter in HeLa cells, thereby reducing c-*myc* mRNA levels. *Proc. Natl Acad. Sci. USA* **88**, 8227–8231

Povsic TJ and Dervan PB. (1989) Triple helix formation by oligonucleotides on DNA extended to the physiological pH range. *J. Am. Chem. Soc.* **111**, 3059–3061.

Povsic TJ, Strobel SA and Dervan PB. (1992) Sequence-specific double-strand alkylation and cleavage of DNA mediated by triple-helix formation. *J. Am. Chem. Soc.* **114**, 5934–5941.

Radhakrishnan I and Patel DJ. (1994) Solution and hydration patterns of a pyrimidine·purine·pyrimidine DNA triplex containing a novel T·CG base triple. *J. Mol. Biol.* **241**, 600–619.

Radhakrishnan I, Patel DJ, Veal JM and Gao X. (1992) Solution conformation of a G·TA triple in a intramolecular pyrimidine·purine·pyrimidine DNA triplex. *J. Am. Chem. Soc.* **114**, 6913–1915.

Raha M, Wang G, Seidman MM and Glazer PM. (1996) Mutagenesis by third-strand-directed psoralen adducts in repair-deficient human cells: high frequency and altered spectrum in a xeroderma pigmentosum variant. *Proc. Natl Acad. Sci. USA* **93**, 2941–2946.

Richmond TJ, Finch JT, Rushton B, Rhodes D and Klug A. (1984) Structure of the nucleosome core particle at 7Å resolution. *Nature* **311**, 532–537.

Roy C. (1993) Inhibition of gene transcription by purine rich triplex forming oligonucleotides. *Nucleic Acids Res.* **21** 2845–2852.

Roy C. (1994) Triple-helix formation interferes with the transcription and hinged DNA structure of the interferon-inducible 6-16 gene promoter. *Eur. J. Biochem.* **220**, 493–503.

Shimizu M, Inoue H and Ohtsuka E. (1994) Detailed study of sequence-specific DNA cleavage of triplex-forming oligonucleotides linked to 1,10-phenanthroline. *Biochemistry* **33**, 606–613.

Singleton SF and Dervan PB. (1992) Influence of pH on the equilibrium association constants for oligodeoxyribonucleotide-directed triple helix formation. *Biochemistry* **31**, 10995–11003.

Soyfer VN and Potaman VN. (1996) *Triple-Helical Nucleic Acids.* Springer, Germany.

Strobel SA and Dervan PB. (1991) Single-site enzymatic cleavage of yeast genomic DNA mediated by triple helix formation. *Nature* **350**, 172–174.

Strobel SA, Doucette-Stamm LA, Riba L, Housman DE and Dervan PB. (1991) Site-specific cleavage of human chromosome 4 mediated by triple-helix formation. *Science* **254**, 1639–1642.

Sun JC, Giovannangeli C, Francois JC, Kurfurst K, Monteny-Garestier T, Asseline U, Saison-Behmoaras T, Thuong NT and Hélène C. (1991) Triple-helix formation by α-oligodeoxynucleotides and α-oligodeoxynucleotide–intercalator conjugates. *Proc. Natl. Acad. Sci. USA* **88**, 6023-6027.

Sun J-S, Francois J-C, Montenay-Garestier T, Saison-Behmoaras T, Roig V, Thuong NT and Hélène C. (1989) Sequence specific intercalating agents: intercalation at specific sequences on duplex DNA via major groove recognition by oligonucleotide–intercalator conjugates. *Proc. Natl Acad. Sci. USA* **86**, 9198–9202.

Svinarchuk F, Bertrand J-R and Malvy C. (1994) A short purine oligonucleotide forms a highly stable triple helix with the promoter of the murine c-*pim*-1 proto-oncogene. *Nucleic Acids Res.* **22,** 3742–3747.

Svinarchuk F, Cherny D, Debin A, Delain E and Malvy C. (1996) A new approach to overcome potassium-mediated inhibition of triplex formation. *Nucleic Acids Res.* **24**, 3858–3865.

Takasugi M, Guendouz A, Chassignol M, Decout JL, Lhomme J, Thuong NT and Hélène C. (1991) Sequence-specific photo-induced cross-linking of the two strands of double-helical DNA by a psoralen covalently linked to a triple helix-forming oligonucleotide. *Proc. Natl Acad. Sci. USA* **88,** 5602–5605.

Thuong NT and Hélène C. (1993) Sequence-specific recognition and modification of double-helical DNA by oligonucleotides. *Angew. Chem.* **32**, 666–690.

Washbrook E and Fox KR. (1994a) Alternate-strand DNA triple helix formation using short acridine-linked oligonucleotides. *Biochem. J.* **301**, 569–575.

Washbrook E and Fox KR. (1994b) Comparison of antiparallel A·AT and T·AT triplets within an alternate strand DNA triple helix. *Nucleic Acids Res.* **22**, 3977–3982.

Westin L, Blomquist P, Milligan JF and Wrange O. (1995) Triple helix DNA alters nucleosomal histone–DNA interactions and acts as a nucleosome barrier. *Nucleic Acids Res.* **23**, 2184–2191.

Wilson WD, Tanious FA, Mizan S, Yao S, Kiselyov AS, Zon G and Strekowski L. (1993) DNA triple-helix specific intercalators as antigene enhancers: unfused aromatic cations. *Biochemistry* **32**, 10614–10621.

Yoon K, Hobbs CA, Koch J, Sardaro M, Kutny R and Weis A. (1992) Elucidation of the sequence-specific third-strand recognition of four Watson–Crick base pairs in a pyrimidine triple-helix motif: T·T, C·GC, T·CG, and G·TA. *Proc Natl Acad. Sci. USA* **89**, 3840–3844.

Young SL, Krawczyk SH, Matteucci MD and Toole JJ. (1991) Triple helix formation inhibits transcription elongation *in vitro*. *Proc. Natl Acad. Sci. USA* **88**, 10023–10026.

14

Protein engineering of therapeutic antibodies: use of antibodies for immunosuppression and treatment of leukaemias

Mike Clark

14.1 Introduction

As a result of the steady and continued accumulation of knowledge concerning the precise relationships between immunoglobulin structure and function, and also through advances in the applications of molecular biological techniques, it has become almost routine practice to attempt to engineer the perfect antibody for any particular therapy (Clark, 1997a; Routledge *et al.*, 1993). This chapter will illustrate this advance in applications of the new technologies with respect to examples of antibodies targeted to human lymphocyte cell surface antigens, in particular CD52, CD3 and CD4. These antibodies can be used for immunosuppression as well as for treatment of lymphoid malignancies. However, it should be remembered that the principles will apply to a wider spectrum of uses of antibodies with many different specificities.

The protein engineering of antibodies is concerned principally with manipulating three physical properties of the antibody: (i) the binding to antigen so as to alter affinity and/or specificity of the antibody; (ii) attempting to reduce the immunogenicity of the antibody in the recipient; (iii) manipulation of the biological properties or effector functions triggered by the antibody (Clark, 1997a).

Genetics of Common Diseases: future therapeutic and diagnostic possibilities,
edited by I. Day and S. Humphries.

Interestingly, the enthusiasm of clinicians, pharmaceutical companies and biotechnology companies for the use of and development of antibody-based therapies seems to oscillate wildly and regularly between optimism and pessimism. An important starting point is to state the obvious in defence of an optimistic view of the applications of antibodies: that antibodies have evolved within the immune system to play a key role in the body's defence against infectious disease (Janeway and Travers, 1996; Male *et al.*, 1996). Millions of years of evolution have selected antibodies which have the appropriate properties to deal with a whole range of antigens and infectious agents. The success of this is evidenced by the fact that immunoglobulins are found within the immune systems of most higher vertebrates, including fishes, reptiles, birds and mammals (Warr *et al.*, 1995). During evolution, the immunoglobulin genes have become duplicated and altered to give rise to multiple isotypes or classes and subclasses of immunoglobulins within a single species, and this allows antibodies with different functional properties based on the different isotypes to be selected. Through a greater understanding of how antibodies function in the course of a natural immune response, we may learn how best to make and use them therapeutically.

The IgG class is an important class which has evolved within mammals, and it has a number of interesting and highly relevant properties which make it the principle antibody type used in therapy (Clark, 1995, 1997b).

(i) It possesses a hinge region between the Cγ1 and Cγ2 domain of the heavy chain which offers a great degree of flexibility with respect to allowing the two Fab arms to orientate independently and to bind to two antigens, while also allowing the Fc region of the antibody to reorientate with respect to binding the effector systems. This makes the antibody readily bivalent for many antigens, and the resulting co-operativity for binding of the two Fabs gives the antibody a higher avidity for antigen.

(ii) IgG antibodies can activate the complement cascade and are also capable of binding to specialized Fcγ receptors present on a range of cell types, uniting the humoral and cellular effector systems. In some species, such as rabbit, a single IgG class has all of these functions, whereas, in other species such as man, mouse and rat, several IgG subclasses have evolved with different effector function profiles (Clark, 1997b).

(iii) An important but frequently overlooked property of IgG is an unusually long biological half-life compared with other plasma proteins, including other immunoglobulin classes (Waldmann and

Strober, 1969). This long half-life ensures that a high level of IgG-based immunity can persist for many months after infection without the need for continuous production by plasma cells, which are relatively short lived.

(iv) The IgG class in mammals is involved in protection of the neonate by active uptake of maternal IgG by a receptor called FcRn (Simister and Mostov, 1989; Story *et al.*, 1994). In some species, such as man, this uptake is across the placenta, whereas in others, such as mice, rats and ruminants, it is transported from the gut in the colostrum.

A key feature of the protein engineering of antibodies is that the final immunoglobulin is encoded by an artificial gene construct which is usually expressed in a mammalian cell culture system from which the antibody can be purified from the culture supernatant. For this reason, there are many different starting points which can all easily lead to the construction of a similar recombinant antibody, so it is difficult to suggest a best strategy for all situations. In essence, the common starting point for an antibody engineering project is a knowledge of the sequence of an antibody of interest (Clark, 1995; Routledge *et al.*, 1993).

14.2 Choice and specificity of antibodies

There are three main sources of antibodies for use in human therapy. The first and major source are monoclonal antibodies made according to the Kohler and Milstein (1975) strategy of generating hybridomas from immunized mice and rats. Strategies for their production are well worked out, and they have been used to define many of the human cell surface antigens which we now use to define different cell subpopulations as a result of grouping the antibodies into the so-called 'cluster of differentiation' or CD groups (Barclay *et al.*, 1993). The rat antibodies CAMPATH-1 against CD52 (Hale *et al.*, 1983) and YTH12.5 against CD3 (Clark and Waldmann, 1987; Cobbold and Waldmann, 1984) were made, for example, by immunizing rats with human T lymphocytes and then fusing the spleen cells with the rat myeloma cell line Y3Ag1.2.3 (Galfre *et al.*, 1981). Hybridomas making anti-lymphocyte antibodies were cloned out from the original culture wells, and then the antibodies were further characterized.

Another source of antibodies of potential use in human therapy is to derive human antibodies from immune donors either by immortalizing the B cells [e.g. with Epstein-Barr virus (EBV)], or by cloning out the variable (V)-genes usage in phage display libraries, and then selecting these phage libraries on antigen (Clark, 1995; Parren and Burton, 1997).

These strategies are dependent upon access to suitable donors who are already immune or who can be ethically immunized, for example in the case of antibodies to blood group antigens such as RhD. In addition, the other problem which can arise is whether the appropriate B cells are present in the material which can be obtained from the donor – usually peripheral blood, but sometimes bone marrow or a lymphoid organ following a biopsy or splenectomy or even occasionally an autopsy (Parren and Burton, 1997). A third major potential source of antibodies for therapy comes from the screening of artificial phage antibody libraries generated using mutated or randomized V gene fragments to try to mimic the natural antibody repertoire (Griffiths and Hoogenboom, 1993).

Independently of how the original antibody was prepared, the specificity and affinity for antigen is essentially a property of the precise combination of the heavy and light chain V region sequences. In particular, the complementarity-determining regions (CDRs) from the heavy and light chain (three from each) generally form the principle interaction with antigen, although this can be further influenced by interactions of antigen with framework residues (FwR) or indirectly through interactions between FwR and CDR residues of the antibody. It may be possible to make changes to the sequence of the CDRs or FwRs of the antibody to increase the affinity for antigen (Routledge *et al.*, 1993). This could be achieved either through computer modelling or knowledge of the structure of the antibody–antigen complex or more often through selecting mutated or randomized versions of the antibody on antigen in an empirical fashion (Griffiths and Hoogenboom, 1993). However, it should be noted that high affinity may not be as crucial as some believe and, for example, the CAMPATH-1G antibody has what might be considered a relatively modest avidity of about 3×10^{-8} M (dissociation constant). The biological effector systems of complement and FcγR have evolved to work in the presence of an excess plasma IgG concentration of about 10 g l^{-1} (approximately 10^{-6} M). In many respects it is the specificity of the antibody rather than the affinity which is most crucial, and in this case the two need not be directly related, as specificity is a function of the particular epitope recognized as well as the degree with which the antibody has a cross-reacting affinity for other antigens. For example, it is possible to have a high affinity and low specificity for a given antigen or even the opposite – a low affinity yet high specificity for a given antigen.

14.3 Immunogenicity

Early therapeutic strategies involved treating patients with rodent monoclonal antibodies. These therapies often were quite successful and, for example, the mouse IgG2a, CD3-specific antibody OKT3 (Ortho Multicentre Transplant Study Group, 1985) and the rat IgG2b antibody CAMPATH-1G (Dyer *et al.*, 1989; Dyer *et al.*, 1990) were able to achieve profound effects *in vivo* in terms of immunosuppression and leukaemia cell clearance. However, a major problem encountered is the patient's anti-globulin response to the rodent-specific residues, within both the V region and the constant region of the antibody. This problem was initially tackled by making chimaeric antibody constructs from the monoclonal antibodies using the rodent variable and human constant region gene segments (Boulianne *et al.*, 1984; Neuberger *et al.*, 1985). However, we also know that the structure of the V region is such that the main antigen-binding residues are found in the CDR loops at the ends of more conserved framework residues making up the β-barrel of the characteristic immunoglobulin fold. The region segments fall into families based on sequence homologies of their frameworks, but within a species there are likely to be FwR sequences which are characteristic of the species (Kabat *et al.*, 1991). Thus came about the concept of transplanting the CDR loops from a rodent sequence on to the FwRs of a human sequence and creating a fully reshaped, humanized antibody which, when compared with sequence databases, would appear to be human derived (Riechmann *et al.*, 1988; Routledge *et al.*, 1993).

In humanizing an antibody, the rodent-derived CDRs from the monoclonal antibody of choice provide the specificity, so these residues are automatically included in the design of the reshaped V region. On the other hand, FwRs in theory can be derived from any human V region and the resultant designed V region would be considered reshaped (Routledge *et al.*, 1993). The first antibody reshaping experiments were performed on anti-nitrophenacetyl (anti-NP) (Jones *et al.*, 1986) and then subsequently on anti-lysozyme (Verhoeyen *et al.*, 1988) and on CAMPATH (Riechmann *et al.*, 1988) antibodies. In the case of NP and lysozyme, the expression of rodent-derived CDRs, along with the framework sequences of the human V_H region NEW, was successful. However, when the same human framework sequences were used in the reshaping of a therapeutic CD52 antibody CAMPATH–1G, the affinity of the antibody for antigen was unacceptable until the serine residue at position 27 had been replaced with phenylalanine, the corresponding residue in the rodent framework. Similar framework changes have been required for other antibodies (Riechmann *et al.*, 1988).

As a result of the problems encountered with the original strategy for reshaping, the so-called 'best fit' strategies for framework selection and design were developed (Gorman *et al.*, 1991; Routledge *et al.*, 1991, 1993).

In the 'best fit' strategy the human framework sequence used for antibody reshaping is derived from the human V region that is most homologous or similar to the rodent-derived V region. It may be assumed that the fewer the changes which need to be made to the selected human framework sequences, the more likely it will be that the affinity will be retained in the reshaped antibody.

At its simplest level, the 'best fit' strategy involves comparing the donor rodent V region with all known human V region amino acid sequences, and then selecting the most homologous to provide the acceptor framework regions for the humanization exercises. In reality, there are several other factors which should be considered, and which may influence the final selection of acceptor FwRs (Routledge *et al.*, 1993). Data has been steadily accumulating on the total number of V region segments available in the human repertoire and also on those which are only present in some individuals represented as allelic genes. This directory of sequences can be used to refine further the choice of frameworks to be used in a reshaping exercise. Logically, the framework of choice should be present at a high frequency in the population (i.e. nearly all individuals) as well as having high homology with the rodent sequence.

14.4 Biological properties

As mentioned above, the exact number of IgG subclasses varies between species. Thus, for example, in species such as the rabbit the single IgG class of antibody seems able to activate complement as well as to bind to Fcγ receptors; however, in several other species, gene duplication events have led to multiple IgG subclasses (e.g. human, rat and mouse having four IgG subclasses each), and then, through changes in sequence, some of these subclasses have lost the ability to activate complement or to bind to some FcγR types (reviewed in Clark, 1997b). Although IgG antibodies and their associated effector systems are obviously conserved and homologous between species, the numbers of IgG subclasses vary and many of the duplications seem to have occurred after speciation. This suggests that there may not be direct functional equivalence between subclasses in each species.

A striking feature of the human IgG subclasses is the very high sequence homology between them. Within the constant region domains, C_H1, C_H2 and C_H3, this is greater than 90% at the amino acid level, whilst the biggest differences are within the hinge region, which differ in both sequence and length between subclasses (reviewed in Clark, 1997b). The hinge of the subclass IgG3 is actually encoded by four repeats of an exon homologous to the hinge region of IgG1. Clearly the functional differences observed in comparisons of the human IgG subclasses must be dependent on some or all of these sequence differences, and thus a knowledge of the crucial sequences might reveal ways in which novel effector function properties could be introduced into recombinant antibodies such that they could be exploited in therapeutic strategies.

With regard to the four human IgG subclassess, a logical starting place for these investigations concerned investigation of the role of the hinge region in complement activation. The site of interaction of C1q with IgG was thought to be within the C_H2 domain, and a C1q-binding motif had been identified in mouse IgG2b compared with mouse IgG1. However, the four human IgG subclasses are identical within the region of this motif and are very homologous for the rest of the C_H2 (Duncan and Winter, 1988). This led to the proposal that the difference between IgG3 and IgG4 was therefore dependent upon the former having a long and very flexible hinge region, whilst the latter has a short and more rigid hinge. However, hinge swap mutants of IgG3 and IgG4 anti-DNS (5-dimethylamino-l-naphtalene sulphonyl dansyl chloride) antibodies, and further experiments with C_H domain swap mutants involving IgG1, 3 and 4 with specificity for the hapten DNS, and then domain swaps involving IgG1 and IgG4 with specificity for the lymphocyte antigen CD52 confirmed that the genetic hinge region has only a marginal effect on complement activation and that the crucial differences between IgG4 and IgG1 are in the COOH-terminal half of the C_H2 domain (Greenwood *et al.*, 1993; Tan *et al.*, 1990; Tao *et al.*, 1991). It was shown that a substitution of serine at position 331 in IgG4 for proline (as in IgG1, 2 and 3) endows IgG4 with the ability to activate complement (Brekke *et al.*, 1994; Tao *et al.*, 1993).

Further evidence has recently cast more doubt on the applicability of the described mouse C1q-binding motif for human IgG subclasses. Using a human IgG1 antibody specific for human MHC class II, HLA-DR antigen, mutations were introduced to attempt to alter complement and FcγR binding. Surprisingly, two results with this human IgG1 antibody show a big difference from earlier studies using mouse IgG2b antibodies specific for NP (Morgan *et al.*, 1995). Firstly, a change in the residue 320, previously reported from experiments with mouse IgG2b as being a crucial residue for C1q binding, from Lys to Ala had no effect on complement. In contrast, a change in the residue Leu235 to Glu, which had previously been implicated in FcγRI binding but not in complement activation using the mouse IgG2b, abolished complement lysis by the human IgG1. Equally the hinge disulphides, as well as hinge length, and also the disulphide bonding of heavy and light chains does seem to have some role to play in determining antibody function, at least for some human IgG subclasses specific for some antigens (reviewed in Clark, 1997b).

In understanding the interactions of IgG antibodies with Fc receptors, it is necessary to take account of the different classes of receptor. In humans there are three identified classes of Fc receptors for human IgG (FcγR) (reviewed in Deo *et al.*, 1997; Ravetch, 1994; van de Winkel and Capel, 1993).

(i) Human FcγRI (CD64) can bind monomeric IgG with high affinity and is expressed constitutively on macrophages and monocytes and can be induced on neutrophils and eosinophils.
(ii) Human FcγRII (CD32) binds IgG only in complexed or polymeric forms and is widely expressed on a range of cell types, including monocytes, macrophages, basophils, eosinophils, Langerhans cells, B cells and platelets.
(iii) Human FcγRIII (CD16) is also a medium to low affinity receptor and is expressed on macrophages, some large granular lymphocytes (LGL), killer cells (K-cell), some natural killer cells (NK-cells) and neutrophils and can be induced on eosinophils and monocytes.

For several reasons, the expression of these three receptor classes on different cell types is complex.

(i) Each of these three classes of receptor are encoded by a cluster of closely related genes, *FcγRIA, B* and *C, FcγRIIA, B* and *C* and *FcγRIIIA* and *B*.
(ii) Most of these genes encode transmembrane receptors which either have cytoplasmic domains capable of signal transduction or which associate with signalling co-receptor complexes. For example, the FcγRIIIa form is a transmembrane receptor expressed in conjunction with γ, ζ and β chains, and is found on K/NK-cells, monocytes and macrophages.
(iii) In contrast to the other Fcγ receptors, the human (but not the mouse) FcγRIIIB, found on neutrophils and eosinophils, is not transmembrane but is instead a glycosyl-phosphatidylinositol (GPI)-anchored receptor.
(iv) Additional complexity arises because some of these genes then give rise to multiple transcripts (e.g. *FcγRIIb1, b2* and *b3).*
(v) Finally some of these genes exist within the population in different polymorphic forms e.g. *FcγRIIa-R131*/*FcγRIIa-H131* and *FcγRIIIbNA1*/*FcγRIIIbNA2*).

Human IgG1 and IgG3 bind FcγRI with the highest affinity (K_d 10^{-8} – 10^{-9} M) followed by IgG4 which is about 10-fold weaker in its interaction while IgG2 does not readily bind to the receptor (Deo *et al.*, 1997; van de Winkel and Capel, 1993). Rat IgG2b and mouse IgG2a also bind human FcγRI readily whilst mouse IgG2b antibodies were found to be poor at binding. A sequence comparison of IgG classes indicated that residues in the IgG lower hinge region (encoded by the 5′ end of the Cγ2 exon) might be crucial, and changing the residue Glu235 to Leu, using site-directed mutagenesis, improved the affinity of mouse IgG2b for FcγRI by 100-fold (Duncan *et al.*, 1988). It was later shown that a region spanning 233–238 (GluLeuLeuGlyGlyPro in IgG1) was critical for binding, and the introduction of the whole of this sequence into IgG2 produced an antibody which bound with higher affinity than IgG1 (Chappel *et al.*, 1991). Investigation of the residues responsible for the lower binding affinity of IgG4 compared with IgG1 and 3 showed that a change of Phe 234 in IgG4 to Leu improved the affinity. In addition, it was found that in IgG3 the change of Pro331 for Ser as found in IgG4 decreased the affinity of IgG3 for FcγRI (Canfield *et al.*, 1991). The introduction of point mutations into NP-specific IgG3 antibodies, in which individual residues were changed to alanine, indicated the critical role of the residues Leu234, Leu235 and Gly237 (Jefferis *et al.*, 1990; Sarmay *et al.*, 1992).

Studies of the direct binding of IgG to the low affinity FcγRII are more difficult. CD3-specific antibodies are able to trigger a mitogenic response from T cells if the antibody Fc aggregates upon binding to Fc receptors on accessory cells, and this can also be used to assay binding function to low affinity receptors (Chappel *et al.*, 1991; Haagen *et al.*, 1995). However, most experimental systems employ some form of complexed or aggregated IgG, such as the rosetting of IgG-sensitized red cells by FcγRII-bearing leukocytes or alternatively IgG aggregated into small multimeric complexes using antigen or F(ab)2 fragments of anti-light chain-specific antibodies mixed in a one-to-one ratio with the IgG (Bredius *et al.*, 1993; Huizinga *et al.*, 1989). Generally, the observations that IgG1 and IgG3 bound well to FcγRII, but IgG4 and IgG2 did not, again suggested that the sequence differences in the hinge and lower hinge regions might be critical. Using rosette formation between red cells sensitized by point-mutated IgG3 antibodies with the cell lines Daudi and K562 has indicated that two of the crucial residues for binding are Leu234 and Leu237 (Huizinga *et al.*, 1989; Sarmay *et al.*, 1992).

As mentioned above, human *FcγRIIa* exists as two alleles, and this polymorphism leads to a functional difference in the ability of the receptor to discriminate between different IgG isotypes. This was originally identified for mitogenic responses with murine CD3 antibodies, but has been found to affect binding of rat and human isotypes (Haagen *et al.*, 1995; Tax *et al.*, 1984). Thus FcγRIIa-R131 binds mouse IgG1 but not human IgG2, while FcγRIIa-H131 does not bind mouse IgG1 but does bind human IgG2. Both forms of the receptor bind human IgG1 but do not bind human IgG4. Only a single amino acid change in the receptor (arginine for histidine at position 131) is responsible for this remarkable ability to discriminate between the different isotypes. Further studies have also indicated that the polymorphism affects binding of rat IgG2b (Haagen *et al.*, 1995). Rat IgG2b behaves in a similar way to human IgG2 and opposite to mouse IgG1; however, unlike human IgG2, the rat IgG2b also binds to the high affinity FcγRI receptor. Recent results suggest that functionally the *FcγRIIa* polymorphism may have important consequences with regard to resistance to certain infections (Deo *et al.*, 1997).

The FcγRIII receptor exists in two different forms in humans, either as a transmembrane receptor (FcγRIIIa) found on K- and NK-cells as well as activated macrophages or as a GPI-anchored molecule (FcγRIIIb) on cells such as neutrophils. Most studies have concentrated on functional assays of antibody-dependent cell-mediated cytotoxicity (ADCC) using effector cells expressing the transmembrane FcγRIIIa (reviewed in Clark, 1997b). The rat IgG2b and the human IgG1 antibodies to NIP and to CD52 were particular potent at triggering ADCC with human peripheral blood mononuclear cells and activated lymphocytes as the effectors, whereas human IgG4 and rat IgG2a were poor. A series of domain swap mutants between human IgG1 and human IgG4 were constructed to identify residues responsible for the observed functional differences (Greenwood *et al.*, 1993). It was verified that all of the critical differences between these two isotypes lie in the Cγ2 domain and secondly that they were in the COOH-terminal end of the Cγ2 domain, a similar result to complement activation as described above and involving four possible amino acid changes (296Tyr→Phe, 327Ala→Gly, 330Ala→Ser, 331Pro→Ser) with, in particular, the residue change Pro331 in IgG1 for Ser in IgG4 prominent. In contrast, using point mutations of residues in the lower hinge region of IgG3, the residues 235 and 237 were identified as critical for ADCC (reviewed in Clark, 1997b). The results obtained with the domain swap mutants of IgG1 and IgG4 antibodies were, however, further complicated as it was also found that the results obtained in this system were dependent upon the donor of the lymphocyte effectors. With some donors IgG1 was effective whilst IgG4 was ineffective, while in others it was found that the four isotypes IgG1, IgG2, IgG3 and IgG4 gave very similar levels of activity (Greenwood *et al.*, 1993 reviewed in Clark, 1997b).

IgG molecules have a highly conserved *N*-linked glycosylation site within the Cγ2 domain at Asp297 which has been found to be critical for complement-mediated lysis as well as binding to and activation of all three FcγR classes of receptor. Antibodies produced without carbohydrate, either through use of metabolic inhibitors, endoglycosidases by site-directed mutation of the attachment site, or produced in bacterial expression systems, all show greatly reduced biological functions (reviewed in Jefferis and Lund, 1997). An aglycosylated human IgG antibody with specificity for the mouse CD8 antigen was found not to deplete mouse CD8 lymphocytes *in vivo,* whereas the glycosylated human IgG subclasses were very effective at depleting cells (Isaacs *et al.*, 1992). This property has been exploited in the production of an aglycosylated form of the humanized IgG1 CD3 antibody which can block T-cell functions without depleting the cells or triggering cytokine release, thus eliminating some of the severe side effects of CD3 antibody therapy (Bolt *et al.*, 1993). While glycosylation of antibody is important for function, the precise structures attached to the IgG are complex and can vary from one cell line to another depending upon the glycosyl transferases present. There is some suggestion that the precise carbohydrate structure present on an antibody might have some influence over the biological activity of the antibody in complement activation and FcγR binding, although further investigation is revealing an important role for the precise oligosaccharide sequences and their interactions with the protein sequences found in each isotype (reviewed in Jefferis and Lund, 1997). Two functions of IgG antibody which do not seem to be strongly dependent upon glycosylation are neonatal transport and the catabolic rate of IgG (Ward and Ghetie, 1995; Wawrzynzak *et al.*, 1992).

For some considerable time there has been a debate as to the nature of the receptor responsible for transport of human IgG across the placenta. In the rat an Fc receptor has been described, FcRn, which is responsible for the neonatal transport of IgG across the intestinal epithelium, and this receptor shows homology to MHC class I, with a heavy chain associated with a β2 microglobulin light chain (Simister and Mostov, 1989). A crystal structure of the rat FcRn has been published, along with a second crystal structure of the receptor complexed with rat IgG in which the receptor interacts with the IgG through contacts with the C_H2 and C_H3 interface (Burmeister *et al.*, 1994, see *Figure 14.1*). A number of histidine residues are involved, and these are likely to be responsible for the observed pH dependence of the transport of IgG by this receptor (Raghaven *et al.*, 1995). A human homologue of the rat FcRn from a placental cDNA library has been reported, and this receptor shows a similar pH-dependency for binding of IgG (Story *et al.*, 1994). The human FcRn seems to be expressed in the syncitiotrophoblast cells and, from the high homology of this human receptor to the rat FcRn, it seems reasonable that it interacts with IgG in a similar structural way as determined for the crystal structure (Kristoffersen and Matre, 1996; Leach *et al.*, 1996; Simister *et al.*, 1996). This should facilitate the modelling of the sites of interaction of human IgG with hFcRn, and the present data would indicate that this interaction is likely to involve sites different from the residues identified as important in the association with the three FcγRs as detailed above (reviewed in Clark, 1997b).

The other likely role for the FcRn molecule is in the maintenance of the longer half-life for IgG. A model proposed by Brambell was that a receptor would recycle IgG antibody, destined for degradation along with other plasma proteins, back to the plasma, thus rescuing it and prolonging its half-life (Brambell *et al.*, 1964). Experiments in mice using IgG antibodies which had been mutated for their interaction with FcRn demonstrated that not only could they not be transported to the neonate but also that they had a greatly reduced biological half-life (Kim *et al.*, 1994). In addition to this observation, mice which lacked the FcRn were unable to transport IgG as neonates, and they had much lower IgG levels and a higher catabolic rate (Ghetie *et al.*, 1996; Junghans and Anderson, 1996).

There are a number of diseases which involve maternal IgG specific for paternal alloantigens; for example, haemolytic disease of the newborn (HDN). The IgG is transported across the placenta where it causes damage in the fetus or newborn through activation of effector mechanisms as described above. It is interesting to speculate as to whether novel future therapeutic strategies could be developed based on a detailed knowledge of all of the interactions with FcR and complement discussed above.

14.5 Clinical use

The CAMPATH-1 antibody was originally developed as an IgM antibody which was very effective in lysing lymphocytes *in vitro* using human serum as a complement source (Hale *et al.*, 1983). The antigen has now been fully identified as CD52, which is a small GPI-anchored glycoprotein expressed by lymphocytes and monocytes but not by

Table 14.1. Clinical uses of CAMPATH-1 (CD52) antibodies[a]

Disease/treatment	Number of patients	Antibody	Reference
Bone marrow transplantation	951	CAMPATH-1M and -1G	Hale and Waldmann, 1994
Bone marrow transplantation	>1000	CAMPATH-1M and -1G	Hale and Waldmann, 1996
Non-Hodgkin lymphoma	2	CAMPATH-1H	Hale *et al.*, 1988
Lymphoid malignancies	31	CAMPATH-1G and -1H	Dyer *et al.*, 1990
Chronic lymphocytic leukaemia	9	CAMPATH-1H	Osterborg *et al.*, 1996
Autoimmune thrombocytopenic purpura	6	CAMPATH-1H	Lim *et al.*, 1993
Systemic vasculitis	1	CAMPATH-1H	Mathieson *et al.*, 1990
Systemic vasculitis	4	CAMPATH-1H	Lockwood *et al.*, 1993
Multiple sclerosis	7	CAMPATH-1H	Moreau *et al.*, 1994
Multiple sclerosis	14	CAMPATH-1H	Moreau *et al.*, 1996
Rheumatoid arthritis	8	CAMPATH-1H	Isaacs *et al.*, 1992
Rheumatoid arthritis	30	CAMPATH-1H	Matteson *et al.*, 1995
Rheumatoid arthritis	41	CAMPATH-1H	Weinblatt *et al.*, 1995
Rheumatoid arthritis	41	CAMPATH-1H	Isaacs *et al.*, 1996a
Rheumatoid arthritis	48	CAMPATH-1H	Brett *et al.*, 1996
Immune-mediated corneal graft destruction	1	CAMPATH-1H	Newman *et al.*, 1995
Cutaneous scleroderma	1	CAMPATH-1H	Isaacs *et al.*, 1996b

[a]CAMPATH-1 (CD52) antibodies, the rat IgM, CAMPATH-1M, the rat IgG2b, CAMPATH-1G, and the humanized IgG1, CAMPATH-1H have been used in a range of clinical therapies. Some selected examples of these uses are reported above together with an indication of the number of patients for which data is presented.

haemopioetic stem cells (Xia *et al.,* 1991). It represents an exceptionally good target for complement lysis. An original therapeutic use for the IgM antibody was to remove lymphocytes from donor bone marrow prior to engraftment to prevent graft-versus-host disease (see *Table 14.1*). The IgM antibody and, more recently, the rat IgG2b antibody has been used regularly by a large number of bone marrow transplantation centres world-wide for this purpose (Hale and Waldmann, 1996).

Although the rat IgM and also the rat IgG2a CAMPATH-1 (CD52) antibodies worked well for lysing lymphocytes *in vitro,* early attempts to treat CD52-positive lymphomas/leukaemias proved unsuccessful (Dyer *et al.,* 1990). However, *in vitro* studies had indicated that rat IgG2b antibodies might be able to activate human FcγR-mediated effector functions, in particular ADCC through human FcγRIII K-cells. A rat IgG2b class-switch variant of the rat IgG2a CAMPATH-1 antibody was selected and this was tried in patients in which the IgM or IgG2a had failed to clear their CD52-tumour cells. The rat IgG2b antibody CAMPATH-1G was found to be highly efficient in clearing CD52-positive lymphocytes *in vivo,* indicating the importance of FcγR-mediated mechanisms over complement for *in vivo* cell clearance. The CAMPATH-1G went on to be used for both lymphoma/leukaemia therapy as well as for immunosuppression in organ transplantation (Dyer *et al.,* 1990). However, the major complication in the use of CAMPATH–1G was a rapid onset of a rat-specific anti-globulin response in a majority of patients treated. This anti-globulin response tended to restrict the course of treatment with the antibody to one course of antibody of about 10 days' duration (Dyer *et al.,* 1990).

The success with the humanization of the heavy chain V region of an antibody with specificity to the hapten NP prompted investigation of whether this could be generalized to both the heavy and light chains of any antibody including those binding complex cell surface antigens such as CD52. As described above, the antibody was successfully humanized although some framework changes were needed in addition to the CDR grafting (Riechmann *et al.,* 1988). The humanized antibody expressed as a human IgG1 turned out to be effective in depleting leukaemic cells and inducing remission in patients (Dyer *et al.,* 1990; Hale *et al.,* 1988).

Following the successful use of the humanized antibody CAMPATH-1H in lymphoma/leukaemia therapy, the antibody began to be used in a number of other disorders where immunosuppression was the desired outcome (see *Table 14.1*). CAMPATH-1H has been used in the treatment of patients with a number of diseases with autoimmune involvement, including refractory rheumatoid arthritis, as well as with patients with systemic vasculitis and also multiple sclerosis (Isaacs *et al.,* 1992; Lockwood *et al.,* 1993; Moreau *et al.,* 1996). In each case, efficacy of the antibody has been established, but perhaps of more general importance has been the observation that the CAMPATH-1H is much less immunogenic than the rat IgG2b CAMPATH-1G (Isaacs *et al.,* 1992). In a series of patients treated with CAMPATH-1H for rheumatoid arthritis none of them made a detectable anti-globulin response following a first course of treatment. Even after a second course of therapy, only some of the patients started to make a response which could be shown to be anti-idiotypic, and this was of low magnitude and did not prevent the course of treatment from achieving a clinical effect. Similar results were obtained in the vasculitis patients (Isaacs *et al.,* 1992; Matteson *et al.,* 1995).

Repeated therapy with a humanized antibody is thus possible, and, compared with rodent antibodies, they are much less likely to provoke an anti-globulin reponse. It seems that, following repeated use of a humanized antibody in patients, some will eventually mount a detectable anti-idiotype response. This may eventually prevent the antibody from achieving any clinical benefit. There may be some practical approaches to circumventing this problem.

(i) It is not clear to what extent this residual immunogenicity might relate to uncharacteristic sequences remaining in the humanized antibody. When CAMPATH-1H was humanized very little information existed on the sequences of human germline V region genes or their distribution in the human population. It now seems that the NEW and REI framework sequences used for the humanization of CAMPATH-1H represent myeloma proteins which have accumulated a number of mutations in the frameworks compared with the closest germline genes now identified (reviewed in Routledge *et al.*, 1993). Thus the humanized antibodies based on these structures could be considered as possessing several FwRs which are not characteristic of human immunoglobulins, and these might be responsible for enhancing the immunogenicity of the antibody.

(ii) Even if the patient mounts an anti-idiotype response to one humanized antibody, it should be possible to use a second or third antibody with different idiotypes in further attempts to treat the same or a different disease. Thus failure with one antibody need not preclude further treatments with another. Clearly this would not be the case for rodent antibodies as the anti-globulin response would most likely cross-react on all rat or mouse antibodies regardless of specificity.

(iii) It may be possible to develop strategies to tolerize patients against the therapeutically used antibody. In mouse models, it has proved possible to tolerize animals using immunosuppressive antibodies to the T-cell antigen CD4 (Cobbold *et al.*, 1990). Once tolerized, it is possible to give repeated doses of several different antibodies, including CD4, without any anti-globulin response arising. The rat and humanized CD4 antibodies described above have been used in experimental human therapies of this type either alone or in combination with CAMPATH-1H (Lockwood *et al.*, 1993).

Antibody therapy is not without side effects. Two principle complications have arisen in the use of anti-lymphocyte antibodies such as CD52 and also CD3.

(i) The first is an acute reaction in patients usually seen during the first

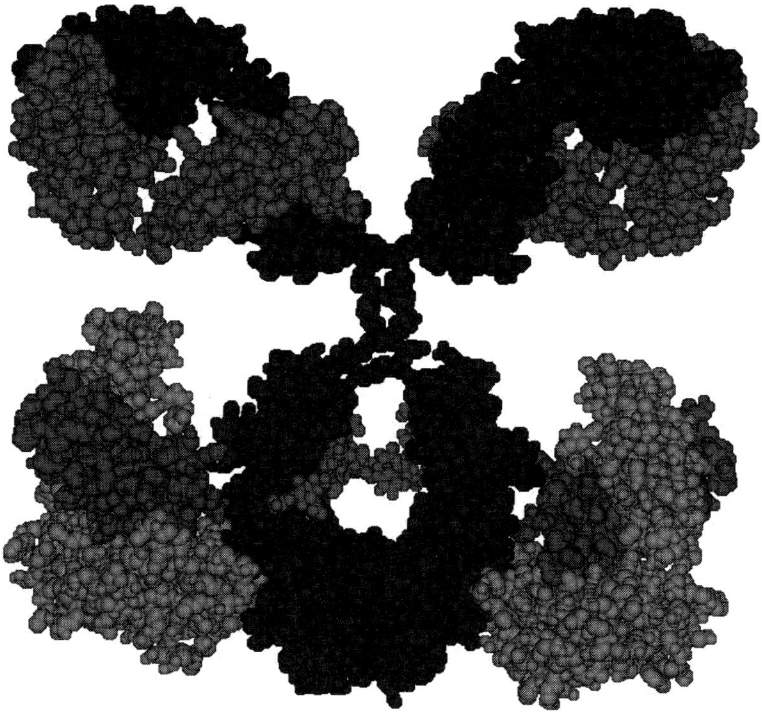

Figure 14.1. Shown are raster space filling models of a human IgG1 antibody interacting with two molecules of FcRn. The models were created from PDB structure files in the Brookaven Database for a human Fab, a human IgG1 Fc and a human IgG1–rat FcRn complex (Clark, 1997b). Certain residues, in particular parts of the hinge, are missing from the crystal structures although attempts have been made to computer model these features. For the model shown here, a peptide for the missing residues in the lower hinge region was created and inserted between the Fab and Fc structures. The final models were then aligned, superimposed and energy minimalized for the protein backbones using the computer software packages Quanta and Charmm, from Biosym Technologies Inc. USA, running on a Silicon Graphics Iris Indigo workstation.

or early doses of antibody. Investigation seems to show that the acute reaction is as a result of the ability of the antibodies to induce cytokine release from lymphocytes and other leukocytes during the first round of treatment (Isaacs *et al.*, 1992). For the CD3 antibodies it has been established in animal models that this side effect is Fc receptor dependent and so antibodies which have altered FcR binding have been generated. An aglycosylated CD3 antibody has been generated which seems to be as immunosuppressive as a wild-type human IgG1 or rat IgG2b. However, the immunosuppression achieved with the aglycosyl CD3 antibody seems not to be associated with side effects due to cytokine release and also does not involve lymphocyte depletion to any extent (Bolt *et al.*, 1993).

(ii) The second side effect is actually the result of profound and sometimes prolonged immunosuppression, which can be achieved using these antibodies. During the early stages of immunosuppression and particularly in patients with other risk factors or receiving additional immunosuppressive drugs, the patients seem at greater risk from infections, particularly viral infections such as cytomegalovirus. There also seems to be a slight risk of a lymphoma or leukaemia arising following anti-lymphocyte therapy. This may be linked to cytotoxic T-cell removal of regulation of EBV-transformed B cells in these patients. CAMPATH-1H- treated rheumatoid arthritis patients seem to remain lymphopenic for some considerable time and also may exhibit long-term alteration of the CD4/CD8 ratio, although they do seem able to mount T-cell responses (Isaacs *et al.*, 1992; Matteson *et al.*, 1995; Weinblatt *et al.*, 1995). Currently there is some hope that immunosuppression without depletion may be possible, and the aglycosylated CD3 antibody, as well as an aglycosylated humanized CD4 antibody, is currently being investigated clinically.

Acknowledgements

I would like to thank Professor Herman Waldmann and Drs Stephen Cobbold, Scott Gorman, Geoffrey Hale and Edward Routledge for helpful discussions. I would also like to acknowledge the support of the Wellcome Trust (grant reference 034817/Z/91) for funding my own studies on recombinant IgG effector functions.

References

Barclay AN, Birkeland ML, Brown MH, Beyers AD, Davis SJ, Somoza C and Williams AF. (1993) *The Leucocyte Antigen Facts Book*. Academic Press.

Bolt S, Routledge E, Lloyd I, Chatenoud L, Pope H, Gorman SD, Clark M and Waldmann H. (1993) The generation of a humanised, non-mitogenic CD3 monoclonal antibody which retains *in vitro* immunosuppressive properties. *Eur. J. Immunol.* **23**, 403–411.

Boulianne GL, Hozumi N and Schulman MJ. (1984) Production of functional chimaeric mouse/human antibody. *Nature* **312**, 643–646.

Brambell FWR, Hemmings WA and Morris IG. (1964) A theoretical model of gamma-globulin catabolism. *Nature* **203**, 1352–1355.

Bredius RGM, de Vries CEE, Troelstra A, van Alphen L, Weening RS, van de Winkel JGJ and Out TA. (1993) Phagocytosis of *Staphylococcus aureus* and *Haemophilus influenzae* type B opsonised by polyclonal human IgG1 and IgG2 antibodies: functional hFcγRIIa polymorphism to IgG2. *J. Immunol.* **151**, 1463–1472.

Brekke OH, Michaelsen TE, Aase A, Sin RH and Slie I. (1994) Human IgG isotype-specific amino acid residues affecting complement-mediated cell lysis and phagocytosis. *Eur. J. Immunol.* **24**, 2542–2547.

Brett S, Baxter G, Cooper H, Johnston JM, Tite J and Rapson N. (1996) Repopulation of blood lymphocyte sub-populations in rheumatoid arthritis patients treated with the depleting humanized monoclonal antibody, CAMPATH-1H. *Immunology* **88**, 13–19.

Burmeister WP, Huber AH and Bjorkman PJ. (1994) Crystal structure of the complex of rat neonatal Fc receptor with Fc. *Nature* **372**, 379–383.

Canfield SM and Morrison SL. (1991) The binding affinity of human IgG for its high affinity Fc receptor is determined by multiple amino acids in the CH2 domain and is modulated by the hinge region. *J. Exp. Med.* **173**, 1483–1491.

Chappel MS, Isenman DE, Everett M, Xu Y, Dorrington KJ and Klein MH. (1991) Identification of the Fc gamma receptor class I binding site in human IgG through the use of recombinat IgG1/IgG2 hybrid and point-mutated antibodies. *Proc. Natl Acad. Sci. USA* **88**, 9036–9040.

Clark M. (1995) General introduction. In: *Monoclonal Antibodies: Principles and Applications* (eds JS Birch and E Lennox), pp. 1–43. John Wiley and Sons, New York.

Clark M. (1997a) Unconjugated antibodies as therapeutics. In: *Antibody Therapeutics* (eds WF Harris and J Adair), pp. 3–31. CRC Press.

Clark M. (1997b) IgG effector mechanisms. *Chem. Immunol.* **65**, 88–110.

Clark M and Waldmann H. (1987) T-cell killing of target cells induced by hybrid antibodies: Comparison of two bispecific monoclonal antibodies. *J. Natl Cancer Inst.* **79**, 1393–1401.

Cobbold SP and Waldmann H. (1984) Therapeutic potential of monovalent monoclonal antibodies. *Nature* **308**, 460–462.

Cobbold SP, Martin G and Waldmann H. (1990) The induction of skin-graft tolerance in major histocompatibility complex-mismatched or primed recipients: primed T-cells can be tolerized in the periphery with anti-CD4 and anti-CD8 antibodies. *Eur. J. Immunol.* **12**, 2747–2755.

Deo YM, Graziano RF, Repp R and van de Winkel JGJ. (1997) Clinical significance of IgG Fc receptors and FcgR-directed immunotherapies. *Immunol. Today* **18**, 127–134.

Duncan AR and Winter G. (1988) The binding site for C1q on antibodies. *Nature* **332**, 738–740.

Duncan AR, Woof JM, Partridge LJ, Burton DR and Winter G. (1988) Localization of the binding site for the human high-affinity Fc receptor on IgG. *Nature* **332**, 563–564.

Dyer MJS, Hale G, Hayhoe FGJ and Waldmann H. (1989) Effects of CAMPATH-1 antibodies *in vivo* in patients with lymphoid malignancies: influence of antibody isotype. *Blood* **73**, 1431–1439.

Dyer MJS, Hale G, Marcus R and Waldmann H. (1990) Remission induction in patients with lymphoid malignancies using unconjugated CAMPATH-1 monoclonal antibodies. *Leukaemia Lymphoma* **2**, 179–193.

Galfre G and Milstein C. (1981) Preparation of monoclonal antibodies: strategies and procedures. *Methods Enzymol.* **73**, 3–46.

Ghetie V, Hubbard JG, Kim JK, Tsen MF, Lee Y and Ward ES. (1996) Abnormally short serum half-lives of IgG in beta-2-microglobulin-deficient mice. *Eur. J. Immunol.* **26**, 690–696.

Gorman SD, Clark MR, Routledge EG, Cobbold SP and Waldmann H. (1991) Reshaping a therapeutic CD4 antibody. *Proc. Natl Acad. Sci. USA* **88**, 4181–4185.

Greenwood J, Clark M and Waldmann H. (1993) Structural motifs involved in human IgG antibody effector functions. *Eur. J. Immunol.* **23**, 1098–1104.

Griffiths AD and Hoogenboom HR. (1993) Building an *in vitro* immune system: human antibodies from phage display libraries, In: *Protein Engineering of Antibody Molecules for Prophylactic and Therapeutic Applications in Man* (ed. M Clark), pp. 45-64. Academic Titles, Nottingham, UK.

Haagen I-A, Geerars AJG, Clark MR and van de Winkel JGJ. (1995) Interaction of human monocyte Fcγ receptors with rat IgG2b: a new indicator for the FcγRIIA (R-H131) polymorphism. *J. Immunol.* **154**, 1852–1860.

Hale G and Waldmann H. (1994) CAMPATH-1 monoclonal antibodies in bone marrow transplantation. *J. Hematother.* **3**, 15–31.

Hale G and Waldmann H. (1996) Recent results using CAMPATH-1 antibodies to control GVHD and graft rejection. *Bone Marrow Transplant*, **17**, 305–308.

Hale G, Bright S, Chumbley G, Hoang T, Metcalf D, Munro AJ and Waldmann H. (1983) Removal of T cells from bone marrow for transplantation: a monoclonal antilymphocyte antibody that fixes human complement. *Blood* **62**, 873–882.

Hale G, Dyer MJS, Clark MR, Phillips JM, Marcus R, Riechmann L, Winter G and Waldmann H. (1988) Remission induction in non-Hodgkin lymphoma with reshaped human monoclonal antibody CAMPATH-1H. *Lancet* **2**, 1394–1399.

Huizinga TWJ, Kerst M, Nuyens JH, Vlug A, Von Dem Borne AEGK , Roos D and Tetteroo PAT. (1989) Binding characteristics of dimeric IgG subclass complexes to human-neutrophils. *J. Immunol.* **142**, 2359–2364.

Isaacs JD, Clark MR, Greenwood J and Waldmann H. (1992) Therapy with monoclonal antibodies – an *in vivo* model for the assessment of therapeutic potential. *J. Immunol.* **148**, 3062–3071.

Isaacs JD, Manna VK, Rapson N, Bulpitt KJ, Hazleman BL, Matteson EL, St.-Clair EW, Schnitzer TJ and Johnston JM. (1996a) CAMPATH-1H in rheumatoid arthritis – an intravenous dose-ranging study. *Br. J. Rheumatol.* **35**, 231–240.

Isaacs JD, Hazelman BL, Chakravarty K, Grant JW, Hale H and Waldmann H. (1996b) Monoclonal antibody therapy of diffuse cutaneous scleroderma with CAMPATH-1H. *J. Rheumatol.* **23**, 1103–1106.

Janeway CA, Jr and Travers P. (1996) *Immunobiology: The Immune System in Health and Disease*, 2nd edn. Blackwell Scientific Publications, Oxford.

Jefferis R and Lund J. (1997) Glycosylation of antibody molecules: structural and functional significance. *Chem. Immunol.* **65**, 111–128.

Jefferis R, Lund J and Pound J. (1990) Molecular definition of interaction sites on human IgG for Fc receptors (huFcγR). *Mol. Immunol.* **27**, 1237–1240.

Jones PT, Dear PH, Foote J, Neuberger MS and Winter, G. (1986) Replacing the complementarity-determining regions in a human antibody with those from a mouse. *Nature* **321**, 522–525.

Junghans RP and Anderson CL. (1996) The protection receptor for IgG catabolism is the beta-2-microglobulin-containing neonatal intestinal transport receptor. *Proc. Natl Acad. Sci. USA,* **93**, 5512–5516.

Kabat EA, Wu TT, Perry HM, Gottesman KS and Foeller C. (1991) *Sequences of Proteins of Immunological Interest.* US Department of Health and Human Services, NIH.

Kim JK, Tsen MF, Ghetie V and Ward ES. (1994) Identifying amino-acid-residues that influence plasma-clearance of murine IgG1 fragments by site-directed mutagenesis. *Eur. J. Immunol.* **24**, 542–548.

Kohler G and Milstein C. (1975) Continuous cultures of fused cells secreting antibody of predefined specificity. *Nature* **256**, 495–497.

Kristoffersen EK and Matre M. (1996) Co-localization of beta 2-microglobulin and IgG in human placental syncytiotrophoblasts. *Eur. J. Immunol.* **26**, 505–507.

Leach JL, Sedmak DD, Osborne JM, Rahill B, Lairmore MD and Anderson CL. (1996) Isolation from human placenta of the IgG transporter, FcRn, and localization to the syncytiotrophoblast: implications for maternal–fetal antibody transport. *J. Immunol.* **157**, 3317–3322.

Lim SH, Hale G, Marcus RE, Waldmann H and Baglin TP. (1993) CAMPATH-1 monoclonal antibody therapy in severe refractory autoimmune thrombocytopenic purpura. *Br. J. Haematol.* **84**, 542–544.

Lockwood CM, Thiru S, Isaacs JD, Hale G and Waldmann H. (1993) Long-term remission of intractable systemic vasculitis with monoclonal antibody therapy. *Lancet* **341**, 1620–1622.

Male D, Cooke A, Owen M, Trowsdale J and Champion B. (1996) *Advanced Immunology*, 3rd Edn. Times Mirror International Publishers Ltd, London.

Mathieson PW, Cobbold SP, Hale G, Clark MR, Oliveira DBG, Lockwood CM and Waldmann H. (1990) Monoclonal antibody therapy in systemic vasculitis. *New Engl. J. Med.* **323**, 250-254.

Matteson EL, Yocum DE, St-Clair EW, Achkar AA, Thakor MS, Jacobs MR, Hays AE, Heitman CK and Johnston JM. (1995) Treatment of active refractory rheumatoid arthritis with humanized monoclonal antibody CAMPATH-1H administered by daily subcutaneous injection. *Arth. Rheum.* **38**, 1187–1193.

Morgan A, Jones ND, Nesbitt AM, Chaplin L, Bodmer MW and Emtage JS. (1995) The N-terminal end of the CH2 domain of chimeric human IgG1 anti-HLA DR is necessary for C1q, FcγRI and FcγRIII binding. *Immunology* **86**, 319–324.

Moreau T, Thorpe J, Miller D *et al.* (1994) Preliminary evidence from magnetic resonance imaging for reduction in disease activity after lymphocyte depletion in multiple sclerosis. *Lancet* **344,** 298–301.

Moreau T, Coles A, Wing M, Isaacs J, Hale G, Waldmann H and Compston A. (1996) Transient increase in symptoms associated with cytokine release in patients with multiple sclerosis. *Brain* **119**, 225–237.

Neuberger MS, Williams GT, Mitchell EB, Jouhal SS, Flanagan JG and Rabbitts TH. (1985) A hapten-specific chimeric immunoglobulin E antibody which exhibits human physiological effector function. *Nature* **314**, 268–271.

Newman DK, Isaacs JD, Watson PG, Meyer PA, Hale G and Waldmann H. (1995) Prevention of immune-mediated corneal graft destruction with the anti-lymphocyte monoclonal antibody, CAMPATH-1H. *Eye* **9**, 564–569.

Ortho Multicentre Transplant Study Group (1985) A randomized clinical trial of OKT3 monoclonal antibody for acute rejection of cadaveric renal transplants. *New Engl. J. Med.* **313**, 337–342.

Osterborg A, Fassas AS, Anagnostopoulos A, Dyer MJ, Catovsky D and Mellstedt H. (1996) Humanized CD52 monoclonal antibody Campath-1H as first-line treatment in chronic lymphocytic leukaemia. *Br. J. Haematol.* **93**, 151–153.

Parren PWHI and Burton DR. (1997) Antibodies against HIV-1 from phage display libraries: mapping of an immune response and progress towards antiviral immunotherapy. *Chem. Immunol.* **65**, 18–56.

Raghavan M, Bonagura VR, Morrison SL and Bjorkman PJ. (1995) Analysis of the pH dependence of the neonatal Fc receptor/immunoglobulin G interaction using antibody and receptor variants. *Biochemistry* **34**, 14649–14657.

Ravetch JV. (1994) Fc receptors: rubor redux. *Cell* **78**, 553–560.

Riechmann L, Clark MR, Waldmann H and Winter G. (1988) Reshaping human antibodies for therapy. *Nature* **332**, 323–327.

Routledge EG, Lloyd I, Gorman S, Clark M and Waldmann H. (1991) A humanized monovalent CD3 antibody which can activate homologous complement. *Eur. J. Immunol.* **21**, 2717–2725.

Routledge EG, Gorman SD and Clark M. (1993) Reshaping antibodies for therapy. In: *Protein Engineering of Antibody Molecules for Prophylactic and Therapeutic Applications in Man* (ed. M Clark), pp. 13–44. Academic Titles, Nottingham, UK.

Sarmay G, Lund J, Rozsnyay Z, Gergely J and Jefferis R. (1992) Mapping and comparison of the interaction sites on the Fc region of the IgG responsible for the triggering antibody dependent cellular cytotoxicity (ADCC) through different types of human Fcγ receptor. *Mol. Immunol.* **29**, 633–639.

Simister NE and Mostov KE. (1989) An Fc receptor structurally related to MHC class-I antigens. *Nature* **337**, 184–187.

Simister NE, Story CM, Chen HL and Hunt JS. (1996) An IgG-transporting Fc receptor expressed in the syncytiotrophoblast of human placenta. *Eur. J. Immunol.* **26**, 1527–1531.

Story CM, Mikulska JE and Simister NE. (1994) A major histocompatibility complex class I-like Fc receptor cloned from human placenta: possible role in transfer of immunoglobulin G from mother to fetus. *J. Exp. Med.* **180,** 2377–2381.

Tan LK, Shopes RJ, Oi VT and Morrison SL. (1990) Influence of the hinge region on complement activation, C1q binding, and segmental flexibility in chimeric human immunoglobulins. *Proc. Natl Acad. Sci. USA* **87**, 162–166.

Tao M, Canfield SM and Morrison SL. (1991) The differential ability of human IgG1 and IgG4 to activate complement is determined by the COOH-terminal sequence of the CH2 domain. *J. Exp. Med.* **173**, 1025–1028.

Tau MH, Smith RI and Morrison SL. (1993) Structural features of human IgG that determine isotype specific differences in complement activation. *J. Exp. Med.* **178,** 661–667.

Tax WJM, Hermes FFM, Willems RW, Capel PJA and Koene RAP. (1984) Fc receptors for mouse IgG1 on human monocytes: polymorphism and role in antibody-induced T cell proliferation. *J. Immunol.* **133,** 1185–1189.

van de Winkel JGJ and Capel PJA. (1993) Human IgG Fc receptor heterogeneity: molecular aspects and clinical implications. *Immunol. Today* **14,** 215–221.

Verhoeyen M, Milstein C and Winter G. (1988) Reshaping human antibodies: grafting an antilysozyme activity. *Science* **239,** 1534–1536.

Waldman TA and Strober W. (1969) Metabolism of immunoglobulins. *Prog. Allergy* **13,** 1–110.

Ward ES and Ghetie V. (1995) The effector functions of immunoglobulins: implications for therapy. *Ther. Immunol.* **2,** 77–94.

Wawrzynczak EJ, Denham S, Parnell GD, Cumber AJ, Jones P and Winter G. (1992) Recombinant mouse monoclonal antibodies with single amino acid substitutions affecting C1q and high affinity Fc receptor binding have identical serum half-lives in the BALB/c mouse. *Mol. Immunol.* **29,** 221–227.

Weinblatt ME, Maddison PJ and Bulpitt KJ. (1995) CAMPATH-1H, a humanized monoclonal antibody, in refractory rheumatoid arthritis. An intravenous dose-escalation study. *Arth. Rheum.,* **38,** 1589–1594.

Xia MQ, Tone M, Packman L, Hale G and Waldmann H. (1991) Characterization of the CAMPATH-1 (CDw52) antigen: biochemical analysis and cDNA cloning reveal an unusually small peptide backbone. *Eur. J. Immunol.* **21,** 1677–1684.

15

Getting to know your genes privately

Paul G. Debenham

15.1 Who knows about the genetics revolution?

Molecular biologists and human geneticists are leading a revolution in human healthcare, but unfortunately nobody else knows about it! The public at large has little or no knowledge of what is now, or will be, available to it as a consequence of the latest genetic discoveries, and has only just started to recognize the tremendous impact genetics could have on the lives of each individual.

Genetic tests are not just a new form of health examination, assessing the patient's health that day and relevant only for a brief period, these tests will forecast disease that does not presently exist and may not be due for tens of years, as well as diseases that may exist in children yet to be conceived. Providers of genetic diagnostics must therefore proceed cautiously as to how they explain the impact and consequences of the tests they will provide.

Historically the expectation within the general UK population has been that the National Health Service (NHS) will be the source of diagnostics, and that any important new diagnostic will be introduced to the public via their general practitioner (GP). But will genetic diagnostics be available to all, on request by GP referral? Already, in these days of constantly stretched national health resources, supermarket shelves are the home for self-testing kits for

Genetics of Common Diseases: future therapeutic and diagnostic possibilities,
edited by I. Day and S. Humphries.

pregnancy; there is no longer a norm for the source of any new diagnostic test.

Pharmaceutical sales staff work full time to achieve awareness of their product by physicians, building on a pre-existing knowledge base to educate GPs in the latest advances in the treatment of symptoms regularly seen by the doctors. With medical genetics, however, any educational pitch cannot presume that doctors are, in general, any better informed than the public at large. An informed decision can be made only once the subject is fully comprehended, thus a doctor's evaluation of a genetic diagnostic service on behalf of his/her patients may be very difficult for many doctors. Thus when it comes to genetic diagnostics perhaps an informed member of the public could equally be competent to decide as to the efficacy of such a test in their personal circumstances as would be their GP.

The appropriateness of patient-directed, or requested, diagnosis is both a philosophical and a political issue that will not be readily resolved; yet all the time genetic research ploughs forward producing its harvest of discoveries, with over 15 000 genes already identified. Presently, as a consequence of the widely differing knowledge base amongst health care professionals, one can observe a very patchy introduction of genetic diagnostics within the NHS, as entrepreneurial clinical geneticists and enlightened GPs seek to offer services they consider meritorious and affordable or for which they have gained grant support. In this environment, it seems probable that private provision of genetic diagnostics will be sought where the NHS cannot deliver.

The prospect of private provision of genetic diagnostic services seems to have raised considerable concerns amongst those health care professionals who are sufficiently aware of the potential impact such diagnoses could have. These touch on all aspects of such a service, from pre-test information, the genetics involved, the quality assurance of the laboratory involved, through to the post-test counselling provided (Harper, 1995). Nonetheless, the perception of private genetic diagnostics from the public's position may differ as more discernible priorities of prompt service, privacy of information and the impact of such results on life insurance premiums are favourably attributed to the private sector and not to the NHS. To what extent then is the intervention of the private sector in medical genetic diagnostics both logical and inevitable?

15.2 Commercial genetic diagnostics is already with us

If the public are aware of DNA and genetic analysis, I would expect that this is predominantly due to its role in forensic science and paternity case work following the discovery of genetic fingerprinting in 1985. This may be particularly true for the UK population, as this was a British discovery and pioneered here with the world following our lead. Now in the mid-1990s the UK is again pioneering genetic identification applications with the national DNA database. Since April 1995, approximately 120 000 samples from scenes of crimes and from suspects have been genetically profiled and recorded; eventually the database may hold data on 5 million samples and individuals (Gill *et al.*, 1997). If you are arrested for an indictable crime, even if it has no biological perspective, for instance fraud, then you may be required to provide a sample for typing. Thus genetics is actively encroaching on the public directly and is not only in the realms of television police dramas. Behind the scenes private DNA diagnostic companies have played an active role in the development and ongoing application of genetic services to forensic case work.

DNA analysis has been legally recognized since 1989 as the definitive means of determining the paternity of a child. This application has found its way into public awareness via high profile disputed paternity cases of pop stars and politicians, as well as being a subject linked to the Child Support Agency as it pursues fathers to pay child maintenance. Interestingly the application of DNA tests for the determination of paternity has been very much the preserve of the private laboratories, as has been the government's use, by contract, of DNA-typing to resolve disputed claims of relationships for immigration purposes (Home Office, 1988). To provide these services, the private sector laboratories have established a customer service infrastructure across the UK and internationally. These services are integrated into our court systems, and every day touch the lives of the public.

The issues above affect the lives of adults and children as dramatically as any medical diagnosis, yet it is not a matter of controversy that these genetic diagnostic services are provided privately. Whilst the calibre of scientists who provide this service has to be vetted by a review of their experience, the government has not found it necessary to impose guidelines or establish a Code of Practice. Without requiring regulation, the competitive aspect of private enterprise has made quality of service every bit as much an important issue as is price, accuracy and speed of service. The key laboratories in this field operate to an international quality standard ISO 9002 for genetic diagnostics whilst NHS genetics laboratories do not.

The lifelong consequences of paternity test results are not provided with options for consultation, or follow up, yet it is only occasionally that distress is reported. It is as if the receipt of the results is taken as a matter of fact, and it is assumed that the recipients adjust their lives and get on with the consequences of their restructured family. The public should perhaps be given more credit for their resilience in coping with these life-changing consequences. Further, the fees for these tests run into hundreds of pounds, much of which often has to paid by the individuals involved; there is no expectation that the state will necessarily provide the service nor that it will be free of charge.

Across the UK therefore are families who have some awareness and experience of DNA diagnostics. For completely separate reasons there are also many families with a very real awareness of the impact of inheritance and genetics as they, or their relatives, are affected by disabling or life-limiting diseases. The NHS regional genetics laboratories provide a thorough service to the immediate family and relatives who suffer from rare disorders ranging from muscular dystrophy to Huntingdon's chorea (Department of Health, 1993). Rare as each case may be, the collective sum of these disorders already occupies the attention of 29 NHS regional genetics centres. Thus the care cannot extend without limit, and certainly not to unrelated friends and neighbours who by contact are sensitized to these disorders and might wish to seek reassurance or knowledge of their own status with respect to the disorder in question. While it may be unrealistic to suppose such contacts would have weighty concerns about rare disorders, particularly where a family history is well known, it is not unrealistic to expect that they may seriously wish to know whether they carry the genetics of a common disorder such as CF once learnt of at close hand.

15.3 Medical genetic diagnostics: the perception of private tests

A child with CF is born nearly every day to parents who, in the majority, had no detectable family history nor had any personal indication that they harboured this disorder. What impact will this unsuspected event have on their friends and neighbours with respect to their own child-bearing plans? The NHS does not provide a CF carrier screening service for the public at large, so where do these people go? The NHS does provide a service to the immediate family but intriguingly even some close relatives have sought private diagnosis of CF because of extensive waiting times. Thus the distinction between a private service and that from the NHS, at least on the most obvious grounds of payment, may not be most important to the public with respect to obtaining a genetic diagnosis.

As there is no template for the structure of a private genetic diagnostic service the definition of an appropriate service would appear to be a matter of view point. Importantly it will be highly dependent on the type of genetic defect being diagnosed. This dependence between the requirements for any genetic diagnostic service and the actual disorder it relates to has taken some time to be appreciated.

The media have had to face the challenge of distilling the issue of the private provision of genetic diagnostics to its readership. Unfortunately the initial portrayal of this issue was predicated by the need to have a catchy story, and thus controversy, with powerful quotes and sound bites, yet with a subject that is impenetrable in the few paragraphs or limited seconds of a TV slot that were available. Hence, initial presentations of the private service for CF carrier detection launched by University Diagnostics Ltd (UDL), the first of its kind in the UK, made 'good reading'. Texts emphasized a mail order aspect implying, by stereotypical association, something unsavoury or of downmarket quality. There was also the implication, by juxtaposition of CF with other disorders, that the service for CF carrier screening would be that which would also be offered for such radically different disorders as breast cancer and Alzheimer's disease (Rogers, 1995). The preliminary opinion put forward in these portrayals was that medical genetics was best left to the NHS. However, the balancing considerations that mail order means privacy and lack of sampling appointments have now started to gain a hearing as has the fact that the CF carrier service was designed specifically for the needs of that diagnosis and no other. There has also started to be a recognition that the private sector may be addressing a latent need unmet through the NHS (Bull, 1996).

Along with the public at large, the primary health care sector appears to be largely ill-informed concerning genetic diagnostics. Understandably GPs rarely encounter incidents of inherited disorders and thus prioritize their attention to the ever prevalent problems they encounter. Yet it would be GPs from whom the public might be expected to ask advice in such matters. The historic route of referrals would direct the enquiry to the regional NHS genetic laboratories which traditionally focus on the assessment of presented symptoms and family history. However, with asymptomatic states, such as that of a CF carrier, this route is not available.

Thus the impetus for seeking advice and sourcing a test for a common carrier status or even predisposition testing for a common life-affecting genetic trait may inevitably be driven by self-motivated patients and pioneering GPs and not by the NHS. In some senses therefore it falls upon the private sector to promote their services and at least establish a public awareness of what is available to enable individuals to make an informed choice as to whether they want such a test or not. For its part, if a private laboratory wishes to offer a medical genetic testing service then it must accept that the service will be subjected to the scrutiny of the media and healthcare professionals. This was certainly the case with the CF carrier screening service from UDL. As mentioned above, the detail of this service is now under reasoned scrutiny following the initial reviews which focused more on the principle of private sector involvement and only sketchily on the actual process.

15.4 CF carrier testing service: the component parts

In 1994 the long-heralded first genetic diagnostic test kit for laboratory use for an inherited disease was launched by Kodak Clinical Diagnostics (becoming Johnson & Johnson Diagnostics then Ortho Diagnostics Ltd). This kit detects the four most common mutations in the *CFTR* gene which cause CF in the UK white heritage population (Ferrie *et al.*, 1992). The validation of this technology has been formidable, being trialled on over 40 000 people. This test is built around amplification refractory mutation system polymerase chain reaction (ARMS–PCR) methodologies and has been designed to have minimal false positive and false negative results given the detailed knowledge available in the gene sequences involved. It has now already been superceded by tests that extend the repertoire of mutations that can be detected. In 1997, it is expected that at least two manufacturers will have tests on the market that can detect tens of mutations in the *CFTR* gene. To some extent there is an ever-decreasing incremental growth of genetic definition with each additional test, as the mutations detected become rarer and rarer in incident. However, the coverage of genetic mutations does extend the use of the test to a variety of ethnic groups which have specific subsets of mutations within their populations. Thus, the evolution of the test technology will broaden the application of CF carrier screening even though it is only marginally increasing its coverage of the several hundred possible, and increasingly rare, genetic causes of CF. The original test can detect approximately 85–90% of the CF mutations occurring in the population, and thus could reduce the carrier risk of person testing negatively from a 4% chance of carrying a CF mutation to less than 0.5% after the test. At UDL this initial level of definition was felt to be valuable for the public to take advantage of and so the private test service was established around this genetic technology.

In part, UDL's ongoing role as a paternity and immigration case work laboratory has predisposed the company to be aware of the need to address the specific sensitivities of the clients involved. Thus UDL was grateful that the CF Trust was prepared to comment on all aspects and texts of the proposed service, so that no advert, literature, information, result format or advice has been developed without CF Trust comment. Further, to date, all the advice given by the Trust's representatives has been taken on board. The CF Trust cannot advocate any service from any private organization and its role is strictly advisory as the most appropriate voice with respect to CF. It is expected that reference to appropriate patient bodies will be sought for any new genetic diagnostic test. This will ensure that such services are tailor-made to the specific consequences of the disorder and not just a re-hash of the CF service.

A private genetic diagnostic service is not legally required to be obtained by referral from a GP; in fact GPs do not normally have the financial structures, or funds, to pay for services outside of those contracted to the NHS. Thus it is the private purse that must pay if someone wishes to establish their CF status and is not a relative of an affected individual. A private contract will be established between the customer and the supplier, and it does not necessarily need to involve the individual's doctor at all. However, given the central role of the GP in an individual's healthcare as well as historically his/her role in communicating medical information, it remains appropriate to offer the client, or bring to their attention, the possibility of involving their GP as a conduit for this private diagnostic test. This option has therefore been built into the structure of the service, with the individual indicating his choice by the type of registration form completed. The GP would need to consent to this role and complete the relevant section in the registration form. Intriguingly, to date, most clients have opted to be tested without the involvement of their GPs. This may reflect the self-selected clientele to date who, by the sheer fact of participating in the test, are not perhaps representative of the interests and preferences of the public at large.

The decision as to whether to involve the GP may also hinge on whether carrier status for CF has any life insurance ramifications. This is because a GP may be required to certify that a patient has, or has not, undergone any genetic tests and, if so, has then to divulge the results. If a patient undertakes a test totally privately, this disclosure will happen only at his/her discretion. Fortunately the genetics advisor to the Association of British Insurers has now advised that CF carrier status should not influence premiums. Therefore this aspect should no longer be contentious.

UDL will not offer CF carrier testing to pregnant women without the consent of their GP or their private physician in charge of their maternity care.

Once the appropriate registration is completed and returned with payment, the customer is sent out a simple mouthwash to swirl around the mouth (to gather mouth cells) and to spit back into a collection tube. This tube is then posted back to the laboratory in the return postage-paid packaging provided.

Perhaps surprisingly the mouthwash has turned out to be a very reliable sampling method, with just one or two failures to date. Despite the relatively uncontrolled nature of the mouthwash process and the equally varied nature of the content of our mouths, this is a robust technique which UDL has applied to other applications of human identification with equal success. In terms of quality assurance, the proof in this case is in the accumulated experience and not in adherence to a specific protocol.

The need for quality assurance in the laboratory process is often put forward as a matter of major concern for those who seek assurance that an unregulated private sector will not include 'home brew' technology run out of garages. There is absolutely no question that this assurance is critical to the success of the private sector, yet without regulation this rogue activity possibility cannot be discounted. In forthcoming guidelines (see below) for the provision of private services direct to the public, the recommendation will be that any such service should be certified to a recognized quality standard, such as ISO 9002, and that the laboratories should participate in a relevant quality assessment programme. Unfortunately, at the time of writing, there is no structured quality assessment scheme for CF diagnostics available. Currently UDL runs its own internal controlled analysis of samples using only industrially manufactured gene test technology to rigorously monitor and maintain standards in their service. To date there has not been the appearance of a 'rogue' operator in this field, and such a hypothetical beast may not realistically exist.

The actual laboratory process from receipt to reporting is concluded within 5 working days currently. The report package has been structured to meet the differing needs associated with receiving a CF carrier negative and positive finding. In the UDL service, the CF negative result has to be presented with a clear indication that there remains a small residual risk that the client may still have a rare mutation that can cause CF, and provide a numerical example of what that means in practice. For the client learning that they carry a CF mutation, a more comprehensive explanation is provided to cover the potential consequences for future children and for their relatives. This package includes an excellent booklet prepared by the CF Trust which comprehensively covers all the relevant issues. The information also sets out to reassure the client about their own health, indicating amongst other things that they are not alone as their status is shared with an estimated 2 million other individuals in the UK, and that their health prospects are unaffected. Whatever the outcome of the test, the client is provided with the contact details of the clinical genetics consultant to discuss the findings by telephone or by face to face appointment. The CF Trust has also agreed to have its contact details provided in the information sent.

15.5 Guidelines for future commercial genetic tests available directly to the public

There are many aspects of the current CF carrier screening service as a private genetic testing service that are not detailed in the brief summary above. No detailed guidelines exist for this service, and the private sector must correctly gauge the fine detail. Components of any such service include:

- the nature and placement of advertising materials
- the need and level of any pre-test counselling
- the pre-test information and to what extent it covers the nature of the disorder being screened for
- the accuracy of the test
- the quality assurance programme necessary
- the method of sampling
- the medical consequences of the test
- the relevance of the information to relatives of the tested individual
- the racial variation in the mutations examined
- the appropriate limitations on who should be tested
- the involvement of GPs and private physicians
- the insurance consequences of the diagnosis
- the speed of obtaining results
- the presentation of the results
- the post-test information
- the provision of post-test counselling.

The Department of Health has established an Advisory Committee on Genetic Testing (ACGT) to set guidelines for the commercial provision of genetic tests direct to the public, but their remit does not extend to those offered under medical supervision, that is by GP referral. The ACGT has, at the time of writing, drafted a proposed set of guidelines to cover the broad issues of such a service (ACGT, 1996). This Code of Practice will cover:

- standards for equipment and reagents
- the staffing and quality assurance accreditation of a testing laboratory
- the confidentiality and storage of customer records
- what tests may be offered
- the information and consultation requirements
- the involvement of GPs
- the testing of children.

The service provided by UDL conforms to the majority of the elements of these guidelines, and where it does not it can adapt the service if appropriate. At this time it is important to wait until the final Code of Practice is issued before assessing the value of these guidelines when it comes to defining the actual working details of a service.

The ACGT has drawn a line between the genetic nature of a disorder and its suitability as a subject for a private service to be offered direct to the public. Genetic diagnostics investigating recessive disorders in which the carrier state has minimal discernible impact on the life of the tested individual are considered to be appropriate for such a service, and the Committee is proposing to request only notification of what is intended by the testing laboratory. This Code of Practice will not extend to dominantly, or X-linked, inherited disorders, and at this time any proposal of this nature is asked to be brought to the attention of the ACGT for comment.

Importantly, therefore, there is no extrapolation from, say, CF testing to breast cancer predisposition testing. Each genetic diagnostic will be evaluated on its merits.

15.6 What genetic diagnostics will be provided next?

How many genetic disorders, or predispositions, will be the subject of DNA tests offered commercially direct to the public in the foreseeable future? CF has come to the fore because it is the most commonly inherited disorder in the western white heritage, and tests are not available through the NHS. For the private sector to offer further tests for other disorders the service must make commercial sense. Assuming most services have a strong price-sensitive element nowadays and that competition will drive down margins, the business perspective will only be met through high-volume testing. The opportunity for high profit tests in a competitive industry will be achieved only if competition is limited, perhaps by patent rights to tests. To achieve high numbers of tests, the tests must not be available through the NHS and must be viewed as valuable to the purchaser; no-one buys anything just because it is available, and particularly not a genetic test. Therefore, there is a combination of commercial criteria, let alone medical relevance and potential Code of Practice requirements, that will not be readily met by a lot of genetic disorders and predispositions. Even the track record of CF carrier screening uptake indicates that genetic diagnostics will not be a pre-eminent aspect of population health care for some years to come.

It is therefore questionable whether there will be a rush to offer the public at large diagnoses for monogenic disorders such as predisposition to certain cancers, or complex diagnostics for multigenic traits such as cardiovascular disease. It is particularly important though to distinguish tests that may be offered directly to the public commercially from those that may be offered to the public commercially via private health care programme and clinics. The latter category may well be the focus for substantial growth in the next few years.

Whilst genetic screening within families directly at risk of inherited disorders, such as breast cancer, may be available through the NHS, there will be a growing group of potentially affected individuals who would wish to ameliorate the difficult process of gaining the knowledge within the benefits of a private treatment environment. Within this potentially directly affected population, the issue of cost may be of reduced significance when the possibility of private counselling and rapid service are at stake. In parallel, providers of private health screening programmes wishing to maximize the value of their service to a patient may seek genetic tests to help define the presented condition. Thus cardiovascular disease-associated genetic analysis, while still an incomplete science, might usefully aid a consultant in his advice to patients with the early signs of heart disease. It may be the case that such diagnostics are not actually available through the NHS apart from for those presenting with symptoms indicating increased risk. Thus private health screening programmes may be the only access to these tests for the pre-symptomatic public. The availability of such tests directly to the public will depend on the tests being worthwhile in the absence of an accompanying structured well-person programme or follow-up service, and it is too early to make that judgement scientifically. It may well be that such a service would evolve after the assessment of the value of genetic information as an adjunct to a health screening programme.

The involvement of financial reward with respect to motivation of scientists and health care professionals is mostly regarded with concern. However, it could be the incentive for the research scientist to push his discoveries from being just an academic analysis of, for example, the genetics of a cardiovascular trait in which his involvement with this disease would thus finish when the research is published, to his active proselytizing of the diagnostic benefits of his finding to the medical profession. The cardiovascular area is a good example where lots of genes interplay with respect to the various elements of lipid metabolism, blood vessel wall integrity and plaque structure, and the progression of thrombosis. There are already indications that knowledge of the specific genotypes of the FH, ApoE, fibrinogen, stromelysin and other genes could valuably aid clinicians in dietary and exercise regimes for their patients (see Chapter 7). It may well be that the introduction of this knowledge into public awareness will come about by the media attention paid to the private sector as it introduces these diagnostics as part of a commercial service.

The genetic bases of a growing number of important medical disorders have been reported steadily over the last few years, yet despite this only the CF carrier test is presently available privately. It is important to recognize that the development of commercial services cannot be divorced from the expected media attention accorded to those services. Thus genetic tests to complement cardiovascular disease diagnosis may be expected to be well received and, as long as the insurance implications are acceptable, they therefore might be introduced. Conversely, the genetic determination of increased risk for the early onset of Alzheimer's disease associated with the ε4 allele of the *APOE* gene (see Chapter 6), which could be offered today, is not offered – for the very reason that such a service would be roundly condemned, even if some people really wanted to know and would pay to have the test done.

The logic behind the discussion so far is that private genetic tests in the medical sphere will be purchased as the diagnosis becomes viewed by the patient as something they wish to know about their life and health. However, one can imagine a possible growth of privately commissioned tests as an element of employment safety where the person participating in the test is not necessarily going to benefit from the genetic diagnosis obtained. For instance, airline companies may wish to assess the cardiovascular genetics of their pilots to augment their conventional clinical assessment, or security firms may equally wish not to employ guards who are genetically at risk of cardiovascular disease. Such genetic screening would be very much a private sector enterprise if it were to occur.

Throughout this chapter I have considered genetic testing as a laboratory-based enterprise with a sample being despatched to a specialized facility where the genetics are analysed and from which results are sent. However, the skills of genetic research hold out the prospect of technologies which could enable self-testing kits to be a reality. Whilst this technology may not have a public application for some years, the fact that people are now choosing to learn their CF carrier status by taking totally independent steps indicates that like-minded individuals in the future might avail themselves of this technology. It is a radical concept and its success will depend fundamentally on whether, in the main, the general public really do want to know more about their genetic make-up so that its analysis becomes more than just a specialist service.

Could the genetics revolution be embraced by the public, not through the advances in the diagnosis of life-limiting disorders and predispositions, but through a range of non-threatening discoveries? For instance, it is not inconceivable that there is a genetic basis for whether we should be left or right handed; would curiosity create a market for such inconsequential knowledge? Perhaps genetic research will define human characteristics that could be totally non-controversial as private diagnostics, such as a musical ear, neat handwriting or an affinity for animals? Could whimsy be commercial in genetics ? The answer may lie in the skill of the advertising campaign. In fact, the next private genetic diagnostic service to find a demand in the UK could be determined in part by the ability to communicate its value to the public irrespective of mode of availability or nature of its findings.

15.7 Presenting the private possibilities

At present, advertising materials for CF carrier testing have been very cautious in their approach. The texts have sought to both educate the reader with regard to the disorder and its frequency of occurrence, as well as indicate aspects of the ease and availability of the test and the support provided. Inevitably the impact, and thus equally the potential commercial reward, is diminished by advertisers who do not tell their story in an instant catchy idea. However, any light-hearted or shock tactics concerning genetic inheritance to promote genetic testing would be roundly condemned. The advertising campaign by the government some years ago to promote human immunodeficiency virus/acquired immune deficiency syndrome (HIV/AIDS) awareness involved stark images of icebergs and, although eventually very effective in achieving its goal, was ridiculed at the time. The progression in advertising style will therefore probably cautiously balance the commercial imperative with essential sensitivity to the subject of concern. Intriguingly, the response to current advertising of UDL's CF carrier testing service for pre-conceptual adults appears to continue for many months after the magazine or paper carrying the advertisement was sold. Clearly, genetic diagnostics as currently presented is being considered thoughtfully and put aside until it becomes appropriate for an individual to follow it up.

In many respects, present advertising is timid in addressing CF when compared with the stronger images of some diseases and handicaps that are used by charitable organizations seeking public awareness and donations. However, as awareness of genetics increases so will an acclimatization to the impact of the public presentation of the issues surrounding inherited disorders. Inevitably, the advertising materials will become more direct and persuasive, and then the skills of the advertisers will start to make their mark.

However, it is unlikely that the advertising of diagnostic tests will be too intrusive in our lives. The advertisements will have to comply with the good taste and honesty imposed by the Advertising Standards Authority. Further, these services cannot be advertised on the radio and TV because they do not provide the opportunity to reflect on the services offered; only clinics are allowed to offer their services in this manner (Radio Authority Advertising and Sponsorship code, 1997).

15.8 Genetics for diagnostics or for what?

Is the impetus for genetic research the betterment of mankind or the advancement of the scientists who discover the genetics of a life-threatening disorder? It may well be a happy fusion of these two, although there is no doubting the kudos associated with being the first one to print and patent the discovery. That being so, the actual genetic discovery once known must somehow find its place in clinical practice. There is an obvious synergism between diagnostics and therapeutics, yet the ideological and commercial value of the cure far outweighs the use of the genetic research for a diagnostic. Thus genetic diagnostics will rarely be the goal of genetic research, and the complexity and number of changes that can alter a coding gene sequence do not lend themselves to be translated easily into meaningful tests. The future progression of genetic research into diagnostics will in part benefit mass markets, and thus the establishment of private sector test providers can only encourage the future availability of the fruits of current genetic research.

To my knowledge, the picture of provision of genetic tests internationally essentially distils down to its provision through the established structure of health care provision (either state or private insurance reimbursement) in each country. Thus the launch of CF carrier screening directly available to the public in the UK is in its own small way a minor revolution in genetics.

It is therefore a very open question as to exactly how much of the human genome project will actually become available as diagnostic tests and how readily we will partake of such opportunities to learn more about ourselves from our genes.

References

ACGT (1996) *Draft Code of Practice for Genetic Testing Offered Commercially Direct to the Public.* Department of Health, London.

Bull S. (1996) Mail order gene testing. *Bull. Med. Eth.* (February 1996), 20–21.

Department of Health (1993) *Population Needs and Genetic Services: an Outline Guide.* Department of Health, London.

Ferrie RM, Schwartz MJ, Robertson NH, Vaudin S, Super M, Malone G and Little S. (1992) Development, multiplexing, and application of ARMS tests for common mutations in the CFTR gene. *Am J. Hum. Genet.* **51**, 251–262.

Gill P, Urquhart A, Millican E, Oldroyd N, Watson S, Sparkes R and Kimpton CP. (1997) A new method of STR interpretation using inferential logic – development of a criminal intelligence database. *Forensic Sci. Int.* (In press).

Harper PS. (1995) Direct marketing of cystic fibrosis carrier screening: commercial push or population need? *J. Med. Genet.* **32**, 249–250.

Home Office (1988) *DNA Profiling in Immigration Casework.* HMSO, London.

Radio Authority Advertising and Sponsorship Code, UK (March 1997) *Diagnosis, Prescription or Treatment by Correspondence* (appendix 4, rule 17). Radio Authority, London.

Rogers L. (1995) Mail-order genetic tests set alarm bells ringing at Westminster. *The Sunday Times* (16 July), p. 9.

Index

ABO blood group, 1
Adenovirus, 212, 227, 229, 231
Admixture test, 8
Affected, 11
AIDS, 245, 319
Aldosterone synthase, 188
Allele, 12
  frequency, 34
  multi, 14
  -sharing, 30
  two, 14
Allele-specific hybridization, 51, 57
Allele-specific oligonucleotide, 56
Alkylation, 36
α_1-antitrypsin, 204, 209, 210, 211, 213, 214
Alternate strand recognition, 268
Alzheimer's disease, 76, 129, 185, 186, 192, 318
Amiloride, 188
Amnion, 20
Amphipathic, 134
Amplimer, 54
  radio-labelled, 55
Amyloid, 146
Ancestral, 8
Angina pectoris, 163
Angiography, 163
Angiotensinogen, 189
Antibodies, 30, 295
  GAD, 30
  insulin, 30
  islet-cell, 30
Antichymotrypsin, 205
Anticipation, 83, 84
Antigen, 1, 287
Antigene, 264
ApoE
ARMS (amplification refractory mutation system), 51, 57, 312
Arthritis, 185, 192
Ascertainment, 25, 26, 92
Aspergillus fumigatus, 247
Association, 14
Asthma, 93, 185, 186, 249
Astrocytes, 134, 139
Ataxia telangiectasia, 223
Atherogenesis, 165
Atheroma, 165
Atherosclerosis, 164, 185, 192
Autoantibodies, 171

β-amyloid, 146
β-cell-dysfunction, 29
Bacteriophage, 37, 49
Basophil, 292
B cell, 292
Behavioural studies, 22
Beta-blockers, 242
Blood group
  ABO, 1
  Rhesus, 8
Bottleneck, 40
BRCA1, 8, 249, 254
BRCA2, 111
Breast cancer, 9, 77
Bronchiectasis, 203
Bronchitis, 203
Buccal wash, 132

CAAT box, 153
Cancer, 139, 185, 191, 192
Cases, 7
Cathepsin O, 247
Cell-lines, 4, 111
Centromeres, 44
CEPH, 4
Cerebrospinal fluid, 146
Cerebrovascular, 91
CCM (chemical cleavage of mismatch), 50, 53
cDNA, 111
  libraries, 111
CDGE (constant denaturant gel electrophoresis), 51, 56

CFTR (CF transmembrane Conductance regulator), 213, 224
Charcot–Marie–tooth disease type 1A, 223, 225
Chimaeras, 224
Chip, 13, 51, 59
Cholera, 41
Chorion, 20
Chromosome, 1, 5, 7, 8, 9, 33
Chronic granulomatous disease, 3
Chylomicron, 133, 134
Cirrhosis, 205
Cis-acting, 100
Classical twin study, 21
Cloning, 94
Clustering, 10
COAD (chronic obstructive airway disease), 203, 208, 210, 211
Co-adaptation, 46
Coding, 12
Collagenase, 161
Concordance, 24, 25, 29, 92
Concordance rate, 26
  casewise, 26
  pairwise, 26
Consanguineous, 9
Coronary heart disease, 28, 151
CpG
  islands, 36
  sites, 35
Cushing's syndrome, 94
Cystic fibrosis, 4, 8, 10, 15, 186, 191, 254, 310, 311, 312, 313, 314, 315, 320
Cytochrome, 252
Cytokines, 157, 158, 300
Cytotoxic, 272

Database
  GenBank, 247
  Human Genome Sciences, 247
  IMAGE, 247
  Incyte, 247
Defibrination, 155
Dementia, 139
Deoxyribonucleic acid (DNA), 4, 5
Descendants, 8
Dexamethasone, 188
DGGE (denaturant gradient gel electrophoresis), 51, 53, 112
  horizontal, 60
  transverse, 60
Diabetes, 14, 136, 185, 186, 192
  gestational, 100
  mitochondrial, 93
  type I, 81
  type II, 249
Dinucleotide, 4
Discordant, 21, 25
Dominant-negative, 38
Dopamine-β-hydroxylase, 6
Dopaminergic, 81, 221
  D1, 81
  D2, 81
Dot blot, 57
Duchenne muscular dystrophy, 15
Duplication, 45

ECM (enzymatic cleavage of mismatches), 50
Electrophoresis, 50, 102
  isoenzyme, 50
  protein, 50
Electrophoretic analysis, 130
Elliptocytosis, 8
Emphysema, 203
EMSA (electrophoretic mobility shift assay), 164
Endonucleases, 49
Enhancer, 44
Enhancer binding protein, 158
Eosinophil, 292
Epigenetic, 47
Epistasis, 136
Epistatic, 47
EST (expressed sequence tags), 110
Exon, 44
Extinct, 41
Extragenic, 77

False-positives, 3
Familial, 10, 92, 185
Fecundity, 178
Felty's syndrome, 172, 175
Fertilization, 20, 23
  in vitro, 21
Fibrinogen, 151, 152, 155, 159, 318
  a, 155
  b, 155
  g, 155
Fitness, 39

Fluorescent, 102
genotyping, 4
labelling, 5, 13
Founder effect, 7, 8, 40, 103, 109

Gain-of-function, 38
Gangliosidoses, 221
Gelatinase B, 161
Gene, 1, 2, 5, 8, 11, 13
amplification, 44
candidate, 7, 8, 9, 10, 12, 77, 96, 114
co-evolution, 46
cytotoxic, 232
developmental, 47
expression, 43
frequency, 33, 34
HIV-1 protease, 254
housekeeping, 46
nested, 44
PMP-22, 225
regulatory, 45
reporter, 231
single, 15
therapy, 119
variability, 152
GENEHUNTER, 102
Generation, 8
Genetic
drift, 40
forward, 96,
linkage, 1
maps, mapping, 4, 5, 45
markers, 1
reverse, 96
Genome, 4, 5, 11, 13
Genotype relative risk, 10
Glucokinase, 100, 114
G6PD, 16
Guillain-Barre, 231
Glycation, 136
GRA (glucocorticoid remediable aldosteronism), 194

H2-antagonist, 242
Haploinsufficiency, 38
Haplotype, 102, 145, 178
Hepatectomy, 214
Hepatocyte, 206
Hepatocyte nuclear factor 1-α, 100
Heritability, 27, 28, 31, 92
Herpes-simplex virus 1, 227, 231
Heterodimer, 154
Heteroduplex analysis, 50, 53
Heterogeneity, 8, 9, 29, 76, 77, 100, 103, 203, 255
Heterozygous, 12
Histochemical, 155
HLA, 11, 29
HNF, 101
Homodimer, 154
Homogeneous, 4
HOMOG, 9
Homozygosity, 9
HPRT (hypoxanthine–guanine phosphoribosyl transferase), 224
HRR (haplotype relative risk), 116
Human Genome Diversity Project, 47
Human Genome Project, 7, 47, 111
Huntington's chorea, 310
Huntington's disease, 8, 84, 223, 226
Hybridized, 13
Hybridomas, 287
Hybrid-vigour, 39
Hypercholesterolaemia, 185, 197, 198, 318
Hyperglycaemia, 93, 95, 97, 193
Hyperinsulinaemia, 111
Hyperlipidaemias, 132
type III, 132, 194
Hyperphosphorylated, 139
Hypertension, 93, 185, 188, 189
Hypervariability, 37
Hypokalaemia, 188
Hypothyroidism, 133, 136

Incretins, 98
IDDM, 93
Identity-by-descent, 11, 77
IL-6 element, 153
Immunoblotting, 135
Immunoglobulin, 286
Immunological, 155
Immunosuppression, 300, 301
Implantation, 20
Imprinting, 47
Inheritance, 1, 7, 15
oligogenic, 76
polygenes, 76
Insertion, 36
Insulin, 11
Insulin-dependent diabetes mellitus, 23, 28

Insulin resistance, 29
Integrase, 263
Interference, 6
Intima, 155
Intra-uterine, 23
Intron, 44
Invertons, 36
Isoelectric focussing, 133, 134
Isomorphous, 266, 270
Isoproteins, 138
Isozymes, 48

λ_{sib}, 102
LADA, 101
Langerhan cells, 292
Lesch–Nyhan syndrome, 224
Leucine zipper, 158
Leukaemia, 297, 298, 301
Leukodystrophies, 221
Libraries, 287
Liddle's syndrome, 188, 193
Ligases, 49
Likelihood, 11
Linkage, 2, 3, 5, 7, 9, 10, 11, 15, 30
  disequilibrium, 7, 8, 12, 14, 47, 103, 110, 116, 117, 164, 177
Liposomes, 119, 212, 227
Loci, 11, 15, 16, 35, 93, 101
Locus, 8, 9, 10, 11, 108
LOD score, 3, 5, 10
Lymphomas, 297, 298, 301
Lymphopenic, 301
Lysosomal, 176, 225, 226
Lysozyme, 289

Macrophages, 134, 169, 206, 207, 292
Macrovascular, 91
MADGE, 60, 129, 130
Malaria, 16, 41
Manic depressive psychosis, 28
Mapping, 4, 9
Marker, 2, 4, 5, 11, 14, 28, 30, 44, 102, 173, 175
  cellular, 47
Matrilysin, 161
Maximum likelihood, 2
MEDPED (Make Early Diagnosis and Prevent Early Deaths), 193, 196, 197, 198
MELAS, 101
Mendelian, 3, 10, 15
Methylation, 35, 52
Microglial, 221, 226
Microplates, 130
Microsatellites, 4, 44, 100, 173
Microvascular, 91
Migration, 41
Mini-satellite, 36, 44
Mitochondrial, 13, 101
Mitogenic, 155
MMP (metalloproteinase), 161, 164
MODY (maturity onset diabetes of the young), 100
  MODY 1, 100
  MODY 2, 100
  MODY 3, 100
Monoclonal, 134, 287
Monocyte, 206, 207, 292
Monogenic, 8, 9, 10
Morbidity, 91
Mortality, 23, 91
MTHFR (methylene tetrahydrofolate reductase), 190
Mucopolysaccharidoses, 221
Multifactorial, 10, 19, 46
Multigenerational, 94, 105
Multimer, 154
Multiple endocrine neoplasia, 195
Multiple regression, 28
Multiple sclerosis, 186, 249, 298,
Multiplexing, 102,
Multipoint linkage analysis, 11
Muscular dystrophy, 310
Mutation, 4, 7, 8, 14, 33, 39, 254
  adaptive, 37
  de novo, 34, 35
  directive, 37
  gain-of-function, 118
  hot spots, 35
  loss-of-function, 118
  neutral, 38
  pressure, 38
  rate, 33
  silent, 38
  somatic, 35
  stochastic, 34
Myocardial infarction, 156
Myotonic dystrophy, 84

Nail-patella syndrome, 3

Neocortical, 231
Neo-Darwinian, 39
Nephropathy, 91
Neurodegeneration, 226
Neurofibrillary tangles, 139, 146
Neuroleptics, 82
Neuropathy, 91
Neuropeptide, 229
Neutralism, 40, 41
Neutrophils, 204, 292
NIDDM, 77, 91
Nigro-striatal, 221
Non-parametric test, 80
Non-random, 2
Non-transcribed, 35
Northern blotting, 50
Nucleases, 53
Nulliparous, 178

Obesity, 93, 136
Offspring, 12
Oligomers, 13
Oligonucleotides, 13, 264, 267
Oligopurine, 270
Osteoporosis, 185, 186, 192

Paired helical filaments, 139
Pancreatitis, 93
Pandemic, 41
Pannus, 171
Parametric test, 80
Parkinson's disease, 223, 226
PDGF (platelet derived growth factor), 160
Pedigree, 3
Penetrance, 8, 94, 100, 203
  incomplete, 9, 77
Penetrant, 4
Peripheral vascular disease, 91, 156
Phenocopies, 4, 9
Phenotype intermediate, 95,
Phenylketonuria, 40
Phospholipase C, 247
Phosphorylation, 154
Placenta, 20
Plaques, 155, 160
Plasmid, 37, 49, 227
Platelets, 155, 292
Pleiotropy, 45, 110, 139, 154
PMD, 60
PODGE, 60
Polyacrylamide, 130
Polygenes, 46
Polymerase, 49
Polymerase chain reaction (PCR), 4, 43, 49
  liquid phase, 131
  long, 50
  reverse transcriptase, 50
Polymorphic, 4, 39, 154
Polymorphism, 8, 12, 14, 96, 114, 117, 129, 175, 178, 254
Polyneuritis, 231
Population, 12
Positional cloning, 3, 15, 101
Post-partum, 178
Post-prandial, 99
Prednisone, 188
Pre-eclampsia, 189
Proband, 26, 75
Probe, 13, 43, 114, 165
Promoter, 44, 45, 212
Prophylactic, 15
Protease, 263
Proteinase, 161
Proteolysis, 176
Provirus, 37
Pseudogenes, 44, 134
Pseudohypoaldosteronism, 9
Psoriasis, 186
Psychosis, 77

Rat
  Goto-Kakizaki, 99
  sand, 99
  Zucker diabetic fatty, 99
Recessive, 7
Recombination, 3, 5, 7
  double, 6
  hot-spots, 36
  non-homologous, 33, 44
  unequal, 36, 37
RED (repeat expansion detection), 84
Regulon, 45
Relative fitness, 41
Relative risk, 173
Replication, 33
Repressor, 153
Restriction analysis, 50
RFLP (restriction fragment length polymorphism), 49
Retinitis pigmentosa, 47

Retinopathy, 91
Retrotransposons, 36
Retroviral vectors, 212
Reverse transcriptase, 263
Rheumatoid arthritis, 171, 298, 301
Rheumatoid factors, 171
Rous sarcoma virus, 231
RT-PCR, 114

Satellite, 44
Scanning, 5
Schizophrenia, 28, 93, 111, 112, 186, 192
Screening, 8
Secretase, 145
Segregation, 2, 3, 33, 92, 95
Sequencers, 5
Sequencing, 13, 43
Sex-linked, 3
Sib-pair, 1, 11, 77, 79, 81, 105
  multipoint, 80
Sickle cell anaemia, 41
Signal sequences, 44
Sjögren's syndrome, 172
Skipped generation, 77
Slippage, 36, 37, 44
Southern blot, 49, 114
SP-1 box, 153
Spironolactone, 188
Splicing sequences, 44
SSCP (single-strand conformational polymorphism), 51, 54, 112, 162
Statin, 186, 187, 190
Steroid 11-beta-hydroxylase, 188
Stratification, 12, 117
Stroke, 185, 249,
Stromelysin-1, 161, 162, 164, 318
Structural equation modelling, 28
Susceptibility, 12, 14, 28, 95, 102, 108
Synctiotrophoblast, 296
Systemic vasculitis, 198

TATA, 153, 208
TAU, 139, 146
Tay-Sach, 15
Telomeres, 44
Template, 132
TGGE (temperature gradient gel electrophoresis), 50, 54
Thalassaemia, 15, 154
Therapeutic, 15
Thermogenesis, 98
T lymphocytes, 287
Trait, 1, 3, 10, 11, 15, 23, 42, 102
  discontinuous, 24
Transcribed, 35
Transcription, 153, 156, 207
Transcription factor, 254
Transduction, 292
Transgenic, 224
Transition, 35
Transmission disequilibrium test (TDT), 79, 116
Transmission distortion test (TDT), 12, 14
Transposons, 36
Transversion, 35
$tRNA^{Leu(UUR)}$, 92, 101
Tuberous sclerosis, 5, 9
Twins
  monozygotic, 20
  dizygotic, 20

Ultracentrifugation, 136
Unlinked, 2, 9

Variance, 23, 92, 114
Variation, 13, 14
Varicella zoster, 229
Vasculitis, 175
Venesector, 132
VNTR, 14, 36, 44

Western blotting, 50
Wild-type, 39

X inactivation, 20
X-linked, 316

Yeast chromosomes, 49

Zygosity, 22, 23, 28